DINOSAURS,

SPITFIRES, and

SEA DRAGONS

DINOSAURS, SPITFIRES, and SEA DRAGONS

Christopher McGowan

Harvard University Press · Cambridge, Massachusetts · London, England · 1991

This is a completely revised and updated version of *The Successful Dragons: A Natural History of Extinct Reptiles,* published by Samuel Stevens and Company, Toronto, in 1983.

This book is printed on acid-free paper, and its binding materials have been chosen for strength and durability.

Library of Congress Cataloging in Publication Data

McGowan, Christopher.
   Dinosaurs, spitfires, and sea dragons / Christopher McGowan.
      p. cm.
   "Completely revised and updated version of The successful dragons : a natural history of extinct reptiles. 1983.
   Includes bibliographical references and index.
   ISBN 0-674-20769-6 (alk. paper)
   1. Reptiles, Fossil. 2. Paleontology—Mesozoic. 3. Evolution.
I. McGowan, Christopher. Successful dragons. II. Title.
QE861.M38 1991
567.9—dc20   90-41552
                                                CIP

Designed by Gwen Frankfeldt

To Liz, with love
You make everything worthwhile

Contents

Acknowledgments

PARTS of this book were more challenging to write than others, but trying to find adequate words to express my gratitude for the generous help I received during its preparation is the most difficult. Perhaps I should simply say "thank you" and trust that those concerned will know that it comes from the heart.

I have many people to thank, people who have given so generously of their time. I will begin with those who have critically read parts of the original manuscript. For their valuable comments I thank Andrew A. Biewener, Walter Coombs, Philip J. Currie, Peter Dodson, James O. Farlow, Andrew J. Forester, Robert C. Goode, Grant Hazelhurst, Trevor Hearn, James A. Hopson, Nicholas Hotton III, Farish A. Jenkins, Judy A. Massare, Charles B. Officer, John H. Ostrom, Kevin Padian, Michael J. Plyley, Jeremy M. V. Rayner, Anthony P. Russell, Dale A. Russell, Martin Sander, Jeffrey J. Thomason, André L. Vallerand, Paul W. Webb, Daniel Weihs, Daphne Weihs, David B. Weishampel, and Peter Wellnhofer. For their valuable discussions I thank Walter Coombs, Peter Dodson, Andy Higgins, Jack Horner, Kay Mehren, Dale Russell, Geoffrey Sherwood, Paul Webb, and Rupert Wild.

Most of the illustrations drawn for the book were prepared by Julian Mulock. Working with Julian was a pleasure, as always, and his magnificent drawings lend immeasurable support to the written word. Marg Sansom's restorations of Mesozoic life add a dimension of reality to the remote past. I thank her for her vision and sensitivity for her subjects.

Catherine Hollett helped me in so many ways—from looking after copyright requests to unscrambling reference files. For her good-humored assistance I am eternally grateful. For helping with proofreading I thank Sue Baker, Joan Ludlow, and Liz McGowan.

Working with the professionals at Harvard University Press has been a thoroughly enjoyable and rewarding experience. I thank them all for their enthusiasm, their help, and their warm hospitality. Howard Boyer initiated the project, saw it through its various stages, and gave encouragement when most it was needed, for which I am sincerely grateful. Kate Schmit practiced her editorial skills with meticulous care, great patience, and good cheer. It was a privilege and a joy working with Kate, and I will always be in her debt. Special thanks also to Paul Adams, Claire Silvers, A. J. Sullivan, and Gwen Frankfeldt.

Most of what I know about Mesozoic reptiles was learned at the Royal Ontario Museum. I am happy to acknowledge my thanks to that institution for providing a supportive and stimulating environment in which to work. Many of my colleagues at the ROM have helped with this project, as they have helped me over the years. I would like to thank them all, with special thanks to Brian Iwama for countless trips to the library and Xerox machine, Brian Boyle and Alan McColl for photography, and Pat Trunks for interlibrary loans. I would also like to thank the Natural Sciences and Engineering Research Council of Canada, which has generously supported my researches over the years.

This book arose out of a similar book, *The Successful Dragons,* published by Samuel Stevens. I am particularly grateful to Alan Samuel, not only for his encouraging me to write in the first place, but also for his continued support of the present undertaking.

Claire McGowan typed parts of an earlier draft, for which I thank her. Liz McGowan provided a peaceful environment in which to write, gave constructive criticism when asked, and dispensed encouragement all the time—as always.

DINOSAURS,

SPITFIRES, and

SEA DRAGONS

Prologue

IGHT is a golden age. An age of make-believe and wonder. Our Camelot days. Little wonder that eight-year-olds are so enthralled by dinosaurs. They know their names, collect their pictures, and plan dinosaur projects at school. But as they grow older and become wiser to the ways of the world, other interests compete for their attention. They outgrow the dinosaur phase.

Sometimes the muse returns. If the current flood of magazine articles and television specials on dinosaurs is any indicator, she has been unusually busy lately. Not only have dinosaurs become news on the popular front but there has also been a veritable explosion of interest within paleontological circles. When I attended my first meeting of the Society of Vertebrate Paleontology in 1969, the number of North American scientists working on dinosaurs and other Mesozoic reptiles could have been counted on one hand. Now it seems that everyone is working on them—almost a quarter of the two hundred papers presented at the 1989 meeting were on Mesozoic reptiles.

Peter Dodson, a dinosaur specialist himself, estimates that about 40 percent of all the known dinosaurs have been discovered during the last twenty years or so. This statistic is all the more remarkable when it is remembered that the first dinosaur was discovered in the early 1820s. Part of the reason for the rebirth of interest is attributable to renewed collecting activities—primarily in Mongolia, China, Alberta, the western United States, and Argentina and Brazil. Discussions of warm-bloodedness, parental care, and extinction have sparked considerable interest, but some of the accounts have been misleadingly speculative. People often ask me at the Royal Ontario Museum how we know the color of dinosaurs, what sounds they made, how fast they ran, how long they lived, and so forth. When I reply that we simply do not know these things, I sometimes get some odd looks. Surely we know . . . some paleontologist said so on the television . . . I read it in a dinosaur book . . . there was this article in a science magazine.

The point I always make is that we do not know everything about living animals, so how can we possibly know so much about animals that disappeared over sixty million years ago? How can we have a complete picture of creatures that left little more than their bones behind? Questions about how fast zebras run, how much elephants weigh, and for how long crocodiles live may sound simple enough, but when we try looking for answers we are often surprised at how few pieces of data are available. Zebras are remarkably uncooperative at running in straight lines for biologists to clock their speeds, and crocodiles are no more conscientious when it comes to keeping family records! Our knowledge of the living world is built on shaky foundations, so why should we be so ready to build dinosaurian sand castles in the air?

All we have are bones, but we usually don't have *all* the bones. When people see a particular dinosaur—say the genus *Diplodocus* or *Tyrannosaurus*—in a museum, they may get the impression that they are seeing a complete skeleton, with many more like it in other museums. Sometimes this is true, but more often than not they are looking at a plaster cast of an original skeleton or at an incomplete skeleton that has been largely restored in plaster. According to Dodson, almost half of the 540 genera of dinosaurs are represented by single specimens and very few specimens comprise complete skeletons.

Firm in the conviction that we can begin to understand extinct animals only by understanding living ones, I have chosen to present my interpretations of Mesozoic life by drawing analogies with relevant examples from the living world. This explains why so much space has been devoted to modern animals. I must also serve warning that I stand at the conservative end of a broad spectrum of opinion on how much can be said of life in the past. The words *may, perhaps,* and *possibly* therefore appear frequently, as I make every attempt to avoid straying beyond the data, but this does not mean that we will not take off on some flights of fancy together, nor avoid having some fun.

The Mesozoic Era, which began about 245 million years ago and ended 65 million years before the present, is often referred to as the Age of Reptiles because they were the predominant land animals. That is not to say that the other major groups of vertebrates were not present at the time. The fishes and amphibians had appeared long before the first dinosaurs, the mammals only a little later; the birds did not appear on the scene until about the middle of the era. Reptiles roamed the land, swam in the sea, and flew in the air. Judged in terms of their diversity, their numbers, and their long tenure on the Earth—over 150 million years—they were a phenomenal success. Modern reptiles—the crocodiles, turtles, lizards, snakes, and *Sphenodon* (sole representative of an ancient group related to lizards)—though successful in their own right, pale into insignificance when compared with reptiles of that former age. Why were the Mesozoic reptiles so successful? Was there anything special about them? The answer to the last question is probably no. The reptiles probably just happened to be at the right place at the right time. Today, it seems to be the turn of man, and tomorrow . . . who knows?

Mesozoic reptiles faced similar problems in their world that modern animals face today, and they often resolved them in similar ways. Some of their solutions were unique, though, which is one of the reasons why they are such a fascinating group to study. Little wonder that so many books have been written about them. The purpose of this book is to explore how Mesozoic reptiles lived and functioned and, in so doing, to gain some insights into the underlying reasons of their success. Extinct animals, like living ones, are dominated and constrained by the same physical laws that govern the rest of the world. The same forces that determined the shape of the wing of a Spitfire, or the I-beam used in construction, were operational during the evolution of the pterosaur wing and the bones of sauropods. We can therefore gain insights into the functional significance of biological structures by drawing from the world of the engineer. This not only gives us a better understanding of the mechanics of animals, but also of the properties of the materials from which they are constructed. We will therefore try to view our subjects with the eyes of an engineer. But our knowledge of Mesozoic reptiles is only as good as the fossils upon which they are based. The fossil record, we shall see, has certain pitfalls and shortcomings, and we have to be constantly aware of these in the interpretations we make. No attempt will be made at a comprehensive treatment of all the reptiles of that time—that would require several books. Instead attention will be focused on selected groups and on topics of particular interest. Sauropod

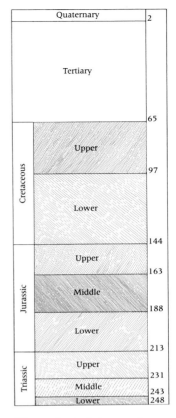

The geological time scale, measured in million years before the present, for the Mesozoic Era—the Age of Reptiles. (Based upon data from Harland et al., 1982.)

("lizard-footed") dinosaurs, for example, will be used to illustrate the problems of being a giant, and a wide range of Mesozoic reptiles will be used to discuss the great reptilian extinction that occurred at the close of the Cretaceous.

The biological phenomena we encounter will be interpreted within the framework of Darwinian evolution, the unifying concept of the biological sciences. It is true that biologists may disagree on certain aspects of the mechanisms of evolution, and that we are revising our ideas on such problems as the evolution of higher taxa (macroevolution) and the role of natural selection ("survival of the fittest") on large-scale evolutionary change, but few would doubt that evolution has occurred and continues to occur. According to the theory of evolution, organisms that are alive today are similar to, but not identical with, organisms that once lived on the Earth. Today's plants and animals are the modified descendants of organisms that once inhabited the world. The modern horse, for example, evolved from a smaller, horse-like animal that had three toes instead of the single toe that characterizes living horses.

Darwin's theory of evolution by means of natural selection provided a mechanism to explain how evolution may have taken place. His theory is elegant in its simplicity. Offspring are similar to, but not identical with, their parents. Therefore no two individuals are exactly alike. Each species produces far more offspring than can possibly survive and, as the offspring are not exactly alike, it follows that some individuals will have features that give them an advantage over others. These advantages, however slight, give the individual a better chance of surviving. Because advantaged individuals have a better chance of survival, they tend to leave more offspring and, since these offspring inherit some of their parents' favorable features, they too tend to have improved chances of survival. The action of natural selection, operating over long periods of time, would therefore bring about a modification of the species, eventually leading to the appearance of a new species. When we talk of advantageous features, we mean features that are advantageous to an individual in the environment in which it must live. For dinosaurs, this was the Mesozoic world.

The early part of the Mesozoic Era was a subtropical world, dominated by evergreens—pines, firs, and other needle-bearing plants. Ferns were abundant, some of them growing as tall as trees. So, too, were cycads, which looked like palm trees with stunted trunks. There were no grasses, shrubs, or other flowering

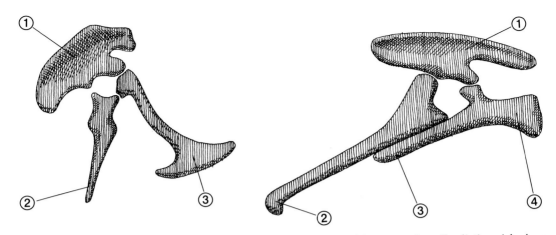

The two major groups of dinosaurs, the Saurischia and Ornithischia, are primarily distinguished on their pelvic girdles. Left, the saurischians have a simple, three-pronged pelvis comprised of three paired bones. Dorsally (above) there is the ilium (1), which is attached to the vertebral column, and ventrally (below) there is the pubis (3), which is anterior (in front), and the ischium (2), which is posterior (behind). Right, the ornithischian pelvis is more complex in that the pubis (3) is directed posteriorly, alongside the ischium (2), and there is an additional anterior process, called the prepubis (4). The ornithischian pelvis is therefore a four-pronged rather than a three-pronged structure.

plants, and ground cover was provided by various types of mosses and by horsetails (*Equisetum*). A flora somewhat similar to this may be seen today in New Zealand, New Guinea, Central and South America, and the Caribbean. It was not until Cretaceous times that flowering plants appeared, giving the landscape an appearance like that of the Gulf coast of North America today.

Some of the major groups of reptiles that lived during the Mesozoic Era, and which appear later in the book, are depicted on pages 6 and 7. There are two major groups of dinosaurs, the Saurischia ("lizard-hipped") and the Ornithischia ("bird-hipped"). The appropriate scheme to be used for the classification of the dinosaurs is a subject of considerable debate among specialists and one which is undergoing rapid change. In the interest of simplicity, the conservative system adopted by Carroll (1988) will be followed here.

Many other reptilian groups shared the Mesozoic world with the dinosaurs, but we will be concerned with only two of these: the pterosaurs ("winged lizards") and the ichthyosaurs ("fish liz-

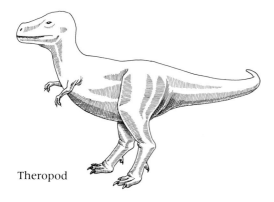

Theropod

Sauropod

There are two main types, or suborders, of saurischian dinosaurs: theropods and sauropods. The theropods were mainly bipedal and carnivorous. The sauropods were mainly quadrupedal, herbivorous, and often of gigantic size. The ornithischians, probably all of them herbivorous, were of five main types: the unarmored ornithopods; the ankylosaurs, which were armored with flat, bony plates; the pachycephalosaurs, which had a thick dome of solid bone on top of their skulls; the stegosaurs, which had a series of vertical plates along the back; and the ceratopsians, which had horns. The dinosaurs were land animals, whereas pterosaurs were creatures of the air and ichthyosaurs and plesiosaurs lived in the seas.

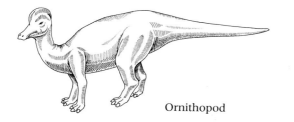

Ornithopod

Ankylosaur

ards"). Passing reference will also be made to a second marine group, the plesiosaurs ("near lizards"), which have been likened to a snake strung through the body of a turtle.

Not all dinosaurs were giants. Some, like *Ornithomimus*, were the size of ostriches, and others were considerably smaller. But most tended toward large size—the size of elephants or bigger. And the sauropods, the largest of them all, were the heaviest animals ever to walk the land.

Pachycephalosaur

Pterosaur

Stegosaur

Ichthyosaur

Ceratopsian

Plesiosaur

A century and a half has passed since Richard Owen coined the name *dinosaur* ("terrible lizard") for these incredible animals. During that time there has been an explosive growth in knowledge, not only about dinosaurs and other animals but in all branches of the sciences. So let us turn the pages and see how the Mesozoic world is viewed from the perspective of the late twentieth century.

ONE

Material Things

B O N E S and teeth are a paleontologist's stock in trade because these are the parts of an animal that are most resistant to decay and are therefore the most likely to be preserved. Fossil bones were once part of the skeleton of a living, moving creature; they were the architectural framework upon which the whole animal was constructed, and we can learn a great deal about an animal and its way of life by studying its skeleton. But there is more to a skeleton than just an assortment of bones. There are the ligaments that help hold the individual bones together, the tendons that connect muscles to bones, and the cartilage that forms the surfaces of joints. We may not see all of these components in a given fossil—ligaments and tendons are seldom preserved, nor is cartilage—but we need to know about them if we are to understand how a living skeleton functions. Similarly, we may learn about an animal's habits and its diet by studying its teeth.

Like any other architectural structure, the skeleton is largely built of a series of beams and columns. And just as certain building materials are used for particular functions—concrete for supporting columns, aluminium for window frames—so do certain tissues have particular functions in the skeleton. We therefore need to know something of the physical properties of the various skeletal materials to see how they are suited to their particular roles. A knowledge of their microscopic structure and chemical composition is also useful, especially in understanding the changes that take place in them during fossilization. The purpose of this chapter is to look at skeletal structures and the materials from which they are built from an engineering perspective. Attention will naturally be focused on bone because most of the fossils dealt with in the chapters that follow are bony.

The subject of materials science, no longer the prerogative of the engineer, is as useful to biologists for answering questions such as why tendons are able to store so much energy during locomotion, as it is to the aircraft designer for understanding why turbine blades break. The limitations of space do not permit an in-depth treatment of the subject, but I hope to cover enough of the basics to provide important insights into the properties of skeletal structures. I will also touch upon some elementary engineering principles, including simple beam theory, and a little basic physics too, but do not despair—the subject is easy to follow. Many of the terms used, like *elasticity* and *stiffness*, are in everyday usage, but they have precise meanings in engineering science and will therefore have to be redefined.

Some thoughts on beams and columns

Let us imagine that we have two bricks and a stout plank of wood—one about six feet (2 m) long and two inches thick (5 cm) would do very nicely. We can set the plank up as a beam simply by supporting it at either end with a brick. If we now stand in the middle of the beam, we notice that it sags under the force of our body mass.[1] The lower surface is bowed out, and we can visualize how the wood fibers there are being stretched. The upper surface curves the other way, and the wood fibers there are being pushed together rather than pulled apart. When things are pushed together they are said to be in *compression,* and when they are pulled apart they are said to be in *tension.* The top surface of the beam is therefore being loaded in compression, the lower surface in tension.

Engineers are usually more interested in measuring force per unit area than total forces, and this is given the term *stress.* If my wife stepped on my toe with the heel of her running shoe I would feel it but it would not hurt. If she trod on me with a stiletto heel, however, the pain would be intense. The forces are the same in both cases (mass × acceleration due to gravity), but the stress is higher the second time simply because the area of a stiletto heel is only a fraction of that of a running shoe.

Suppose we could insert a probe into the wooden plank and measure the stresses in the wood at different levels.[2] The compressive stress would be highest at the top surface and would diminish as the probe went deeper into the wood. Similarly, the tensile stress would be highest at the bottom surface and lower toward the center. A point would be reached in the middle of the plank where there was no stress at all; this is called the neutral axis of the beam.

Not only do the magnitudes of the stresses change at different depths, they also change along the length of the plank for a given depth. The stresses are highest in the middle of the plank, where the bending is greatest, and decrease toward either end. The maximum, or peak, stresses would therefore be recorded in the middle of the plank. The peak compressive stress occurs at the midpoint of the upper surface, the peak tensile stress at the midpoint of the lower surface. When structures are loaded so that they bend, they always experience higher peak stresses than when there is no bending. They are therefore more likely to be broken under these conditions, as anyone who has broken a bone will attest—breaks usually occur when arms and legs are bent or twisted.

Although our plank makes a perfectly good beam the way we have it set up, it does sag rather a lot. And if we wanted to support the weight of several people, it would probably sag all the way to the floor. A simple solution is to turn the plank on its edge, and although it would require a bit of balancing now because we would have to stand on a two-inch width, the beam would probably not sag noticeably at all. If we could measure the stresses at the top and bottom of the plank in this new position we should find that they were much less than before. But they would still change from compression to tension in passing from the top to the bottom surface, with a neutral axis in the middle. Since the middle of the beam is not under stress, we could drill holes in it and barely diminish its bending strength. Indeed, this is the principle of the steel I-beams used in construction: the steel is concentrated in the areas of maximum stress, the top and the bottom, with relatively little material in the middle.

Instead of using the plank as a beam we could use it as a vertical column. It could, for example, be used as a prop to shore up the sagging roof of a mine shaft. In this situation the plank would be loaded in simple compression. If we were to measure the stresses within the wood we should find that they were compressive from top to bottom, with little, if any, variation from one region to another. We should also find that, for the same load, the stress at any point would be much less than the peak stress when the plank was loaded as a beam.

We could use our wooden plank and bricks to explain elasticity, stiffness, and all the other terms of materials science. But that would be a bit boring, so we shall use skeletal tissue instead, with references to nonliving systems where appropriate.

The nature of bone

Any treatment of bone structure, especially from the perspective of a mechanical engineering problem, is bound to be an oversimplification of the true situation. This is partly because bone is a living tissue, with more roles to play than just that of providing mechanical support, and partly because details of its fine structure are complex and not completely understood. But, at the risk of oversimplification, let us pick up a freshly cleaned bone—the rib of a sheep is suitable—and examine its mechanical properties. A sheep's rib is about as long as a pencil and about as slender, though it is flattened rather than rounded and is curved. Now if we grasp the rib firmly between both hands and try and break it. . . . Well, I can't do it, but I have no difficulties breaking the pencil. One obvious feature we notice when we conduct this simple experiment is that the rib is quite resilient—springy like the blade of a sword. To be more precise in our terminology, fresh bone is elastic. In engineering terms an *elastic* material is one that returns to its original shape when the stress is removed, without suffering any permanent strain (permanent change in shape). A tennis ball is obviously elastic; so too is a saw blade and a diving board. To avoid possible confusion, a distinction needs to be made between the properties of *materials* and of *structures*. When we bend a steel saw blade and see it spring back straight again when we let go, we are actually observing the elasticity of the saw blade, which is a structure. If we wanted to observe the elasticity of the steel—the material of the blade—we would have to use a machine that could exert a large enough force to stretch the blade in a straight pull. Most of the terms we will be using can be applied equally to materials and structures, but there are some exceptions.

In addition to being elastic, bone is also stiff. *Stiff* materials are those in which a large stress is required in order to bring about a small change in shape, or, to be precise, in strain. The stiffness of a material—the relative ease with which it can be pulled or pushed temporarily out of shape—is expressed by a quantity called *Young's modulus,* named after English physicist Thomas Young. Young's modulus is simply the stress applied to a material divided by the strain produced (change in length).[3] Young's modulus applies only to materials, *not* to structures. A small pull on a rubber band produces a large change in length (small stress but large strain), so rubber has a low Young's modulus. Conversely, we would have to pull very hard on a steel wire to stretch it even a tiny bit (enormous stress but minute strain); steel has a high Young's modulus. The stiffer the material, the higher its modulus. Common stiff materials include concrete, glass, pottery, metals, and bone. Common materials that have low values for Young's modulus include neoprene, polythene, sponge, and cork.

Elastic materials behave elastically only if the stress applied to them is below a certain critical level. If the stress exceeds this threshold, which is called the *elastic limit,* then the material does not return to its original shape after the stress is removed. If we flex the blade of a saw it will bow, and provided we do not bend it too far, the blade will return to its original shape and become straight again when we remove the stress. But if we exceed the elastic limit, the blade will suffer permanent strain and be permanently bent out of shape. When we exceed the elastic limit of a material it actually gives or yields and it is then permanently stretched. When the material is being permanently stretched its behavior is said to be *plastic.* Steel does not behave plastically for very long before it breaks. If we took a slab of toffee instead and pulled on it, we would be able to draw it out into a long ribbon. Toffee, then, has a low elastic limit and an extensive plastic phase, during which time it experiences permanent strain—in other words, it is permanently stretched.

The relationship between elastic and plastic behavior is very easily demonstrated with one of those plastic supermarket bags, especially one with writing on it. If you hold the bag in both hands and apply a gentle force, you will see that the bag stretches—it actually feels quite elastic, as if it were made of rubber. Provided you do not pull too hard and exceed the elastic limit of the material, the bag will return to its original shape when you release your grip. If next time you pull a bit harder, you will exceed the elastic limit and the material will become plastic. When you stop pulling this time you find that the bag has been permanently stretched—it has been pulled out of shape and the writing on the bag is distorted. Bags are not as plastic in their behavior as toffee, but they are more plastic than steel. Steel does have a plastic phase, by the way, and that is one of the reasons why car insurance rates are so high. When the door panel of a car is struck by another car the stress is usually sufficiently high to exceed the elastic limit of the steel. There is therefore some permanent strain, and when the dent is sprung out, sometimes simply by pushing from the other side, the door does not quite look the same as it did before. It needs some expensive body work and new paint.

Not all materials pass through a plastic phase before they break. Some materials, like glass and ceramics, cannot be bent out of shape, and they are described as being *brittle*. Windows and dishes, unlike cars, do not get dented, they just get broken. Bone actually does pass through a plastic phase prior to breaking, which contributes to its strength (Burstein et al., 1972), but we can think of bone as being essentially a brittle material. Thus a skier does not bend a bone during a serious mishap on the slopes; the bone breaks.

To return to the sheep's rib: I could not break the rib with my bare hands, nor would it break if I threw it onto a hard floor. But if I were prepared to do some work on it with a hammer, it would probably break. Materials that require relatively large amounts of work to be done on them before they break are described as *tough* materials.[4] Steel is tough, and so is nylon. Glass, in contrast, is easily broken and if I dropped a wine glass onto the floor it would break for sure. Eggshell is not a tough material either, nor is good pastry.

So far we know that bone is elastic because it returns to its original shape when the stress is removed, that it is stiff because a large stress is required to produce a small strain, and that it is tough because much work has to be done to break it. There are two more mechanical properties to consider, its tensile strength and its compressive strength.

Things can be broken by pulling them apart or by compressing them. Materials that stand up well to being pulled apart are said to have a high *tensile strength*, while those that withstand compression well are said to have a high *compressive strength*.[5] There is a subtle but important distinction between tensile and compressive strengths on the one hand and toughness on the other. The former are concerned with the application of a steady force (mass ×

acceleration) to a material, the latter with the work done (force × distance) on the material. A simple way of looking at it is to think that tough materials stand up well to rough treatment, being dropped or struck; materials with high tensile or compressive strengths—*strong* materials—stand up well to being loaded gently. Materials can be strong without being tough. Porcelain, for example, like glass, has a high compressive strength but it is not tough. This was elegantly demonstrated a few years ago by a special display of English china outside a rather expensive shop in Toronto. Four simply splendid English teacups had been set on the sidewalk, and a Rolls Royce was balanced on top of them. Though strong in compression, porcelain is not tough; we can only guess how many cups were broken in the attempt to lower two tons of car onto the teacups gently enough.

Bone is not only tough, it is strong in both tension and in compression. It is also stiff and fairly elastic and has a relatively low density—about twice that of water. Its tensile strength is about half that of mild steel, it is about twenty-five times tougher than brick, and its stiffness exceeds that of wood. How are these remarkable properties achieved?

The answer lies in its composite nature. Bone comprises two main nonliving components, *collagen,* a protein, and *apatite,* which is a crystalline mineral.[6] The nature of each of these components can be demonstrated very easily by removing one or the other of them from a fresh bone. If we took our sheep's rib and left it in an oven for an hour set at about 350° F, the collagen would break down, leaving just the apatite. Apart from being a bit discolored and a little lighter in weight, the bone would look just as it did before treatment, but handling would reveal major changes. The rib would still feel stiff and hard, and it would have retained much of its compressive strength, but it would lack elasticity. Indeed, if we tried bending it, even gently, it would snap in our fingers like a potato chip. We would also notice that it was so brittle it could be chipped with a fingernail and shattered by dropping it onto the floor. The mineral portion of the bone, on its own, is neither very elastic nor tough, but it is stiff and is moderately strong in compression.

If a second bone were taken, this one soaked in acid, the mineral apatite would dissolve, leaving just the collagen. A 10 percent solution of a strong acid such as nitric acid completes the job in a few hours; vinegar also works, but it takes about two weeks.

Again the treated bone would look just as it did before it went into the acid, but it would feel soft and rubbery, not at all stiff; it could even be tied into a knot. Its behavior is elastic. We could flatten it by squeezing it between our fingers, showing that it had lost its compressive strength, and if we tried pulling it apart we would find that it had little tensile strength.

Bone therefore combines the stiffness and compressive strength of apatite with the elasticity of collagen, to form a strong, stiff, composite material. Comparisons are often made between bone and man-made composites like reinforced concrete, but this is probably a gross over-simplification. For one thing, the relationship between collagen and apatite is far more intimate than that between the concrete and steel. Collagen is organized into very fine fibers—far too small to be seen with a light microscope—and the apatite crystals are of a similar small size. There is some uncertainty regarding the precise relationship between the two, but it seems that the microcrystals of apatite are cemented to the collagen fibers, with their long axes oriented along the fibers.[7] In typical man-made composites, like reinforced concrete, it is the fibers that provide most of the tensile strength (concrete, on its own, is very weak in tension). In bone, however, it is the *mineral* portion that provides most of the tensile strength. This has been shown experimentally by progressively dissolving out the apatite from the bone samples, using acid. As more of the mineral is removed from the bone its tensile strength decreases (Burstein et al., 1975).

If we picked up one of the pieces of sheep's rib and examined a broken end we would see that bone is porous. The inner part of the bone looks especially spongy, but the outer layer, which is not very thick, looks quite solid. The inner, spongy region is described as cancellous bone, the outer layer as compact bone. Although the compact bone appears to be nonporous, when it is viewed under the microscope numerous spaces are apparent, but they are relatively small. (In order to examine bone microscopically slices have to be taken from the bone that are thin enough to allow light to pass through them.) Bone is highly vascular and, in life, most of the pores are filled with blood vessels. The most striking feature of the compact bone is the orderly arrangement of the bone matrix into a series of concentric rings looking something like an onion cut in half. Each layer surrounds a central blood canal, which in life contains a blood vessel bathed in tissue

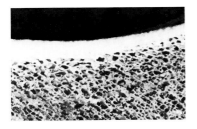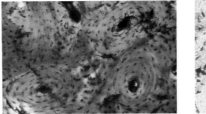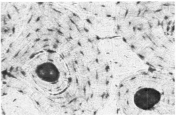

The internal structure of bone is visible in cross sections. Left, a slice cut through a long bone shows the difference between compact bone (white layer at the top) and the spongy cancellous bone. More details are revealed by examining thin slices under a microscope. Middle, the concentric whorls of lacunae (small dark features) are characteristic of Haversian bone. Right, at higher magnification the lacunae can be seen more clearly.

fluid. The whole structure is referred to as a single osteon. This distinctive pattern of organization is referred to as Haversian bone and is found in mammals, birds, dinosaurs, and some other vertebrates. Since the most readily available bone specimens for microscopic examination are mammalian, we tend to get the idea that Haversian bone is common. But if we looked at a wide range of vertebrates we would find that Haversian bone was the exception rather than the rule—it is not found in fishes (the largest vertebrate group) nor in amphibians or most reptiles.

Increasing the magnification of the microscope would reveal a ring of small spaces within each bone layer or lamella. These spaces, called lacunae, house the living bone cells. Bone commences its embryonic development as a loose cluster of cells called osteoblasts ("bone buds"). These begin manufacturing the nonliving bone matrix—the collagen fibers and apatite crystals—and eventually become completely surrounded by it. At this point their work is complete and the osteoblasts mature into osteocytes ("bone cells"). Osteocytes are smaller and flatter than osteoblasts and have a spidery appearance because of the fine threads of cytoplasm that project from the cell bodies. These threads lie in minute tubes, called canaliculi ("small canals"), that penetrate the bone matrix. Since there is some continuity among the canaliculi of adjacent lacunae, and since the lacunae in the innermost lamella are close enough to the central canal to exchange fluids with the blood vessel inside it, the osteocytes are able to obtain nourishment from the blood supply. The role of the osteocytes is somewhat unclear, but they appear to be involved in the exchange of materials between the tissue fluid (in the central canal) and the bone matrix, and they also have some limited capacity for bone manufacture. The life span of osteocytes may extend over many years, but it seems that when they die the bone around them undergoes breakdown. There is a constant turnover in bone tissue, old osteons break down and new ones replace them. We can actually detect areas in which this has happened because the replacement, or secondary, osteons are surrounded by a layer of cement; this shows up in a microscope under polarized light[8] as a clear zone.

The organic portion of the bone matrix, comprising mostly collagen, makes up about 20–25 percent of the dry weight of fresh bone. Collagen, being a protein, is made up of amino acids, and there are some twenty different amino acids. Each particular protein, whether it be a protein in muscles, such as myosin, or a protein in blood, such as albumin, has its own characteristic makeup of amino acids. It is therefore possible to identify a protein by determining what amino acids are present and in what proportions. Collagen, a common protein, is found not only in bones but also in skin, ligaments (the tough, white sinews that connect bones together), and tendons (like ligaments, but connecting muscles to bones). Many of us have actually extracted collagen from various animal parts, probably without even realizing it. When bones are boiled up to make soup the liquid gets thicker because of the extraction of collagen from the bones. When the liquid cools it sets like jelly—indeed it is jelly because the extract is called gelatin and gelatin is the main ingredient of jelly. Gelatin can also be obtained from tendons, as when a stew is made from tough meat. Provided the meat is boiled for long enough, all those white gristly pieces—the tendons which make the meat so tough—are reduced to soft, brown, and rather tasty gelatin.

The amino acid composition of bone collagen does not differ very much from the makeup of collagen in skin or tendon. Nor is there very much difference from one species to another. Not all of the protein in bone is collagen, however, and when all the collagen has been removed the remaining protein has a different amino acid composition from that of collagen. The most significant difference between them is that collagen contains the amino acid hydroxyproline, whereas the noncollagen proteins do not. Hydroxyproline therefore acts as a useful indicator of collagen in fossil bone. Protein molecules, in contrast to inorganic molecules like hydroxyapatite, are not stable over long periods of time and break down into their component amino acids. It is this breakdown of collagen which causes bones to become brittle when they are exposed to weathering, a subject that is dealt with in the next chapter.

Before leaving the subject of bone I must emphasize that it is a living tissue. We have already seen that there is a continual turnover of osteocytes and the bony lamellae in which they lie; furthermore, the rate of bone replacement can be speeded up in response to increased stress. Therefore, if there were a marked increase in the stresses acting on an animal's skeleton, the bones would respond by becoming thicker in the regions of the increase. If an office worker, for example, were to take up manual work, those bones subjected to increased stress would become thickened in the appropriate places. Conversely, a reduction in stress would result in bone reduction. The response to changing stress takes days rather than weeks and can actually be measured. In humans the clavicle (collarbone) lies close to the surface, and an approximate measure of its diameter can be obtained using callipers. One of my students sportingly let me measure his clavicle both before and after he took up weight training. Over the period of the Christmas vacation his clavicle increased in diameter by three-sixteenths of an inch (4.6 mm), an increase of about 25 percent.

Cartilage, ligaments, and tendons

Cartilage, like collagen, is found in many different parts of the body, often in association with collagen, as in the intervertebral discs (see Chapter 3). In its purest form cartilage is a smooth, hard material which looks and feels like white plastic. The articular surfaces of all bones are capped with a thin layer of cartilage, which provides for a smooth, low-friction joint surface—take a look at the ends of the long bones next time you eat chicken. The human external ear (the part called the pinna) is cartilaginous, so too is the tip of our nose. If we apply a small stress to either one of these structures, we notice that they are readily strained (bent out of shape). But when the stress is removed they return to their original shape. Cartilage, then, is an elastic material and is not very stiff, that is, it has a low Young's modulus. Nor is it very tough, as many boxers have discovered, and it has relatively low tensile and compressive strengths.

Ligaments and tendons are resilient—that is, they have a low Young's modulus (but not as low as that of a rubber band)—and they are tough, too. Some tendons are able to stretch more than some ligaments because they have a rubbery protein called, appropriately enough, *elastin*. Ligaments and tendons have a high tensile but a low compressive strength and have a low Young's modulus (relative to bone). Now elastic materials that have a low modulus have a particularly important property that we have not yet considered: an ability to store and absorb large amounts of energy. This property is encountered in everyday life, as we shall see below, and aids locomotion in many animal species, as we shall see in Chapter 3.

Strain energy

Slingshots (catapults) were very popular in my youth, in spite of attempts by schoolteachers and parents alike to banish them from society. They were very easy to make. All that was needed was a Y-shaped piece of wood—usually fashioned from a tree branch—a length of rubber (the thicker the better), and some string to bind the two ends of the rubber to the two prongs of the handle. In the more sophisticated models the rubber was threaded through a scrap of leather, which served to hold the projectile prior to launching, but most of us managed without this elaboration. And it was surprising how much damage could be done with such a simple device, windows being the prime targets. Slingshots are dangerous though. I recall being hit on the temple by a stone fired from a considerable distance, and the impact sent me to the ground.

The reason slingshots are so dangerous is that rubber has such a low Young's modulus. When the slingshot is pulled back prior to firing, energy, strain energy, is stored in the rubber; this energy is transferred to the stone when it is released. The amount of strain energy stored in the rubber is inversely proportional to its Young's modulus,[9] which is why we used elastic in our slingshots rather than piano wire. The same principle is used in the longbow, but here the strain energy is stored in the wood of the flexed bow rather than in the string. Because Young's modulus of wood is much higher than that of rubber, it is much harder to pull back a longbow than a pocket slingshot. Relatively more strain energy is stored, though, which explains why an arrow from a longbow can do so much damage, as demonstrated by English bowmen at the Battle of Agincourt.[10] In these examples the useful end product is the storage and release of energy, but there are other applications where it is the absorption of the energy that is the desired end result, as the following examples show.

Museums often lend specimens to other institutions. A few years ago the Royal Ontario Museum sent a particularly fine skull of an ankylosaurian dinosaur to another paleontologist for study. At that time we were experimenting with a plastic foam for making casts of bones, and somebody had the brilliant idea of using foam to pack the skull. A wooden box was duly built and the skull snugly foamed into place inside it—like an almond in the middle of a chocolate. This particular foam is stiff when it sets, that is, it has a high Young's modulus. This makes it ideal as a casting material, but very poor as a packing material because it absorbs very little energy. If we had used a material with a low Young's modulus, like foam rubber, the energy applied to the outside of the box during transit would have been absorbed by the packing material rather than being transmitted to the skull. Needless to say the skull arrived in many pieces; but it is all together again now.

The earliest cars had solid rubber tires, and since the modulus of rubber is considerably less than that of steel they provided a much smoother ride than did the steel-rimmed wheels used in horse-drawn carriages. But the modulus of compressed air is only a fraction that of rubber, which is why pneumatic tires soon replaced solid tires.

Teeth

Teeth are chemically very similar to bone, but they are quite distinct structurally, so there is usually never any difficulty in distinguishing between them microscopically. Tooth material comprises two separate components, dentine, which forms the bulk of the tooth, and enamel, which forms a hard protective layer on the outside. Both components are formed of the mineral apatite, together with proteins, though there is less protein in enamel than in dentine. There is a central pulp cavity containing blood vessels and nerve fibers, fine branches of which pass through the dentine. Nutrients are able to permeate through the dentine from these finer blood vessels because the dentine is perforated by a dense network of very fine tubules. These tubules run parallel with one another and are so dense that they give the dentine a characteristic fibrous appearance under high magnification. The enamel, in contrast, is not porous and is essentially featureless under microscopic examination.

Dentine is probably stronger in compression than bone, but it may be weaker in tension. Enamel is probably stronger still in compression, but weaker in tension, and may not be as tough as bone or dentine—it is certainly more brittle. But it is a very hard material and is able to withstand considerable amounts of wear, as we will see in a later chapter. Indeed, it is the hardest of all vertebrate materials.

Primary Data

E VERYTHING we know about dinosaurs, pterosaurs, ichthyosaurs—and all those other denizens of the remote past—has been gleaned from studying their fossils and the geological environment in which they were found. This may sound like the unnecessary stating of the obvious, but it is important to remind ourselves of the fact because our interpretations of the past are only as good as the data upon which they are based. We must therefore always be aware of any shortcomings in the data. Can the study of trace elements in fossil bone, for example, really tell us anything about the diets of extinct animals? Does the microscopic structure of dinosaur bone offer any clues to thermal physiology? Does the relative proportion of carnivores to herbivores in a fossil sample give a true representation of the actual proportion of meat eaters to plant eaters in that particular environment at that particular time? Some of the chapters that follow will be concerned with examining certain of these questions in depth. The purpose of the present chapter is to give an overview of fossil data—warts and all! We will begin by looking at the process of preservation to see how, and under what circumstances, fossils are formed. The initial view will be at the macroscopic level to see what happens to a body after death. As we proceed we will see some of the biases that creep into the fossil record. We will then take a microscopic view and ask what chemical and structural changes take place in the bones themselves.

The study of the transformation of the remains of animals from the living world to the fossil record—from biosphere to lithosphere—is called taphonomy. This term, literally the "laws of burial," was coined in 1940 by the Russian paleontologist I. A. Efremov. Although he published a series of important papers on this new branch of paleontology, the subject was largely over-looked until a great resurgence of interest in the late 1960s and 1970s. This renewal is largely attributable to the painstakingly detailed field studies of researchers like Anna Kay Behrensmeyer and others on contemporary mammalian communities in Africa. These neontological studies (studies of living as opposed to extinct organisms) revealed principles that could be applied to the inter-pretation of animal communities of the distant past (Behrens-meyer and Hill, 1980; Dodson, 1980).

Most fossils are the remains of the hard parts of animals and plants; bones, teeth, shells, wood, seeds, spores, and pollen grains being among the most common. But soft parts are sometimes also preserved, usually as impressions in the rock or as natural casts. We have fossils of soft-bodied animals such as worms and jellyfish, natural casts of brains, skin impressions, feather impressions, foot-prints, and more besides. Most of our attention will be focused on fossil bone, though.

Extracting data from fossils is much like gathering evidence in a murder investigation; in both instances the longer the interval that has elapsed since the time of death, generally the less infor-mation is preserved. Thus we may be lucky enough to find natural skin impressions of a 75-million-year-old dinosaur, or a natural brain cast of a 140-million-year-old bird, but these stony replicas are all that remain of the original organs; the organic tissues that once formed them have long since decomposed and disappeared. If we hope to find remnants of the actual organs themselves, we have to look at considerably younger fossils—ones that are thou-sands rather than millions of years old. Mammoths have been found in which remnants of hair, skin, and even muscles have been found (Augusta, 1962) and similar preservation is known for some other fairly recent fossils, including fossils of ground sloths (Woodward, 1899; Sutcliffe, 1985) and moas, large New Zealand birds, now extinct, related to emus (Oliver, 1955). The preservation of soft parts is sometimes quite remarkable. Some human fossils were recently found in a bog in Florida that were so well preserved that the brain, which is a particularly soft organ, retained its normal appearance. These human fossils were only about 8,000 years old and it was possible to see microscopic details of the cells and even to extract DNA (Doran et al., 1986). Whether to describe such specimens as fossils or as subfossils is debatable,

A natural cast of the skin of a hadrosaurian dinosaur, preserved in sandstone. The skin had a pebbled struc-ture, clearly seen in the close-up, and was devoid of scales.

but we will follow the convention of using the term *fossil* in its original sense, from the Latin *fossilus,* for anything that is "dug up." Although these relatively recent fossils fall beyond the purview of this book, some of the techniques involved in their analysis are of sufficient interest to warrant their inclusion here; we will return to them later in this chapter.

A study in black

When an animal dies, its carcass soon begins to break up. The process is accelerated by the action of scavengers, both large and small, by the putrefying action of bacteria, and by weathering. The rate of breakdown is determined by a number of factors, including temperature, moisture, the type of terrain and the types of scavengers. Decomposition occurs much more rapidly in warm, moist conditions than in cold, dry ones. Antarctic explorers sometimes come across the desiccated remains of penguins that may have been dead for hundreds of years. Just recently the bodies of three of the victims of the ill-fated Franklin Expedition to the Arctic were discovered, very well preserved since their burial in 1846 (Beattie and Geiger, 1987). The tropics, in marked contrast, destroy their dead rapidly. During a five-year study of orangutans in the tropical rain forests of central Borneo, field-workers found the remains of dead orangutans only twice (Galdikas, 1978). On the first occasion an old male had been found probably less than twelve hours after its death; wild pigs had already devoured the contents of the body cavity, removed many of the ribs, and nearly finished stripping the shoulders and hips of flesh. While the remains were being examined, one pig returned and began dragging the carcass around with such vigor that bones were wrenched from their sockets and removed. On the second occasion only a fragment of a skull had been discovered and a thorough search of the surrounding area failed to reveal even a scrap of bone. Although large scavengers like the wild pigs can destroy a carcass within days, they are not a necessary part of the process of decay— a body can be completely destroyed by the combined action of bacteria, insects, and the elements.

Droughts are not uncommon in Africa, and the high mortality they wreak upon wild and domestic animal populations provides a unique opportunity to study taphonomic processes on a large scale. During the Kenyan drought of 1970–71, observations were made on the decomposition of some one hundred elephant carcasses in Tsavo National Park (Coe, 1980). Under conditions like these, where there is an excess of carrion, carcasses may escape the attention of large scavengers altogether, but it is still surprising how rapidly they are broken down. It was found that large carcasses would lose all of their soft parts (the internal organs and muscles) within two weeks by the action of bacteria and invertebrates. All that remained then was the skeleton encased in the hide. Within another three weeks the hide and all the ligaments had been eaten away by beetles (of the genus *Dermestes*), at the remarkable rate of eighteen pounds (8 kg) per day, leaving the skeleton exposed to the weathering action of the elements. (Dermestid beetles are so good at this job, in fact, that they are put to work in museums throughout the world to strip the last remnants of soft tissue from bones during the preparation of skeletons.) Under these conditions the bones became very brittle and began cracking and flaking within five weeks. Their breakup was greatly speeded up by the trampling of other animals. Within a year the skeletons were disarticulated and scattered, many of the bones having been broken or lost. Carcasses tracked over a five-year period were found to be scattered over distances of up to 160 feet (50 m) from the site of death. Elephants themselves have been observed carrying tusks and bones for considerable distances.

The main causes of weathering are the daily fluctuations in temperature, from the cool of night to the destructive heat of the sun by day; alternations between wetting and drying also play a part (Behrensmeyer, 1978). The rate of weathering is more rapid for the bones of smaller species (less than 100 pounds or 45 kg body weight) than larger ones, and it is also faster in bones of juveniles than in bones of adults. As a result, the chances of survival are biased against both small species and juveniles. There is also evidence that the skeletons of some species are affected by weathering somewhat differently from those of other species. A study in Uganda showed that carnivore skeletons appeared to remain articulated relatively longer than those of bovids (antelopes, buffalo, cattle, and their relatives; Hill, 1980). The vertebral column of a lion, for example, remained fully articulated, perhaps because active carnivores have stronger ligaments binding their vertebrae together than do herbivores.

The general habitat—grasslands, for example, compared with bush—seems to play little part in the long-term fate of bones, but micro-environments appear to have a significant influence (Behrensmeyer, 1978). Thus bones that have been overgrown by plants and that are partially protected from the harsher surrounding environment often show less evidence of weathering than nearby bones fully exposed to the elements. The lower surfaces of bones are similarly afforded some protection by being subjected to less extreme changes in temperature and moisture than the exposed surfaces. The investigators of the Franklin Expedition to the Arctic, for example, found fragments of a human skull that had been weathering on the surface of the soil for over one hundred and thirty years (Beattie and Geiger, 1987). The upper surface was white and powdery and easily broken by handling, whereas the brown underside was in good condition. Burial, of course, offers more complete protection than do micro-environmental effects. This is well illustrated by examples of partially buried bones; the buried parts frequently show no signs of weathering even though the exposed parts are in the later stages of disintegration.

So far we have seen that when animals die their bodies are quickly broken down, especially if they are scavenged by large animals. Small skeletons, and those of juveniles, are broken down the most rapidly, and the rate and sequence of disarticulation of a skeleton appears to differ from one species to another. The process occurs more rapidly in hot, wet climates than in cold, dry ones. Under tropical conditions a carcass can disappear within a few days, but in the absence of scavengers it may take ten or more years before the last traces of bone are lost to the weathering action of the elements. Destruction is almost inevitable, and the only way of preventing this seems to be by burial.

What are the chances of a carcass becoming buried? The chances vary considerably from one environment to another. In East Africa, for example, field studies have shown that although burial is uncommon, it does happen from time to time. A survey of twenty thousand bones, found weathering out on the surface in Kenya, showed that some 4 percent of them were more than half buried in the soil (Behrensmeyer and Boaz, 1980). But preservation requires complete burial, and most of the partially buried bones are likely to be destroyed before this can happen. Rapid burial, then, is an obvious prerequisite for preservation, but this is a relatively rare event, especially on land. Heavy downpours of rain can bury bones rapidly, so too can floods, dust storms, and trampling by other animals. Animals dying in swamps, lakes, rivers, and by the sea have a chance of being rapidly buried in mud and silt. And those animals that have the misfortune to fall into sinkholes and fissures in the ground stand a good chance of being preserved.

Given all these chancy events in the preservation of animal and plant remains, how are we to interpret the fossil record that does survive? If small skeletons are more readily destroyed than large ones, for example, their chances of survival are smaller. This means that those animals whose bony remains do survive are not likely to be truly representative of the species that actually coexisted. How serious is the discrepancy likely to be? In order to assess this question, Behrensmeyer and Boaz (1980) compared the numbers of species living in a given area of Kenya with the number of species represented by bones lying on the surface. Large carnivores (larger than 15 lb or 7 kg body weight) were found to be completely represented by their bones, with a slightly less (95 percent) complete representation for herbivores of similar size. But when it came to smaller animals (1–15 lb or 0.5–7 kg), only 60 percent of the herbivores and 21 percent of the carnivores were represented among the bones. Similar comparisons were made for bones that had been partially buried and these were found to be even less representative of actual species composition than were the surface bones.

What does all this mean? Although these field studies pertain only to mammals, and only to mammals living in one locality in East Africa, they do clearly demonstrate that the bony remains of animals that are found in a particular environment are not truly representative of the animals that actually live there. Since these bones are potential fossils in the making, these studies have important implications for the fossil record. More studies need to be made, and not only for different environments but also for other kinds of animals as well as mammals. But the implications are clear: the fossil record is not only a fractional sample of the organisms that once lived on the earth, it is also a very biased one. It would therefore appear imprudent to draw any firm conclusions on the relative abundances of species in ancient environments on the basis of their relative abundances in the fossil record, especially if the species concerned are animals of small size.

Another problem of interpretation is the question of contemporaneity. Does the fact that a certain assemblage of bones occur together in a certain area necessarily mean that the species represented were living together there at the same time? Most of the herbivores represented in the bone samples discussed above, including zebra and wildebeest, are seasonal migrants, and only a few species (impala, buffalo, and rhinoceros) spend all their lives in the area. The fact that the bones of a zebra and an impala are found together clearly does not establish that the two animals lived together in the same habitat at the same time. Their paths might have crossed, as when animals come together at water holes, but they need not have shared the same habitat. Another point to remember is that rivers can carry carcasses for many miles. Furthermore, bones that have been buried, perhaps for hundreds of years, can be washed out of river banks and mixed up with more recently deposited bones.

So far we have been considering only the vertebrates, primarily the mammals. But the vertebrates constitute only about 3 percent of the total number of living animal species. The largest group, by far, are the insects, accounting for about 76 percent of the total. And the second-largest group is the molluscs, with more than twice as many species as the vertebrates. This group includes snails and clams and oysters and squids, and other good things to eat. Most molluscs have strong shells, which are resistant to decay; most live in the sea or in freshwater—environments that offer the best chances of burial after death; and many spend their adult lives buried in sediments. The chances of a mollusc being preserved after death are therefore considerably higher than those of most other animals. Anyone who has ever gone digging for clams will appreciate this—some beaches are so chock-full of dead clams that there seem to be more shells than mud. It is therefore not surprising that mollusc shells are among the commonest of all fossils, nor that the fossil record is disproportionately biased in their favor. In contrast, there is only a scant fossil record for birds. Because of the lightness and frequent small size of their bones, bird skeletons are fragile and readily broken up after death, greatly reducing the chances of preservation. Birds are therefore underrepresented in the fossil record.

Aside from its being biased in the species represented, the fossil record is also biased in the nature of its data—it is primarily a record of skeletons. Sometimes we are fortunate enough to catch glimpses of the soft anatomy, as when skin impressions and the like are preserved, but, for the most part, all we have to work with are the skeletal remains of animals that once were. We now turn to the nature of fossils themselves, primarily to fossil bones, especially at the microscopic level.

The nature of fossils

Anyone who has handled a fossil knows that it is usually not exactly like its modern counterpart, the most obvious difference being that it is often much heavier. Fossils often have the quality of stone rather than of organic material, and this has led to the use of the term *petrifaction* ("to bring about rock"). The implication is that bone, and other tissues, have somehow been turned into stone, and this is certainly the explanation given in some texts. But it is a wrong interpretation; fossils are frequently so dense mainly because the pores and other spaces in the bone have become filled with minerals taken up from the surrounding sediments. Some fossil bones have all the interstitial spaces filled with foreign minerals, including the marrow cavity if there is one, while others have taken up but little from their surroundings. Probably all of the minerals deposited within the bone have been recrystallized from solution by the action of water percolating through them. The degree of mineralization appears to be determined by the nature of the environment in which the bone was deposited and not by the antiquity of the bone. For example, the black fossil bones that are so common in many parts of Florida are heavily mineralized, but they are only about 20,000 years old, whereas many of the dinosaur bones from western Canada, which are about 75 million years old, are only partially filled in. Under optimum conditions the process of mineralization probably takes thousands rather than millions of years, perhaps considerably less.

The amount of change that has occurred in fossil bone, even in bone as old as that of dinosaurs, is often remarkably small. We are therefore usually able to see the microscopic structure of the bone, including such fine details as the lacunae where the living bone cells once resided. The natural bone mineral, the hydroxyapatite, is virtually unaltered too—it has the same crystal structure as that of modern bone. Although nothing remains of the original collagen, some of its component amino acids are usually still detectable, together with amino acids of the noncollagen proteins of bone.

There is a great deal of variation in the amino acid composition of fossil bones. This is partly due to the environmental conditions to which the fossils have been exposed, and also to their age. In general, younger fossils have relatively more amino acids per gram than older ones, provided they have been exposed to similar

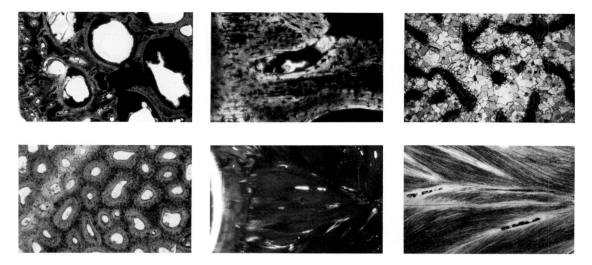

The microscopic structure of fossils. Top row, left to right: *Transverse section through the vertebra of a hadrosaurian dinosaur (Upper Cretaceous); the black areas are pores that have been filled with minerals. At higher magnification small dark features can be seen—the lacunae, which once housed the living bone cells (osteocytes). Often, all pores in fossil bone have been filled with mineral, as in the third section, from the vertebra of the ornithopod dinosaur* CAMPTOSAURUS *(Upper Jurassic); in this section, photographed under polarized light, the mineral crystals (calcite) appear to be arranged in a pattern.* Bottom row, left to right: *Transverse section through the vertebra of an ichthyosaur (Lower Jurassic); notice that the pores are almost devoid of minerals. The next two photographs show transverse sections through the tooth of a hadrosaur. The white layer on the left of the first section is enamel, beneath it is the dentine, and at the far right side is the pulp cavity. At higher magnification blood vessels can be seen radiating through the dentine from the pulp cavity. Notice the dark granules in the blood vessels.*

environmental conditions. The relative amounts of the different amino acids also tend to vary with age, again given similar environmental conditions. It has been found that if the amino acid content of a fossil bone is not less than about 10 percent that of fresh bone (this is only likely to pertain in relatively young fossils), the relative amounts of the different amino acids are similar to those of fresh bone (Hare, 1980).

Modern analytical techniques make it possible to measure the amounts of amino acids and of various elements in fossils with great accuracy. The quantities being analyzed are usually small, so great care has to be taken to prevent contamination. Bone samples for amino acid analysis have to be carefully inspected for molds, lichens, bacterial colonies, and any other foreign proteins. Relatively young, well-preserved fossils are preferred because they are likely to contain relatively larger quantities of amino acids. The remarkably well preserved Pleistocene bones of Rancho La Brea, California, are especially suitable because they have been essentially sealed off from the environment by the oily tar that has impregnated them. Many thousands of years ago this area, now within the city limits of Los Angeles, was a tar seep, and it seems that animals became mired in the soft tar when they came to drink the water that pooled there. Hundreds of specimens have been collected, including saber-toothed cats, bisons, camels, horses, bears, dire wolves, and small mammals like gophers. Many of the bones have amino acid compositions like those of fresh material. The bones of one bison, for example, contained 227.4 milligrams of amino acids per gram, which is similar to the amounts found in fresh bison bone, whereas one of the gophers contained only 6.8 milligrams per gram (Wykoff, 1972), which is probably not more than 5 percent of the original content. We can surmise that the gopher bones had probably been exposed to the elements for a longer time than those of the bison, and that most of the amino acid molecules had been leached away. Even so, the proportions of the various amino acids in the gopher bones were similar to those of the bison. Some of the fossil bones were found to have amino acid compositions somewhat different from that of collagen, but this was attributed to bacterial contamination at the time of death. Although these atypical amino acid compositions may very well have originated from bacteria, it is equally plausible that they originated from the breakdown of the noncollagenous proteins in the fossil bone (after the collagen had decomposed and disappeared) because these proteins appear to be more resistant to decay than collagen (Hare, 1974).

When fossil bone is analyzed for elements, predictably large quantities of calcium and phosphorus are found because hydroxyapatite is a crystalline form of calcium phosphate. Various other elements are usually also found, such as barium, fluorine, iron, manganese, strontium, and yttrium. Some fossil bones also contain radioactive elements, including uranium, sometimes in alarmingly large quantities. Some of these elements may have been naturally present in the skeleton at the time of death, while others may have passed into the bone from the surrounding sediments over the years. Because of leaching between the bone and the sediments, it is virtually impossible to be sure of the origin of any particular element. A possible solution to the problem has been offered by Professor Heinrich Toots and his colleagues at the University of Wyoming, who have been comparing the distribution of certain elements in bone, dentine, and enamel (Toots and Voorhies, 1965; Parker and Toots, 1980).

Dentine, like bone, is porous, though to a considerably lesser extent because the tubules that penetrate the matrix are so very fine. Dentine is therefore less permeable to the passage of materials in solution than is bone. Enamel is not porous and is therefore essentially impermeable. If a given element originates in the sediments, rather than in the skeleton, it should be found in increasingly smaller amounts in these three tissues because of their decreasing permeabilities. The Wyoming researchers found that the levels of flourine in fossils decreased steadily from bone to dentine to enamel. Strontium, on the other hand, was found in similar concentrations in all three tissues. They reasoned that the fluorine had originated in the sediments and leached into the skeleton, whereas the strontium had been present in the skeleton at the time of death. They finally concluded that strontium levels always remain virtually unchanged since the time of death for all vertebrate fossils.

Because strontium found in fossils can be assumed to have been present in the living tissue, and not leached into the fossil from the ground, strontium levels have been used to deduce the diets of fossil vertebrates. Strontium is a metal belonging to the same chemical family as calcium. It occurs naturally in the environment, often together with calcium but in relatively lower concentrations. Plants absorb strontium from the soil and when herbivores eat the plants some of the element is absorbed into their bodies. The strontium passes into all tissues but tends to accumulate in the bones and teeth, along with calcium. Because living organisms have some ability to select against strontium in favor of calcium they act like a filter, and the herbivores probably contain lower levels of strontium than the plants they eat. Carnivores feed on the flesh of herbivores and therefore consume relatively smaller amounts of strontium than did their prey, so relatively lower strontium levels accumulate in their bones. The strontium-calcium ratios are accordingly lower in the bones of carnivores than in herbivores. Measuring this ratio in fossil bones should make it possible to determine whether the fossil in question was a herbivore or a carnivore.

Different soils contain different amounts of strontium, so the plants from one locality may have strontium levels quite different from those found elsewhere. This difference is reflected in the strontium-calcium ratios in the bones of the animals grazing on the plants. The bones of carnivores in high-strontium areas could very well have higher strontium-calcium ratios than those of herbivores living in low-strontium localities. The Wyoming group were well aware of this problem and went to great lengths to emphasize the need to obtain fossil bone samples for animals that were exposed to similar background levels of strontium during their lifetimes. They believed these conditions were met if samples were restricted to the same quarries. However, the fact that two specimens were collected from same quarry does not necessarily mean that they were living contemporaneously—this would be true only if they were taken at exactly the same horizontal level in the quarry. And even if they were living at the same time the fact that they were found together at the same burial site does not necessarily mean that they were living in the same environment, as we have seen. Great caution should therefore be exercised in attempting to use strontium-calcium ratios to cast light on diets of extinct animals. More promising indicators of diet are provided by assessing the concentrations of stable isotopes of carbon and nitrogen in bone collagen, but this methodology is available only for relatively recent fossils, ones that are thousands rather than millions of years old (DeNiro and Weiner, 1988).

A very useful tool for analyzing elemental composition of fossils is the electron probe, which is a modified electron microscope. Samples of bone or rock are cut and polished and placed inside the instrument, where they are bombarded by a fine beam of electrons. When the electrons strike an atom they cause it to become excited, that is, to radiate energy. Each element emits radiation in a characteristic pattern of wavelengths. By measuring the quality and quantity of the emitted radiation we can identify each element and assess its relative abundance. A magnified image of the specimen is displayed on a screen, together with the location of the electron beam, so it is a simple matter to direct the beam to any minute spot and obtain a reading for the concentration of any particular element at that precise point. It is therefore possible to determine whether an element of particular interest resides within the bone matrix itself or in the material filling in the pore spaces. Some years ago I became interested in trying to determine the origin of the iron found in such high concentrations in the hadrosaur skeletons we have at the Royal Ontario Museum (McGowan, 1983). These dinosaurs, all from the Upper Cretaceous of Alberta, were found in a hard sandstone, a sedimentary rock that is frequently streaked with a brown discoloration. Analyses of these discolored areas reveal high levels of iron. The obvious deduction was that the iron in the bones originated from the rock, and similar conclusions have been drawn for other fossil bones (Houston, Toots and Kelley, 1966). When samples of discolored sandstone taken from the vicinity of hadrosaur bones were analyzed, however, it was found that the levels of iron in sandstone near the bones were higher than levels of iron in sandstone some distance away. This suggests that the iron may have originated in the dinosaur rather than in the sediments. Blood is rich in iron, and there is a lot of blood in muscles, but is it conceivable that iron may have leached out from the blood and muscles of a decomposing dinosaur to become deposited in the sandy sediments that buried it? This does not seem very likely, especially considering the porosity of sand, and I would expect that iron, and any other elements that were free to migrate, would have leached out into the surrounding sediments and been lost fairly rapidly. But would this also have applied to the iron contained in the blood within the vascular spaces of the bones and teeth? Electron probe analyses showed that the high iron levels in the bones of the hadrosaurs were concentrated in the dark in-filling material that occupied the vascular spaces. A tooth was sectioned and polished and a similar iron-rich in-filling material was found, in clumps, inside the pulp cavity. Some of the iron-rich material was even found inside the minute blood vessels that radiate through the dentine from the pulp cavity. Did this iron come from the blood of the dinosaur, or did it leach in from the sediments that entombed it? We will probably never know the answer, and I suspect that elemental analyses of such ancient bones will probably never yield reliable information. But the same is not true for the pathological data that sometimes lurk in bones.

Injuries and disease

Abnormalities in fossil bones and teeth are common. The first major work on the subject appeared in 1923, opening up a new branch of paleontology called paleopathology (Moodie, 1923). Interest, naturally enough, has tended to focus on human disease and suffering, and much attention has been given to the autopsies that have been conducted on Egyptian mummies. These studies have revealed all manner of diseases and dietary deficiencies and even parasites that had invaded some of the tissues. During a recent visit to the Dakhleh Oasis in Egypt I had the opportunity to examine some skeletons that were being excavated from the Roman Period. I was surprised at the incidence of arthritic joints, some so chronic that articular surfaces had been obliterated. The resulting wear of bone against bone had caused grooving.

Healed bone fractures, not uncommon in dinosaur skeletons, are revealed by bony swellings, called calluses, at the site of the injury. When a bone is broken the fractured ends are cemented together by a fairly coarse network of bone, the callus serving as a temporary splint. As the fracture heals the callus is reduced in size, but, as anyone who has had a broken bone knows, it takes many years before it disappears completely. Some indication of the amount of healing that has occurred since the time of injury can therefore be obtained from the size of the callus. If the broken ends of the injured bone are not displaced during repair, the bone heals neatly. When this is not the case, repaired bones may sometimes have extensive calluses. A skeleton of the carnivorous dinosaur *Albertosaurus* in the Royal Ontario Museum has such a badly healed fracture of the right fibula. The fibula is the slender bone that runs along the outer edge of the tibia (shin bone). The injured leg appears to have healed and been fully functional, but the large size of the callus suggests that perhaps not many years had elapsed between injury and death.

Sometimes the injuries are so severe that fractured bones puncture the skin, leaving them exposed to invasion by disease-causing agents. Infections contracted in this way may spread throughout the skeleton. There is a horned dinosaur on display at the Royal Ontario Museum that appears to have suffered from such a chronic bone disease. Viewed from the right side the skeleton looks normal, but from the other side one of the ribs appears to be deformed. This rib is about half the length of the others, and it ends in an expanded and roughened bony outgrowth. A close examination of other parts of the skeleton reveals other abnormalities, suggesting something more than a simple injury. Many of the toe bones have been eroded and have various bony outgrowths. Perhaps the rib, which never healed properly, broke the surface of the skin and became infected. The disease may have contributed to the animal's death.

Sometimes dinosaur tails are found in which two or more vertebrae have become joined together. If union involves the fusion or erosion of the articular surfaces as well as outgrowths

Skeletons often provide evidence of bone damage and disease. Top row: *Sometime in life this hadrosaur received a blow to the left side that broke six ribs; note the bony calluses that formed at the sites of repair.* Middle row: *Viewed from the right this horned dinosaur* CHASMOSAURUS *looks normal. From the left side, however, a deformed rib can be seen.* Bottom row: *Close inspection reveals a bony outgrowth and some abnormality in the adjacent rib* (left). *Bony outgrowths from other parts of the skeleton, including some toes* (right), *are indicative of a chronic bone disease.*

bridging across the vertebrae, it was probably a result of injury (Rothschild, 1987). When the only connection is by bony bridges, however, the union is probably not pathological but a response to local high stresses in the skeleton. This condition, common in our own species in later life, is known as DISH (diffuse idiopathic skeletal hyperostosis) and will be referred to again in Chapter 5.

Soft anatomy

The most frequently found evidence of soft anatomy is provided by impressions embedded in the sediments, like footprints. Dinosaur trackways, which have been discovered in many parts of the world, offer us a salutary link with the past. Just to place one's feet in a trackway left all those millions of years ago, to measure one's stride against the beast that once passed that way, is a most thought-provoking experience. But fossil trackways hold more than just fascination, they contain vital clues to posture and locomotion in dinosaurs, as we shall see in later chapters.

Far less common than trackways are impressions of skin made when dinosaurs laid down on the ground. The weight of their bodies pressing into the ground left a natural imprint of their skin. These impressions subsequently filled in with sediments, which eventually consolidated into a stony natural cast of the skin. Skin impressions are often found in isolation, but sometimes they are found in association with skeletal remains, allowing them to be identified. Such positive identifications have been made for hadrosaurs, ceratopsians, and tyrannosaurs; they appear to have lacked scales and to have had a pebbled skin much like that of an elephant.

Skin impressions are not confined to dinosaurs. The most finely preserved impressions have been found in marine sediments, most notably in the Lower Jurassic shales of Holzmaden in southern Germany. These deposits are especially noted for their ichthyosaurs, a number of which have their body outline preserved as a carbonaceous film. Had it not been for the discovery of such perfectly preserved individuals, we would not know that ichthyosaurs had a dorsal fin, nor would we know anything about the shape of the tail flukes. Even more remarkable is the preservation of the details of feathers in *Archaeopteryx* ("ancient wing"), the earliest bird, from the Upper Jurassic (155 million years before the present) of Solnhofen, also in southern Germany. These fossils are preserved in a fine-grained limestone that was laid down in a tropical lagoon close to land. The remarkable state of preservation has been attributed partly to the absence of scavengers and to low oxygen levels in the fine sediments that settled on the floor of the lagoon (Buisonjé, 1985).

Natural casts of internal organs can provide us with some valuable insights into the biology of extinct animals. Casts of the brain have probably been the most instructive of these. The brain, a soft, fatty organ, appears to decompose more rapidly after death than most other structures. Some evidence for this has been provided by autopsies performed on bodies that have remained frozen since the time of death. For example, a forty-thousand-year-old baby mammoth that was found in the permafrost of Siberia in 1977 was in such good condition that blood vessels in the muscles still contained red and white blood cells (Stewart, 1979), but the brain was nothing more than a structureless mass. Similar findings were made during the recent autopsies of some of the victims of the Franklin Expedition to the Canadian Arctic (Beatie and Geiger, 1987). The bodies, frozen in the permafrost for almost one-and-a-half centuries, were generally in a good state of preservation, except for their brains. It is reported that one of the brains, shrunken to about two-thirds of its normal size, had autolyzed (meaning it had been digested by enzymes released from its own cells) and that a second had turned to liquid.

Once a brain has decomposed, and possibly exuded from the foramen magnum (the large hole at the base of the skull), the bony cranium that once housed it may become filled with sediments. If these particles of sediment then consolidate a natural cast is formed, called an endocast. The material forming the endocast is often more resistant to weathering than the bone surrounding it, so it sometimes happens that the skull is destroyed and all that remains is a replica of the brain. This appears to have been the case in the British Museum specimen of *Archaeopteryx*.

When evidence of the soft anatomy is preserved, it is not the original material that remains, only a stony replica of its former structure. Consequently, the internal anatomy of the structure is usually lost forever. We can learn nothing about the structure of brain cells by studying natural endocasts, any more than we can learn about the structure of skin cells by studying casts of dinosaur skin. There are some remarkable exceptions to this, however, such as the preservation of the microscopic structure of the body muscles of a fish that lived during the Jurassic (Wyckoff, 1971). The muscles in this remarkable specimen appear as blocks of material attached to the vertebrae, and a microscopic examination reveals numerous fibers that compare in size with those of modern muscle cells. Each modern muscle cell, or fiber, is made up of a large number of fibrils. Within each fibril is a repeating series of interdigitating filaments, and, because these are in register in adjacent fibrils, the whole muscle fiber has a finely striated appearance when viewed with a light microscope under high magnification. (It is for this reason that skeletal muscle is usually called striated muscle.) When small pieces of the fossilized fish muscle were examined under high magnification, the investigator was amazed to discover that the fibers were striated and that the bands corresponded in width with those of modern muscles. How is such remarkable preservation possible? What we are looking at is not the original muscle material—that has long since rotted away— but a high-resolution replica of its surface features formed in mineral. Replication may have occurred in this case by a process similar to that used by botanists for studying plant cells. If nail polish is lightly coated on the surface of a leaf, allowed to dry, then peeled off and examined under a microscope, all of the details of the superficial cells can be seen. Presumably minerals in solution crystallized onto the surface of the fish muscle soon after death and faithfully conformed to surface details of the cells. Studies on living organisms suggest that this is how skeletal structures are formed in lower plants and animals. It appears that some mechanism exists that causes minerals to become precipitated from solution and then crystallized onto biological surfaces to form the skeletons (Leadbeater and Riding, 1986).

Our survey of fossils has covered much ground, some of which will be revisited in later chapters.

THREE

How the Vertebrate Skeleton Works

T H E biggest problem I had as an undergraduate student of zoology, more years ago now than I care to recall, was keeping track of all the different invertebrate animals. Not only are there so many of them, but they are all so very different from one another. The vertebrates, in contrast, were a joy to study because they were all so similar. This is especially true for the "higher" vertebrates—the amphibians, reptiles, birds, and mammals. Their skeletons, for example, are remarkably similar, for they have undergone relatively little change in basic pattern during their long evolutionary history. A consequence of this skeletal conservatism is that if we have a good knowledge of the skeleton of one particular vertebrate, we have a basic understanding of them all. This idea is featured in most high school texts on biology, often illustrated with a diagram showing that such seemingly diverse structures as the wing of a bird, the flipper of a whale, and the arm of a man all conform to the same basic skeletal plan. In this chapter we will be looking at the skeleton of a camel. The selection of this example is a somewhat arbitrary but reasonable choice. An extant (living) as opposed to an extinct vertebrate was selected because we have knowledge of the muscles and ligaments and other soft parts of living animals. We can also discuss their locomotion. A mammal was chosen because its upright stance was more appropriate to that of a dinosaur than the sprawling, "push-up" posture of modern reptiles. And the camel was chosen just because there is a rather nice old illustration of one in the work of Richard Owen, one of the greatest of anatomists. The camel is a quadruped, meaning that it stands on four feet, in contrast to

Most reptiles, including all living ones, hold their legs out horizontally from the body in pushup fashion. The sinuous movements of their bodies help them to walk.

bipeds, like us, which stand only on two feet. As well as looking at bones and locomotion we will be considering teeth and chewing, but we will begin with the skeleton—specifically, the vertebral column.

The vertebral column, the central supporting structure of the skeleton, comprises a large number of vertebrae joined together to form a stiff but flexible rod. The individual vertebrae vary in shape according to their relative position along the column. It is therefore possible to distinguish between, say, a caudal vertebra, from the tail, and a cervical vertebra, from the neck. Each vertebra consists of a short cylinder of bone, the centrum, surmounted by an arch, the neural arch, through which the spinal cord passes from the brain. The neural arch is usually drawn out dorsally

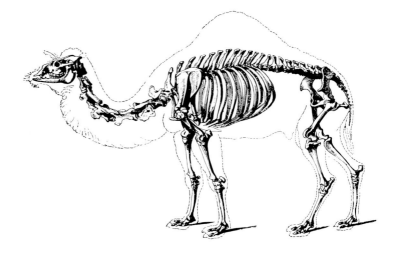

Richard Owen's illustration of the camel, an example of a four-footed animal with an upright posture.

(toward the back of the animal) into a prominent bony process, the neural spine, to which muscles and ligaments are attached. At about the level where the neural arch is joined to the centrum are a pair of processes, the transverse processes, with which the ribs articulate. Sometimes the ribs fuse with the transverse processes, as in the cervical vertebrae of mammals. Adjacent vertebrae articulate with one another by paired processes, fore and aft, called zygapophyses, which lie at the level of the neural arch. The articular surfaces of the anterior zygapophyses face upward and inward and correspond with those of the posterior zygapophyses, which face downward and outward. The anterior and posterior surfaces of the centra also articulate. In many animals, including many dinosaurs, the anterior surface is convex while the posterior surface is concave, so adjacent vertebrae articulate together in a ball-and-socket joint.

Adjacent centra are held together by the intervertebral disks, which are thin pads of fibrocartilage. The fibrous part of the disk is primarily collagen, the remainder is cartilage, so the disk combines the properties of these two materials. Disks are therefore tough and have a high tensile strength—they are actually so strong that very high stresses are more likely to break the individual vertebrae than the discs. And because of their relatively low Young's modulus, they may be able to store and absorb some energy. The corollary of their low modulus is that they lack stiffness, so adjacent vertebrae are not rigidly bound together. The vertebrae are also held together by ligaments—fibrous straps of collagen that bridge across two or more vertebrae at a time. Like the disks, ligaments are tough and have a high tensile strength. Their relatively low Young's modulus gives the ligaments some capacity for storing strain energy. The vertebrae are also held together by muscles. There are, for example, numerous small muscles that crisscross between adjacent neural spines and transverse processes. Like ligaments, muscles have a low Young's modulus and therefore contribute to the storage of strain energy, but they do not have such a high tensile strength. Being contractile, they have the additional property of being able to modify the tensile stresses that they exert on the bones and can therefore effect small changes in posture.

The vertebral column is usually not straight but is arched, like a bow, between the pectoral (shoulder) and pelvic (hip) girdles. This curvature is maintained partly by the shape of the individual vertebrae and the way that they articulate with one another, and partly by the tension in the ligaments and especially in the muscles. The muscles also serve to bring about sideways as well as up-and-down movements of the column. Two major muscle groups control the vertebral column: the epaxial muscles and the hypaxial muscles. The epaxial muscles lie above the level of the transverse processes, and they serve to straighten the arch. There are very many of them, some short, like those spanning from one

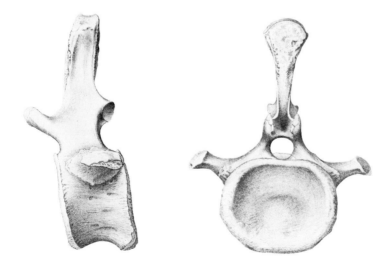

The vertebrae of most quadrupeds have the same general structure. This is a caudal vertebra from the sauropod dinosaur CAMARASAURUS, *viewed from the left side and from the front.*

neural spine to the next, while others are considerably longer. When we have the misfortune to "put out" our back, usually by lifting something awkwardly, we have usually pulled one of these muscles. The epaxial muscles are primarily associated with the vertebral column, but some do spread out over the outside of the ribs. Most of the rib muscles, however, belong to the hypaxial group, which serves to counterbalance the actions of the epaxial muscles. Some of the hypaxial muscles span from one vertebra to another, ventral to (below) the level of the transverse processes, while others span across the ribs, both externally and internally. A major subdivision of the hypaxial muscle group forms the muscles of the abdomen, the most important muscles for counteracting the tendency for the vertebral column to sag between the shoulders and the hips. The abdominal muscles are especially well developed in heavy, long-bodied animals.

The vertebral column therefore provides the body with a stiff but flexible supporting beam that can resist compressive and tensile forces. Its extensive musculature provides for both stability and mobility and contributes to the vertebral column's capacity to absorb and store strain energy. This energy-storage mechanism is an integral part of the running mechanism of many mammals, particularly those that are specialized for galloping (Alexander, Dimery, and Ker, 1985) and the same may have been true for certain dinosaurs.

The weight of the body is supported by the legs, which are connected to the vertebral column through limb girdles—the pelvis for the hindlegs and the pectoral or shoulder girdle for the forelegs. The pelvis has a firm bony union with the vertebral column, usually involving bone-to-bone fusion, but the pectoral girdle is neither fused nor in direct contact with the vertebral column. Instead, the pectoral girdle is attached, primarily by muscles, to the rib cage, often with a good degree of mobility, as in the horse. When a horse is in motion the shoulder girdle swings back and forth, effectively adding another segment to the forelimb and contributing to its speed.

Legs have to do more than just support the body—they have to move the body too, and sometimes compromises have to be made between these two objectives. One compromise involves the angle at which the limb segments are held with respect to the vertical. When the skeleton of a mammal or a bird, or a dinosaur, is viewed from the front or back, the limb segments can be seen to lie in a vertical line. This is described as the erect posture. The erect posture contrasts with the primitive sprawling posture of amphibians and most reptiles (excluding the dinosaurs), whose legs are held out at the sides of the body as if the animal were doing pushups. But when skeletons with the erect posture are viewed from the side, the limb segments do not usually appear to lie along a vertical line. Instead, the humerus (upper arm) slopes backward, the radius and ulna (forearm) slope forward; the reverse holds for the hindlimb—the femur (thigh) slopes forward and the tibia (shin) slopes backward. The knee and elbow therefore point fore and aft, respectively. This, of course, puts added stress on the leg bones because they are being loaded in a way that tends to bend them. The obvious mechanical solution is to keep the segments vertical and to run with stiff legs so that the bones are loaded in simple compression. But having the bones of the legs inclined to one another offers some important locomotory advantages. The inclination of the limb segments is maintained by tension in the muscles and ligaments. This system provides for a natural springing action, which allows for the storage and release of strain energy that is so important during locomotion. Having the limb segments move relative to one another also increases the speed of the leg movement, because the speed of the whole leg is the sum of the speeds of its component parts. Furthermore, the distance through which the foot can be swung is increased, as can easily be demonstrated by some simple experiments with jointed popsicle sticks. The inclination of the limb bones is therefore a compromise between weight support and locomotory performance.

The increased stresses in the inclined leg bones are compensated for by their being hollow rather than solid. This strategy follows a well-known engineering principle that a hollow pipe of a given weight can withstand greater bending stresses than a solid rod of equal weight. This is why steel pipes rather than steel rods are used in construction work.

Now that we have a picture of the basic structure of the skeleton and how it supports the weight of the body, we are ready to give it some movement. But before proceeding we need to define two locomotory terms. *Stride length* is defined as the distance between footfalls of the same foot. To take the example of a human out for a stroll, it is the distance between the point on the ground where the walker's right foot strikes to the point on the ground where it strikes next. The fraction of the stride that a foot is on the ground is called the *duty factor,* and this obviously becomes smaller with increasing speed.

Imagine a circus parade making its way down Main Street. At the lead, moving in line abreast, are an elephant, a horse, and a small dog. The elephant is walking, fairly briskly, the horse is moving at a gentle trot, but the small dog has to gallop to keep up with them. The walk, the trot, and the gallop are three distinct patterns of locomotion, called *gaits,* that quadrupeds use to move at increasingly higher ranges of speed. The three gaits are analogous to the three forward gears of a car. At slow speeds quadrupeds walk, increasing their speed by increasing the number of strides taken per minute—they increase their revs. If a graph were plotted of speed along the horizontal axis and strides per minute along the vertical axis, a straight line sloping upward would be obtained. In other words, speed increases at a steady rate with increasing stride frequency. But a point is eventually reached when a continued increase in speed requires a change in gait from the walk to the trot. The animal "changes gear," but the car analogy is not perfect because there is no drop in revs per minute at this point as there is in a car. Stride frequency continues to increase with increasing speed throughout the trot, but not so rapidly as it did for the walk. The graph would continue as a straight line, but the slope would not be as steep for the trot as it was for the walk.

What is the difference between the walk and the trot? Think of a horse and rider. When the horse is walking the equestrian gets a fairly smooth ride, but as soon as the horse begins to trot the rider begins bouncing in the saddle. The bounce is due to an increase in the vertical oscillation of the horse's center of mass (center of gravity). The horse is also less stable during the trot than the walk, stability being sacrificed for speed; no more than two hooves are on the ground at any one time, and there are instants when none is on the ground (except during slow trots). The horse, as it were, throws more of its body into the trot than the walk. Speed is increased during the trot as during the walk, by increasing the stride frequency, but the point is eventually reached when the transition has to be made to the gallop. This is the point in a riding lesson when the novice really gets thrown about because the horse now launches its whole body into the air during each stride. Galloping is characterized by a leaping movement of the body. Horses have especially stiff backs, but

some mammals, like greyhounds and tigers, have especially flexible backs that are alternately flexed and straightened during a
stride as if they were made of sprung steel.[1] Most back movements
of this sort are brought about by the action of the epaxial and
hypaxial muscles, but the recoil action of the vertebral column
probably plays a role too. Needless to say, galloping is the prerogative of quadrupeds—bipeds cannot gallop.

A singular feature of the gallop is that stride frequency barely
increases with speed, higher speeds being achieved by increasing
the length of the stride. The graph of stride frequency plotted
against speed would therefore depict the gallop as a straight line
with a very slight slope. When an animal changes from a trot to
a gallop it has reached its capacity for making rapid body movements and is essentially running at its maximum sustained stride
frequency.

To return to the circus parade, the elephant, the largest of the
three, is walking; the horse is trotting; while the dog, the smallest,
gallops to keep up. This is because the speeds at which animals
change from one gait to another increases with increasing body
size, as has been demonstrated for a wide variety of animals
(Heglund and Taylor, 1988). The speeds attained at each gait also
increase with increasing body size. Therefore if the dog, the horse,
and the elephant were all three walking along at a comfortable
pace, the elephant would be way ahead of the horse and the dog
would be left far behind. The same would also hold for the trot
and for the gallop, except that elephants cannot gallop, a restric-
tion imposed by their large size. Stride frequency, however, decreases with increasing body size. A walking dog therefore takes
more strides per minute than a walking horse, which in turn takes
more strides per minute than a walking elephant. Smaller animals,
then, are relatively slower and have higher stride frequencies than
larger ones. The fact that speed and stride frequency change in a
regular and predictable fashion with body size provides a means
of comparing locomotory performances of animals of different
sizes. As we saw earlier, the transition from trotting to galloping
essentially marks the upward limit of stride frequency, and it has
been proposed that this transition point serves as a physiologically
similar speed for animals of different sizes (Heglund, Taylor, and
McMahon, 1974). A terrier and a water buffalo running at the
transition point between a trot and a gallop are considered to be
moving at the same equivalent speeds. In terms of miles per hour,
of course, the terrier is moving much more slowly than the buffalo.

Many of the subtleties of animal locomotion are lost to us because they happen so quickly, but we can overcome the problem by looking at slow-motion films of animals in motion. A film of a cursorial (running) animal like a horse trotting or galloping would show the feet snapping backward as they leave the ground. The snap is largely brought about by the springing action of tendons at the back of the leg. These tendons get stretched when the hoof strikes the ground and continue being stretched as the leg flexes under the animal's weight. As we saw in the previous chapter, tendon is elastic and has a relatively low Young's modulus, so it is able to store large amounts of strain energy. This strain energy is released during the push-off phase in which the horse is pushed upward and forward. The residual strain energy causes the hoof to snap back as it leaves the ground. This recoil action allows much of the impact energy of the footfall to be recovered, significantly reducing the cost of locomotion. This mechanism was described for the horse half a century ago (Camp and Smith, 1942) and has been fairly intensively investigated since, not only for the horse but for many other animals too (Alexander, 1984; Dimery, Alexander, and Ker, 1986).

Humans are not very good runners. Although we lack the recoil mechanism seen in cursorial animals, we do have some ability to store strain energy. We store energy in the Achilles tendon of the calf and in the ligaments of the arch of the foot, both of which are flexed during each step. It has been estimated that these two structures can store about 35 percent and 17 percent, respectively, of the total energy turnover during each stride (Ker et al., 1987). This storage capacity explains why the shoes worn by competitive sprinters are so thin. Jogging shoes are well cushioned to absorb impact and reduce injuries, but sprinters do not wear them because they dampen much of the springing action of the arch and Achilles tendon. Before leaving the topic of locomotion, we need to consider the biological significance of the relative placement of the limb muscles.

For many years now it has been recognized that in many cursorial animals—horses, antelopes, ostriches, and the like—the bulk of their leg muscles is concentrated proximally, that is, closest to the body, their action being transmitted to the lower parts of the legs by long tendons. This arrangement has traditionally been interpreted as a strategy for reducing the energy costs of locomotion by reducing the inertia of the legs.

The inertia of an object is its tendency to stay where it is. That it can be reduced by concentrating the mass proximally may be illustrated by a skater performing a fast pirouette on the ice. The spin starts with the arms outstretched, which gives the body the greatest inertia; when the arms are tucked in close to the body the sudden decrease in inertia causes a rapid increase in velocity. How does this work? It all has to do with the conservation of energy. The energy of a moving object, called the kinetic energy, is half the mass of the object multiplied by the square of its velocity ($\frac{1}{2}mv^2$). The outstretched arms of a spinning skater move faster than her body, simply because they have to cover a greater distance with each revolution. Therefore the kinetic energy of her arms is much higher than it would be if her arms were held close to her body. When she tucks her arms in close there is a sudden decrease in their velocity, hence a sudden reduction in their kinetic energy. The excess energy (kinetic energy of outstretched arms minus kinetic energy of tucked-in arms) is not lost but is added to the kinetic energy of her whole body. And since her mass remains constant, her velocity has to increase so that the total kinetic energy of her body remains the same.

Animals with proximally placed leg muscles are therefore expected to expend less energy in running, but experiments by C. R. Taylor and others showed that they performed little better than animals with more distally placed muscles (Taylor et al., 1974). Indeed, the costs of running at the same equivalent speeds are remarkably constant over a wide range of animals, both quadrupeds and bipeds (Heglund and Taylor, 1988). How could these rather unexpected findings be explained?

The energetic costs of locomotion have two major components; there are muscular costs—the costs of activating the muscles and in getting work out of them—and there are mechanical costs— the costs of moving the various parts of the limbs relative to the rest of the body, and of moving the body relative to the ground. The muscular costs are determined primarily by the amount of muscle present, which is obviously related to body mass, whereas the mechanical costs are determined by the anatomy of the body. It has been argued that the muscular costs are the predominant ones (Heglund and Taylor, 1988), so the anatomy of an animal may have less relevance to the energetic costs of locomotion than previously thought. Be that as it may, the conclusion that the relative placement of limb muscles appears to have nothing to do with costs of locomotion were difficult to accept, so other investigators decided to look into the matter.

One of the problems with Taylor's experiments was that the animals being compared (cheetah, goat, and gazelle) differed from one another in more regards than just their limb muscles. What was needed was an experiment where the only variable was the relative placement of the muscle mass. Such an experiment, of course, is not possible, but experiments have been performed on humans where the energetic costs of running were compared for individuals before and after having a load attached to various parts of the body (Myers and Steudel, 1985). In a series of carefully designed experiments, an individual's energetic costs while running were compared without a load and then with a fixed load attached to the waist, thigh, shin, and ankle. The costs increased above the initial condition by 4, 9, 12, and 24 percent, respectively. This clearly demonstrates that the energetic costs of running increase as weight is added to more and more distal portions of the limb. Similar conclusions have been reached by other researchers using an entirely different approach (Hildebrand and Hurley, 1985).

Trained sprinters tuck in their legs during the recovery stroke such that their heels almost touch their buttocks (this can be seen in slow-motion reruns of the Olympics). The marked reduction in inertia of the leg speeds the recovery stroke, thereby increasing the stride rate and hence speed. Runners are not trained to do this consciously but their training includes stretching exercises that increase the mobility of the femur, especially the extent to which it can be retracted (drawn back). The tucking-in motion therefore comes naturally to them.[2] Joggers can simulate this for themselves simply by making a conscious effort to tuck in the lower leg as soon as the foot leaves the ground. If this is done in the middle of normal running an immediate acceleration in stride rate is experienced. This results in a burst of speed but the speed of the power stroke also has to be increased in order to sustain the increase in stride frequency.

So far our attention has been directed solely to the bony parts of the skeleton, but the other hard skeletal structures, the teeth, have much to teach us also. Mammals, unlike most reptiles, have very distinctive teeth, each with a specific pattern of peaks (cusps) and valleys. Indeed the species of many mammals can be identified on the structure of the teeth alone. And because teeth are more resistant to destruction than any other skeletal tissue, most of the fossil record of mammals is represented by teeth. As a consequence, mammalian teeth have been the most extensively studied

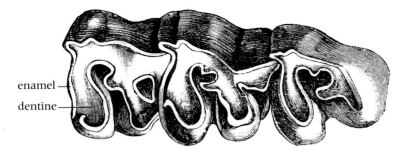

enamel—
dentine—

Three molar teeth from an extinct rhinoceros. The dentine wears much faster than the enamel and therefore becomes hollowed.

of all teeth and the following account is therefore based largely on mammals.

Of the two materials that make up a tooth, dentine and enamel, enamel is much more resistant to wear. For species that do not grind their food, ourselves and carnivores included, the enamel coating usually lasts for the lifetime of the tooth. I can feel wear-facets on the occlusal surfaces (the surfaces that come together) of my incisors (front teeth), but the enamel has not worn down to the dentine. Nor is there any exposed dentine in any of our cat's teeth, but if I checked my friend's horse, I would see exposed areas of dentine in all of its teeth. Herbivores actually exploit the differences in wearing rates of dentine and enamel to form efficient cutting and grinding surfaces. The shape of these areas is partly determined by the specific cusp patterns inherited in the genes for each particular species. It is also determined by the way the animal chews and wears down its teeth, and this again is an inherited characteristic.

When a tooth first erupts in a herbivore, before it starts being worn down by chewing, the various peaks and valleys are all covered with enamel. But as the tooth is used the enamel wears away and the softer, exposed dentine then wears away faster than the enamel. The end result is a series of slightly hollowed dentine islands, each surrounded by a raised edge of enamel, and it is these raised edges of enamel that form the cutting edges so important for grinding up plant material.

Mammalian herbivores usually chew with a side-to-side action of the mandibles, upper and lower teeth on one side coming together first, followed by those of the other side. It is obviously advantageous to have the raised enamel cutting edges of the cheek teeth oriented fore and aft. In this way the top cutting edge meets the bottom cutting edge lengthwise on, guillotine fashion, maximizing the cutting and pulverizing action. And this is just what we find when we examine the teeth of herbivores like cows and sheep that chew with a sideways movement of the teeth—the cutting edges are oriented at right angles to the direction of motion. Animals that chew with a back and forth motion of the mandible, like guinea pigs and elephants, have the cutting edges lined up transversely, again at right angles to the direction of motion.

There is one more clue to the direction of motion of the teeth during chewing, and this is provided by the way in which the dentine wears. As one tooth grinds past another the dentine of its leading edge, being protected by the (harder) enamel that immediately precedes it, wears slowly. The dentine on the trailing edge, in contrast, wears rapidly because it is preceded by dentine rather than by a protective wall of enamel. As a consequence, the dentine at the leading edge of a tooth is essentially flush with the enamel, while that at the trailing edge lies well below the enamel edge and is therefore dished. This differential wearing pattern has been simulated with model teeth (Costa and Greaves, 1981). In this experiment aluminium tubes were filled with plaster, the metal simulating the harder enamel and the plaster simulating the dentine. These "teeth" were then abraded on a grinding wheel, and after some time the worn surfaces were examined. The plaster at the leading edge was found to be flush with the aluminium, while that at the trailing edge was worn down. There were also striations on the plaster, as there often are on teeth, and these too indicate the direction of motion. Because jaw function can be deduced from patterns of tooth wear, we will find these patterns a useful tool when we consider the mechanics of dinosaur jaws.

Reading a Dinosaur Skeleton

P ALEONTOLOGISTS are famous, or infamous, for being able to tell us everything about an animal from the smallest scrap of information. Drawings of complete animals have been constructed from fragmentary remains. I have even read a discussion of the possible diet of an animal that was represented only by its footprints. Baron Georges Cuvier, generally regarded as the father of vertebrate paleontology, was not averse to producing bold reconstructions on the most tenuous evidence, and paleontologists of lesser personage could hardly be criticized for emulating the master. We might forgive our forefathers for their transgressions, but there is no excuse for continuing the tradition of extrapolating beyond the data; the discussion on the possible diet of the animal that left the footprints appeared in a reputable scientific journal in the late 1970s. The best safeguard against falling into this trap is to make constant reference to present-day animals. If it is not possible to draw conclusions regarding a particular feature from the skeleton of a modern animal, then it will not be possible for a fossil skeleton. Paleontologists have no more knowledge of the extinct beasts they study than neontologists have of recent animals, if their conclusions rest solely on skeletal remains. Imagine if elephants had not yet been seen in the flesh. A neontologist would need a vivid imagination indeed to invent the elephant's trunk because it is formed entirely of soft tissue. And who could surmise from those dry dead bones that elephants often spend much of their time in the water?[1]

A wealth of information can be drawn from the detailed study of fossils, but there are also many things we cannot learn from them. At very best we have a complete skeleton to study, but even this is only part of the animal. We have none of the soft tissues, though we may sometimes have clues to these in the form of skin impressions, muscle insertion scars on bones, endocranial casts of the brain, and the like. We cannot watch the fossil move, observe its behavior, measure its temperature, record its heart rate, or measure any of its other body functions. We seldom know its sex, we can rarely determine how old it was when it died, and we cannot always be sure whether it was an adult or a juvenile. If we have several skeletons that look the same we can probably never be sure whether they belonged to the same species, because, with rare exceptions, we have no way of recognizing a biological species in the fossil record. (A biological species is a natural entity in the living world and is recognized by the fact that its members are freely interbreeding.) There are limitations to what we can learn from fossils and we must constantly be aware of these. Forewarned of the pitfalls, we can proceed to examine a dinosaur skeleton and see what it can tell us about the animal that once gave it life.

Dinosaurs appear to have been predominantly terrestrial and were therefore confronted with the same problems of supporting and moving their bodies against the force of gravity as present-day land animals do. Most dinosaurs were herbivores, like many living vertebrates, and were preyed upon by the carnivores. The skeletal adaptions of these two life-styles were quite distinct, and we usually have no more difficulty in distinguishing between herbivorous and carnivorous dinosaurs than we do in distinguishing between a lion and an antelope. Most carnivorous dinosaurs conformed to the same general pattern throughout the Mesozoic Era, whereas the herbivores displayed great diversity. One very successful group of herbivores was the hadrosaurs, which appeared in the latter half of the Cretaceous and flourished until the end of the period, some thirty million years later. They have been known for well over a century and have been found in many parts of the world—Europe, the Far East, even South America—but they are especially well represented in North America (Charig, 1973; Brett-Surman, 1979; Bonaparte et al., 1984). Numerous well-preserved skeletons have been collected from western Canada and the United States, some of which can be seen on display in major museums in both countries. Hadrosaurs are therefore ideal subjects for finding out how skeletons can be interpreted and also for learning something about dinosaurs. And the fact that many of them possess the most extraordinary crests on their heads, the interpretations of which have ranged from the sensible to the ridiculous, is an added bonus. Most of the account that follows will therefore be based on hadrosaurs, but I will also make brief references to some other dinosaurs.

By dinosaurian standards most hadrosaurs ("heavy lizard") were not particularly large. They were about twenty-five feet (8 m) long, the size of the Indian elephant, though some hadrosaurs have been found in Mexico at twice the length (15 m; Morris, 1981). Although there are some differences in their bodily skeletons, most of the features used to distinguish the various species relate to their skulls. Some species, like *Edmontosaurus regalis,* were without crests, while others, like *Corythosaurus casuarius,* had an elaborate hollow crest reminiscent of a Corinthian helmet. *Para-*

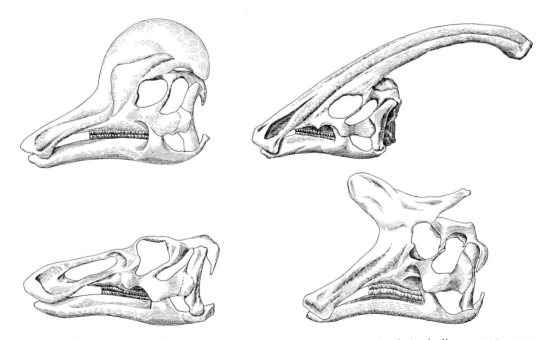

The main differences among the various hadrosaur species are found in their skulls. Top left, CORY-THOSAURUS CASUARIUS, *with an extensive semicircular crest.* Bottom left, *the crestless* EDMONTO-SAURUS REGALIS. Top right, PARASAUROLOPHUS WALKERI, *with an extensive tubular crest.* Bottom right, LAMBEOSAURUS LAMBEI, *with an anterior hollow crest and a small solid crest in-clined backward. All four fossils are from the Upper Cretaceous of western Canada, drawn from specimens in the Royal Ontario Museum.*

saurolophus walkeri was unusual for the greatly elongated and hollow crest that projected backward over the neck. *Lambeosaurus lambei* had two crests, a hollow one pointing forward and a small solid crest pointing backward. There are several other species but, mercifully, not as many as there were just a few years ago. The reason for the previous surfeit of species is that earlier paleontologists had mistakenly identified juveniles as separate species from adults. These errors were put to right by Peter Dodson (1975), who made a detailed study of the changes that take place during the growth of hadrosaurs. He showed that their crests, like those of the cassowary, a large flightless bird that lives in the rain forests of Australia and Borneo, did not develop until later in life. Consequently, juvenile hadrosaurs did not look very much like their parents, which explains why they have been misidentified as separate species. Before attempting to interpret the significance of these curious head structure, we will first look at the skeleton to find out what we can about posture and locomotion.

Posture and locomotion

Most museum collections of dinosaurs were gathered during the latter part of the nineteenth and early part of the twentieth century, and a great deal has been learned about them since those times. Indeed, there have been tremendous strides even within the last twenty years, and it is therefore not surprising that we have to question the posture in which some dinosaur skeletons have been mounted. In the dinosaur gallery at the Royal Ontario Museum, for example, there is a fine skeleton of *Corythosaurus* that was mounted in the 1920s in an upright posture. At that time many paleontologists thought that hadrosaurs walked upright, on their hindlegs. But others, impressed by the position in which many hadrosaur skeletons have been found, believed that they walked on all fours with the tail curving gently to the ground. Close examination of the upright *Corythosaurus* skeleton reveals that the head of the femur articulates with the pelvis at the point where two of the bones, the pubis and ilium, join together, which is a point of weakness. It is therefore unlikely that the body could have been held in an upright position, but some caution is required here because we do not always draw the correct conclusions from the study of skeletons. The skeleton of the elephant, for example, is adapted for supporting a large body mass, and a bipedal posture is inconceivable. Nevertheless, elephants have been observed to stand on their hindlegs in the wild, in order to reach up into trees. Perhaps hadrosaurs could stand, but not walk, with a semi-erect posture.

The hindlegs are considerably larger than the forelegs, not only in length but also in thickness, and it is reasonable to conclude that they carried most of the weight, so hadrosaurs were functionally bipedal. We are bipedal, but we are rather unusual because we hold our vertebral column vertically. Other bipeds, from birds to kangaroos, have an essentially horizontal vertebral column. Perhaps it was our own upright posture that led early paleontologists to reconstruct hadrosaurs the way they did. Peter Galton (1970) made a convincing case that hadrosaurs held their vertebral column horizontally and their tail stiff and fairly straight, rather than curving to the ground. The tail therefore served to counterbalance the weight of the front part of the body about the pivot formed between the head of the femur and the socket, or acetabulum, of the pelvis. Although they walked on all fours, it is believed that they probably lifted their forelimbs off the ground when they ran. Even while running, however, hadrosaurs kept the vertebral column horizontal.

We cannot be sure, but it seems reasonable to surmise that the hadrosaur's vertebral column was arched. Evidence for this conclusion is provided by the curvature that is preserved in the vertebral columns of many skeletons. The tail, in contrast, appears to have been relatively straight when viewed from the side, because of the numerous ossified tendons that crisscross the neural spines in the pelvic region and in the anterior (front) half of the tail. We know that these rodlike structures, which are about as thick as a pencil, are ossified tendons rather than bones because of their microscopic structure. Ossified tendons are also found in some modern animals—those stiff white splints in the leg muscles of the turkey are ossified tendons, but they are used not for stiffening the leg but for transmitting the pull of muscles to bones. They are obviously stiffer than ordinary tendon, and consequently they have a higher Young's modulus (Bennet and Stafford, 1988). In addition to the neural spines, which project dorsally, the caudal vertebrae of most reptiles also have spines, called chevron bones, which project ventrally (downward, toward the belly side of the body). The neural spines and the chevron bones are both relatively long in hadrosaurs, especially in the giant hadrosaurs from Mexico, and consequently the tails were very deep (dorso-ventrally).

The counterbalancing role of the tail was beautifully demon-

The upright posture in which this hadrosaur skeleton (CORYTHO-SAURUS CASUARIUS) was mounted is now believed to be incorrect. Close inspection of the pelvis shows that in this position the head of the femur abuts against the joint between the pubis and ilium, which was probably an area of weakness.

strated when we were mounting a skeleton of *Lambeosaurus* (named for its discoverer) in the dinosaur gallery of the Royal Ontario Museum. The method now used to mount a skeleton is to drill out the bones and thread them onto a steel framework so that the skeleton stands on its own legs, as it did in life. The vertebrae are strung onto a horizontal rod that has been welded to the vertical rods in the legs. During construction of the *Lambeosaurus,* the tail vertebrae were the first to be strung on, and the leverage generated by the tail was so great that a large concrete block had to be attached at the other end to prevent the whole thing from tipping. Once the other vertebrae and the skull had been added, the skeleton was so well balanced that the touch of a finger was enough to make it rock up and down.

Some modern lizards that are able to run on their hindlegs have long tails as balancing organs. Experiments have shown that tail loss in such lizards often causes individuals to stumble and that running speeds are reduced by more than one-third (Ballinger, Nietfeldt, and Krupa, 1979). Birds, of course, lack the long tails of reptiles, but they achieve balance by having their heavy internal organs (primarily the gut) slung between their legs. The body still pivots about the hip joint, as in *Lambeosaurus.* This was demonstrated to me during a visit to Australia, when I had the opportunity to watch some emus (large flightless birds, related to ostriches) at close quarters. Emus are naturally inquisitive, and one of them approached me and was quite happy to let me rub him under the chin. This slight movement was enough to cause his body to rock gently up and down about the hip joint.

The problem of deciding on an appropriate walking posture for our skeleton of *Lambeosaurus* was simplified by the fact that we had a set of tracks in which to mount the skeleton. The trackway had not been made by *Lambeosaurus,* because it was from an earlier geological horizon, but it had obviously been made by a similar dinosaur (probably by an iguanodontid). The impressions of the three toes, broad and blunt, corresponded quite closely with the skeleton of the foot of *Lambeosaurus,* showing that the unknown trackmaker was of about the same size. The trackway was made by a dinosaur walking on all fours, and the imprints of the hindfeet were considerably larger and deeper than those of the forefeet, confirming that most of the weight was carried on the hindlegs. The stride length was six feet (just under 2 m), which is quite modest compared with the length of the skeleton, twenty-five feet (8 m). The short stride indicates that the animal was walking fairly slowly at the time, because the length of the stride increases with speed.

When I am walking slowly my stride length (see page 45) is about three feet (a little less than 1 m), but it increases to fifteen feet (5 m) when I am running. The direct relationship between stride length and speed may be demonstrated by changing speed while travelling on snow and then going back and examining one's trackway. It may also be demonstrated on a dry day by wetting one's shoes before walking and then running on a dry sidewalk, but this method usually worries the neighbors. R. McNeill Alexander, probably one of the most prolific authorities on animal locomotion, derived a relationship between stride length, animal size, and the Froude number, which is a dimensionless quantity used by engineers as a measure of the interaction of inertia and gravity (Alexander, 1976). To test how well this relationship held when applied to actual data, he plotted a graph of relative stride lengths (stride length divided by height of the hip above the ground) against Froude number (u^2/gh, where u = speed, g = acceleration of gravity, and h = height of the hip above the ground) for several different animals. Five species were sampled: man (both walking and running), horse, ostrich, elephant, and a small African rodent called a jird. The data points appeared to conform to a straight line, suggesting that there was a good relationship between relative stride length and Froude number; the relationship promised to be applicable to other animals, too. It was a simple matter to obtain the equation of the straight line, which was then rewritten so as to give an equation for the speed of an animal in terms of its stride length and the height of its hips above the ground. Now stride length can be measured from a trackway, and if an estimate of the hip height of the trackmaker can be made, it is possible to deduce its speed. This raises all sorts of exciting possibilities for fossil trackways, as Alexander demonstrated when he used his method to deduce the speeds of various dinosaurs. Paleontologists seized upon the equation with unrestrained enthusiasm and a large number of publications have since appeared giving running speeds for a wide range of dinosaurs. Most of these estimates were made without a second thought being given to the reliability of the method, but how reliable is it?

One of the obvious problems is trying to estimate the hip height of a dinosaur that left only its footprints behind. Alexander compared the size of the hindfeet with the hip height for some di-

This contemporary mount of the hadrosaur LAMBEOSAURUS LAMBEI *has been given a horizontal vertebral column. Notice that the feet, which are set in a trackway, have the heel raised from the ground in the digitigrade pattern of walking. Hadrosaurs are believed to have run on their hindlegs with their forelegs clear of the ground. The animal that made the trackway shown here was walking rather slowly.*

nosaur skeletons and concluded that hip height could be approximated at four times the length of the footprint. But the reliability of this estimate is unclear and a more direct estimate of hip height is obviously more desirable. If a given trackway can be attributed to a particular type of dinosaur, it is possible to estimate the hip height by reference to known skeletons. But identifying dinosaur tracks is not like identifying the tracks of living animals, when positive identifications can be made because of direct associations between the track and the trackmaker. All a paleontologist can do is identify a trackway with a particular group of dinosaurs, the hadrosaurs, for example, and use the size of a footprint as only a general guide to the size of the actual animal that made it. Estimates for hip height inferred from trackways are therefore only very approximate.

How good is the equation itself? Alexander certainly made a good case that there is a relationship between relative stride length and Froude number, but the fact that only five species were used to derive the equation seriously undermines its predictive value. This was borne out by a later paper in which twice as many (modern) species were sampled (Alexander and Jayes, 1983). There was now so much scatter among the data points that they no longer fell along a single line, so four separate lines were plotted. There was therefore no single equation that could be used for estimating speed from relative stride length. This means that the original equation is inappropriate and that the estimates for the running speeds of dinosaurs that have been derived from it are invalid.

In a more recent study, Alexander (1989) presented a graph based on data from fewer species (seven instead of ten) that had less than half as many data points as before. There appears to be less scattering among the data points (though this might be because fewer were used) and he was able to fit a single line to them. Alexander cautioned that the scatter of points about the line made an estimate of running speed "inevitably a rough one." Nevertheless he still estimated dinosaur speeds to the nearest tenth of a mile per hour (and to the nearest tenth of a meter per second).

So although there may be some value in making statements regarding the *relative* speeds represented by different trackways, I do not think we are able to give any definitive estimates of *absolute* speeds.

Let us return to our example of *Lambeosaurus* and the trackway in which it was mounted. The width across the track, that is, the

An artist's restoration showing a group of lambeosaurs being disturbed and taking flight.

distance between the center of a left footprint and the center of a right one measured across the track, was only eight inches (20 cm). The footprints were therefore not far off from being in a straight line, which means that the feet were placed well beneath the body during locomotion. This is quite common among quadrupeds and can be verified by looking at animal tracks left in the snow. When an animal lifts its feet off the ground during locomotion it becomes unstable. If the feet were placed well to the side, as they are in mice, for example, the body would always be toppling toward the unsupported side. Placing the feet close to the center of the body is therefore a strategy for increasing stability.

The hadrosaur foot has three large toes, all pointing forward and each ending in a small hoof. Instead of walking on the entire foot the hadrosaur raised its heels clear off the ground, so that it walked on its toes. This method of walking is called digitigrade. When we walk we place the sole of the foot on the ground, which is called plantigrade walking, but we become essentially digitigrade when we sprint. The digitigrade pattern adds length to the leg (that is, the "heel" acts as part of the leg); as we shall see, this is a strategy for increasing speed. The part of the foot that is held clear of the ground, the metatarsus, is fairly long in hadrosaurs. Only the tips of the toes can be discerned in footprints and it seems likely that an extensive pad of connective tissue cushioned the hindfoot. Pads are found in most modern animals that run (except hoofed ones, of course), including cats and dogs. They are elastic and serve to absorb energy during locomotion, reducing the shock of impact and damping out the tendency for the foot to bounce (referred to as chattering; Alexander, Bennett and Ker, 1986). It is possible that some of this energy is returned to the foot after impact and gives some rebound to the step—much like a good running shoe does. The front foot is much smaller and more slender than the hind one. The digits do not spread out, and impressions of skin that have been found show that they were connected by a web of skin.

The massive hindlegs of a hadrosaur not only supported most of the body weight but also supplied most of the power for thrusting the body forward. The power was provided by muscles, and although these have long since disappeared, they have left traces of their presence on the surfaces of the bones. If we examine the skeleton of a modern animal we can detect areas where muscles once attached to the bones. These attachment sites, called muscle scars, have various forms, from ridges and depressions to roughened areas. The general plan of muscles is fairly conservative within the vertebrates and consequently we can get some idea of the muscle arrangement of a dinosaur by comparing their muscle scars with those of their modern relatives. However, it must be emphasized that this approach gives a very approximate idea of the locations of particular muscles and reveals little or nothing about their sizes. My concern for caution in interpreting skeletal marks arises from some studies I conducted on the limb muscles of modern animals, specifically two species of birds, the kiwi and the weka (McGowan, 1979a, 1982, 1986b). Surprisingly few of the muscles could be associated with any specific marks on the bones—indeed many of the features that would have been described as muscle scars had nothing to do with muscles at all. It was therefore impossible to reconstruct their muscles by studying their bones. It is possible that the kiwi and weka are atypical among vertebrates, but colleagues have reported similar findings to me for other species. More work needs to be done, but the indications are that muscles cannot be reconstructed from skeletal data. Probably the best that can be hoped for extinct species is to obtain indications of the relative developments of some of their muscles. The numerous muscle reconstructions that have been published for fossils are therefore questionable. I hasten to point out that I too (McGowan, 1973b) am among the guilty!

Most of the hadrosaur's muscular energy was probably used in propelling the body forward by pulling the hindlegs backward (retraction) against the resistance of the ground. The extensive muscles that performed this task originated largely from the posterior half of the pelvis and from the root of the tail and inserted onto the back edge of the femur. Pulling the leg forward (protraction) was accomplished by muscles that originated from the anterior half of the pelvis and inserted onto the anterior edge of the femur. The protractor muscles did not have to work to propel the body forward but they did have to overcome the inertia of the leg. Experiments on modern animals have shown that this requires large amounts of work, especially as speed increases.[2]

The speed of an animal is determined by the length and speed of its stride. Both may be increased by increasing the length of the lower leg segment (below the knee). The horse, for example, is a fast runner and it has a lower leg that is considerably longer than the upper leg; the ratio of the two segments is about 2.5:1. The elephant, in contrast, which is not specialized for running, has a lower leg that is much shorter than the upper leg; the ratio is only 0.7:1. In *Lambeosaurus* the ratio of the lower leg to the upper leg is 1.5:1, which suggests that it may have been capable of running fairly fast. Walter Coombs (1978), who made a detailed study of the adaptations of cursorial animals, concluded that hadrosaurs were probably somewhat better runners than rhinos. In *Lambeosaurus* the metatarsus (the sole of the foot) contributes to the length of the lower leg because of its digitigrade stance. Many modern animals are also digitigrade, including cats and dogs. This stance not only increases the length of the lower leg but also the speed of the stride, because the speed of the leg is determined by the sum of the speeds of its individual segments. Increased speed cannot be accomplished by increasing the length of the upper leg segment because doing so would increase the distance through which the powerful retractor and protractor muscles would have to work, which would decrease the speed of the stride. The inertia of the leg would also be considerably increased if the upper leg

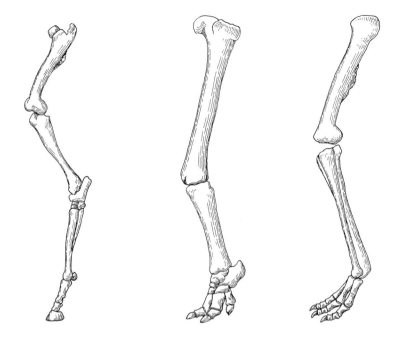

The relative lengths of the upper and lower leg segments are used as measures of running potentials. Left: *In the horse, a fast runner, the lower segment is more than twice the length of the upper one.* Middle: *The elephant, which can neither trot nor gallop, has a lower segment that is shorter than the upper one.* Right: *In* LAMBEOSAURUS, *the lower segment is about one-and-a-half times longer than the upper segment, which suggests an ability to run fast.*

were longer because of the large muscle mass attached to the femur.

From the time of their discovery, hadrosaurs were thought to have been aquatic. Their tails, which are long and deep, were believed to have been used primarily for swimming. Evidence that the digits of the foreleg were webbed was naturally taken as further evidence of their aquatic nature. The belief that hadrosaurs lived in water has colored most interpretations of their anatomy. Although they may well have spent much of their time in the water, there is no reason to believe that they were not also fully adapted to living on the land. The elephant, as mentioned earlier, often spends much time in the water but it would be foolish to attempt to interpret all of its anatomical features in terms of adaptations to the aquatic environment. Perhaps the hadrosaur's deep tail and webbed fingers *were* aquatic adaptations, but this is not necessarily so. It is true that crocodiles, marine iguanas, and other aquatic reptiles have deep tails that are used for swimming, but there are also many terrestrial reptiles, including *Sphenodon* and many lizards, that similarly have deep tails but do not venture into the water. The tails of carnivorous dinosaurs, horned dinosaurs, and plated dinosaurs are also fairly deep, but few paleontologists would consider these animals to have been aquatic. The webbed hand seems an obvious aquatic adaptation, but since it is so small compared with the rest of the body it would probably have been of little assistance in swimming. Furthermore, the fingers of a kangaroo are united by a sheath of skin—so for that matter does my friend's dog—but this is clearly not an aquatic adaptation. Kangaroos use their front limbs to prop themselves when moving slowly or when stopping to feed. We know from the evidence of trackways that hadrosaurs used their forelimbs in a similar fashion. There is also some dietary evidence to support the claim that they spent at least some of their life on the land. This leads us to the next topic of hadrosaur anatomy, their teeth and jaw apparatus.

Teeth and jaws

The skull, and particularly the teeth, offer the most information on the possible feeding habits of extinct animals. Hadrosaurs have as many as one thousand teeth, packed tightly into four separate dental batteries, one in each half of each jaw. There are therefore left and right upper batteries and left and right lower batteries. Each battery has between two hundred and two hundred fifty teeth, and these are arranged in about forty vertical series with between three and six teeth in each series. Only those teeth that are at the free edge of each dental battery have worn surfaces, showing that they alone were in use. Why such a complex arrangement? Reptiles, unlike mammals, continually replace their teeth throughout life. As the hadrosaur wore away its teeth, new ones took their place from lower down in the vertical series. The individual teeth are only about an inch long (3 cm) and have a diamond-shaped cross section. Just to complicate matters further, there are two teeth in wear at any one time in each vertical series. These lie next to each other, one toward the outside edge of the jaw, the other toward the inside edge.

The batteries were arranged so that the lower ones cut inside the upper ones, and the two sets of grinding mills thus formed were extensive. Whenever we see extensive grinding surfaces like this, whether it be in rabbits, horses, or elephants, we can be fairly sure that the animal was a herbivore. Plants need a great deal of

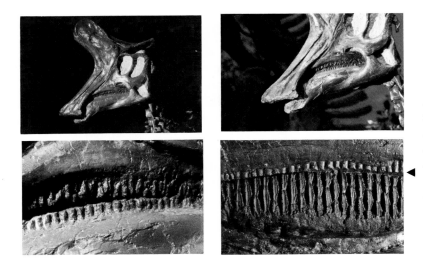

Skull of LAMBEOSAURUS LAMBEI, *showing details of the teeth.* Bottom right, *inside surface of the dental batteries: the individual teeth are diamond shaped, and the arrow marks the upper edge of the lower dental battery.*

chewing because they contain cellulose and other tough materials. In spite of their soft appearance many plants are abrasive, which causes teeth to wear rapidly. Horses resolve the problem by having high-crowned teeth that take a lifetime to wear down. The strategy of hadrosaurs was to replace worn-out teeth continuously.

When food is being chewed it is important that the grinding surfaces come together and remain parallel to one another. If the jaw joint were at the same level as the grinding surfaces, the jaws would close like the blades of a pair of scissors. The back teeth would therefore come together before those at the front, so it would be impossible to have all of the occlusal (grinding) surfaces in play at the same time. In mammals this problem is overcome by the placement of the jaw joint above the level of the teeth. You can check this out easily enough for yourself by moving your jaw while holding a hand to the side of your face, at about ear level. You can also check the way that upper and lower teeth come together when the jaws are closed by feeling with your tongue.

In contrast to the mammals, the hadrosaurs and many other reptiles have jaw joints below the level of the teeth. The joint is formed between the quadrate, a stout, rod-shaped bone that extends vertically downward from the back of the skull, and a shallow depression in the mandible into which the end of the quadrate fits. The depression is larger than the end of the quadrate and is elongated in the direction of the mandible. Its shape suggests that there was a good degree of freedom in the jaw joint and that the mandible may have been capable of moving backward and forward and from side to side as well as up and down.

The unique jaws and teeth of hadrosaurs have attracted considerable interest among paleontologists. There has been much speculation on how they actually functioned. Did the mandibles move from side to side, like those of a cow, up and down, back and forth, or a combination of all three? Three decades ago John Ostrom (1961), in a landmark study on hadrosaur skulls, concluded that the mandibles probably moved back and forth, but a great number of discoveries have been made since that time. The problem has recently been reexamined by David Weishampel (1983, 1984), who arrived at a satisfactory solution to the problem, largely on the evidence of tooth wear.

Unlike the teeth in most other animals, hadrosaur teeth are not completely surrounded by enamel. Instead, the enamel is restricted to the buccal (outer) surface in the upper teeth, and to the lingual (inner) surface in the lower teeth. The enamel cutting edges are therefore arranged fore and aft, which means that the teeth must have moved from side to side and not from front to back, as Ostrom had supposed. This deduction is supported both by the presence of very fine wear striations (revealed by the scanning electron microscope) oriented from side to side and by the way that the dentine has worn away. Recall from Chapter 3 that the dentine that is at the leading edge of a tooth wears down flush with the enamel, whereas that at the trailing edge becomes hollowed. We know from examining hadrosaur skulls that the lower teeth ground inside the upper ones, therefore the outer edges of the upper teeth are at the leading edge. We should therefore expect the dentine to be flush with the enamel at the outer edge but hollowed at the inside edge, which is just what we find. The reverse holds true for the lower teeth.

Mammals chew on one side of their mouth, then on the other. Reptiles seem unable to do this and bite down on both sides of the mouth at the same time (precious few modern reptiles actually chew their food). Weishampel concluded that hadrosaurs chewed in this manner too: as the mandibles were raised and the food was crushed between the teeth, the maxillary bones (the ones at the side of the face bearing the upper dental batteries) were pushed out to the sides by the biting force. While this seems absolutely reasonable it also seems likely that there may have been some side-to-side movements of the mandible as well, as in most modern herbivores. The jaw joint certainly appears to have been sufficiently flexible to have allowed such an action.

Effective grinding requires not only that the food be abraded between the teeth but also that the abrading take place under pressure. The muscles that provided this pressure were located at the back of the skull, just in front of the jaw joint, and ran from the skull to the mandible. The teeth closest to the jaw joint experienced a more powerful bite than those further away. We know this from our experience: nuts are cracked more easily between the back teeth than between the front ones. The pressure on the teeth at the front of the dental batteries was therefore probably much lower than on those at the back, and this may be the reason why the dental batteries are not extended any further forward than they are. Like all other ornithischians (bird-hipped dinosaurs), hadrosaurs have an extra bone, the predentary, capping the tip of the mandible.

The hadrosaurs got their popular name—"duck-billed dinosaurs"—from the fact that the upper and lower jaws are expanded into broad bills. The occlusal edges of the upper and lower bills correspond with one another quite closely, and there is evidence that each may have been invested in a horny sheath, similar to the beak of a bird or a turtle. Natural molds of an upper sheath have been preserved in a few specimens, from which it seems that the original structure was thin and had a fluted internal margin. These features suggested to some paleontologists that the upper bill may have functioned as a filtering device, like the fluted upper bill of the duck, and were used to support the aquatic interpretation (Morris, 1970). Some specialists have even gone so far as to argue that the bill was capable only of cropping soft water plants. Hadrosaurs may have used their bills for dabbling in water for plants, but there is no reason to suppose that they were not eminently suited for cropping land plants too. The considerable wear on the teeth leaves no doubt that they did consume fairly tough plants.

Direct evidence of the food preference of at least one hadrosaur became available in 1922 (Kräusel) with the discovery of the remains of the stomach contents of a particularly well-preserved skeleton, which is now in the Senckenberg Museum in Frankfurt. All of the identifiable remains are of land plants, and conifer needles of the species *Cunninghamites elegans* are particularly abundant. At the time of this discovery the belief that hadrosaurs were aquatic was so deeply entrenched that the author was somewhat hesitant to accept his findings at face value. He therefore made the point that the evidence indicated only that hadrosaurs *could* feed on land plants. The subjugation of evidence that is contrary to accepted beliefs is unfortunately not a rare event in the progress of science.

Plants, being relatively low in nutrients, are consumed in large quantities. As a consequence, herbivores have a much longer digestive tract than carnivores, and this is normally filled with slowly decomposing food. Herbivores therefore tend to have relatively wider bodies than carnivores; horses and cows, for example, are broad across the back, while lions and tigers are narrow. We can visualize that in the hadrosaurs, and in most of the other ornithischian dinosaurs, the large gap between the end of the rib cage and the posteriorly directed pubes and ischia of the pelvis would have been occupied by a large and pendulous belly.

Unravelling the crest

The most striking feature of many hadrosaurs is their cranial crest. Numerous interpretations have been offered for the crest's function, many of which, predictably, are aquatic in nature. Some thoughts on the subject have been more than a little imaginative. The major stumbling block to arriving at a satisfactory account of the crests is that no living animals possess similar structures. Another problem has been the tendency for paleontologists to assume that crests fulfilled a single role; but there is no reason why they should not have had several different functions.

There are two types of crests, hollow and solid. The hollow ones are connected with the external nares (nostrils) to form part of the respiratory apparatus. Some hadrosaurs have both hollow and solid crests, but others have one or the other. The solid crest may be nothing more than a small knob on top of the skull, as in *Prosaurolophus,* or it might be drawn out into a modest horn extending from the back of the skull, as in *Lambeosaurus,* which also has a hollow crest. Or it might have the form of a unicorn's horn, as in the Chinese hadrosaur *Tsintaosaurus.*

The hollow crests, as seen in *Corythosaurus,* appear to be relatively large, but about half of the crest is actually flattened from side to side and only the proximal portion is hollow. The internal anatomy of the hollow crest, which can be seen in only a few

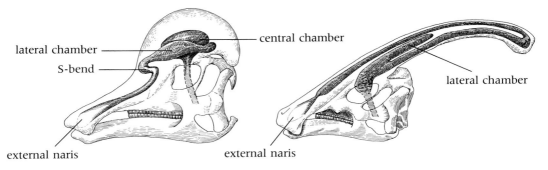

lateral chamber

S-bend

central chamber

lateral chamber

external naris

external naris

The hollow crests of the hadrosaur skulls conducted air from the paired nares to the throat. The air passages are relatively simple in CORYTHOSAURUS *(left), but are doubled back upon themselves in* PARASAURO-LOPHUS *(right).*

specimens, is fairly complex (Ostrom, 1961; Weishampel, 1981). We will start at the external naris and work our way back. The external naris is very large, as in all hadrosaurs, and opens posteriorly into a fairly wide bony canal. At about the level of the orbit (eye socket), each of these two canals (left and right) is thrown into an S-shaped bend, much like the trap in the drain pipe of a sink. Just posterior to the bend each tube widens into a fairly large lateral chamber. Each lateral chamber lays to the left and right of the center line of the skull and opens into a common central chamber. A single bony tube extends downward from the central chamber and connects up to the internal nares, which are located at the back of the throat. When the animal breathed, air was drawn in through the external nares, passed along the paired bony tubes, through the S-bends, into the lateral chambers, then into the central chamber. From there the air passed down the single bony tube to the internal nares, then into the trachea (windpipe), and finally into the lungs. All very complex! The details of the duct system vary from species to species. *Parasaurolophus,* for example, has a large backwardly-projecting crest, and the two bony canals from each naris, which lack an S-bend, continue back to the end of the crest, double back on themselves, and then pass forward again toward the throat. The lateral chambers are very long and extend back about half the length of the crest. The central chamber is relatively small.

Speculation on the possible function of the crests covers a wide range, from the plausible to the ridiculous. In one of the aquatic interpretations, the complex arrangement of tubes, especially with the built-in S-bend, was seen as a trap to prevent water from being taken into the lungs during underwater feeding. But lots of other aquatic animals manage very nicely without such elaborate devices—crocodiles, ducks, and hippos, to name but a few. They avoid drowning themselves by using sphincters (a ring of muscles surrounding an opening) to close off their external nares. Some paleontologists have suggested that the chamber functioned as a reserve supply of air that could be drawn upon during submersion. Aside from the objection that the volume of air involved is so small compared with the volume of the lungs—an estimated 4 percent for *Corythosaurus* (Ostrom, 1961)—it is inconceivable how this air could have been transferred to the lungs when the hadrosaur was underwater. It has even been suggested that the crests functioned as snorkels, but there is no evidence of an opening into the crest in any skull. Franz Baron Nopcsa, a Hungarian paleontologist, proposed that all the crested hadrosaurs were males and all the noncrested ones females. This notion is refuted by the fact that there are other skeletal characteristics that distinguish between the two types; there is also the problem that crested and noncrested hadrosaurs are frequently not found together at the same geological level.

Ostrom (1961) approached the problem by considering which of the several functions of the nasal passages of modern animals best fitted the pattern of elaboration seen in hollow-crested hadrosaurs. The narial passages of living animals function to remove particles, to humidify, and, in the case of warm-blooded animals, to warm the inhaled air before it passes to the lungs. The passages also provide for a sense of smell (olfaction). There is no reason why the crest should not have served several or all of these functions, but after considering each one in turn Ostrom concluded that the most likely function was olfaction.

If an improved sense of smell was the primary function of the hadrosaur crest, how did the flat-headed hadrosaurs like *Edmontosaurus* manage so well without a crest? This is especially puzzling in light of the fact that the flat-headed ones out-survived most of the crested ones. Furthermore, what was the significance of the solid crests? Some of them were just as prominent as the hollow crests, and they could not have served an olfactory function. James Hopson (1975) offered a solution that would resolve all of these problems. He proposed that the crests were organs of display. Living animals have to communicate with one another for a variety of reasons. They need to warn of danger, advertise abundant food supplies, attract mates, communicate with their young, defend territories, recognize members of their own species, and more besides. And they employ a variety of techniques to do this. Birds sing and display their colorful plumage, honeybees dance to tell other bees where to search for pollen, and dogs use scent to mark out their territories. Closely related species often look alike, and it is then very important for individuals to be able to recognize their own kind, to prevent matings between individuals of different species. Many bird species, for example, are virtually indistinguishable from one another, but the males sing different songs so that they do not attract females of another species. There are many species of hadrosaurs, and, at least in terms of their body skeletons, they appear to have been very similar to one another. Hopson therefore proposed that the most likely function of the crest was for species recognition. The solid crests could obviously serve for visual recognition, and the hollow crests, with their extensive bony chambers, could have served as resonating chambers for producing sounds as well. He also suggested that the hollow areas around the external nares, which are especially extensive in the flat-headed species, may have housed an inflatable extension of the fleshy nostrils. These structures may have been brightly colored, like the scarlet throat sac of the frigate bird; they may also have been capable of producing sounds.

Weishampel (1981), working independently on the idea of sound production, made an extensive study of the vocalization potentials of the crests, paying particular attention to *Parasaurolophus*. Little imagination is required to equate its U-shaped tube with a trombone, and by the application of simple acoustical theory he calculated the resonant frequencies of the tube. Long tubes resonate with lower frequencies than shorter ones, and the lowest note that can be obtained from a given tube is called the fundamental frequency. If you pursed your lips and blew fairly gently down an open tube (copper plumbing pipe is ideal!) the first note you would obtain is the fundamental. If you then blew harder you would obtain a note that is twice the frequency of the first. This note, called the first harmonic, is one octave higher than the first. With a suitably long tube, and a little practice, it is possible to obtain at least the first five or six harmonics. Weishampel calculated that the fundamental frequency for *Parasaurolophus walkeri* was forty-eight cycles per second (48 Hz) and that the frequency range for the first five octaves was 48–240 Hz. A second species, *Parasaurolophus cyrtocristatus*, which has a shorter tube, was estimated to have a range of 75–375 Hz for the same harmonics. Higher harmonics are possible for both species. Furthermore, the lateral chambers, being much shorter than the main U-tubes, would produce sounds of higher frequencies. Weishampel did not attempt to calculate the resonant frequencies for other crested hadrosaurs because their acoustics are far more complex than the acoustics of the simple U-tube of *Parasaurolophus*.

We will never know whether hadrosaurs really did use their crests for vocalization, but the idea is very plausible. The fact that they would have made relatively low frequency sounds is very significant (see page 81). And if the crests were for visual and vocal displays, the hadrosaurs would obviously have needed suitably well developed senses of sight and hearing. Is there any evidence that they did? Before we attempt to assess the sensory potentials of animals that lived more than 60 million years ago, we need to know something of the faculties of modern ones. Specifically, we need to know whether there are any features of the skull that can be used as reliable indicators of sensory function.

Taking stock of the senses

An animal's sensory awareness—to sight, sound, smell, taste, and touch—requires both the possession of the relevant sensory organs (eyes, ears, olfactory organs, taste buds, touch receptors) and the development of those particular regions of the brain that convert impulses from these organs into sensations. Therefore we have to investigate both the sensory organs and the brain if we hope to draw conclusions about the dinosaurs' senses.

The vertebrate brain is a conservative organ in which particular regions are usually associated with specific functions. Consequently, the relative development of a certain area provides an indication of the relative development of the function with which it is associated. It is therefore of much interest to see whether the skull reveals any structural information about the brain—but that is the subject of Chapter 7. Here we will restrict ourselves to searching for skeletal clues to the sensory organs themselves.

Some indication of the size of an animal's eye, for example, may be obtained from the size of the orbit. Obviously an eye can be no larger than its orbit, but it may be smaller, and so the orbital diameter indicates only the maximum possible size. Many reptiles and birds and also fishes have a series of bony plates embedded in the white part of the eye. These are joined together to form a bony ring called the sclerotic ring, which gives a fairly precise measure of the actual size of the eye. Sclerotic rings have been preserved in many fossil groups, including dinosaurs, pterosaurs, and ichthyosaurs, so we have a good idea of eye sizes in these particular instances.

Most birds and mammals have large eyes, sight being their predominant sense, but there are exceptions, like the kiwi. This nocturnal bird forages for insects and worms beneath the forest litter using its long, sensitive bill. Sight plays a minor role in its life and its eyes, and thus its orbits, are correspondingly small. The owl is a nocturnal bird too, but sight is its predominant sense and its eyes are relatively huge. Large eyes have more light receptor cells than smaller eyes and are therefore more sensitive, especially under low light conditions. For this reason large eyes are often correlated with the nocturnal habit. An owl's eye is both absolutely and relatively much larger than that of the mouse that it hunts. This probably makes little difference during broad daylight, when the mouse probably sees as much of the world as the owl. But at night the greater light-gathering capacity of the owl's eyes gives visual supremacy over its prey. This principle was well illustrated for me when I once had to decide which of two pairs of binoculars to purchase. They had the same magnification ($\times$7) but the sizes of the objective lenses were different (35 and 50 mm), so they had different light-gathering capacities. There was absolutely no difference in their performance during the day, but as dusk approached the 7 $\times$ 50 binoculars appeared to have the edge. By nightfall I could still pick out details of objects with this pair that I could not even see with the 7 $\times$ 35 glasses.

In most vertebrates the visual angle of the eye, which is essentially the maximum angle of the cone through which light can enter the eye, is about 170 degrees (Walls, 1942). This means that most eyes, our own included, have a field of view that covers an arc of almost half a circle. To demonstrate, just close your left eye and concentrate your attention on an object in front of you. Now stretch out your right arm to the side, just behind the level of your head, and, while waggling your fingers, swing your arm slowly forward until you just catch sight of your fingers at the edge of your field of vision. Try it a few times, making sure to keep your right eye facing forward all the time. The angle between your forward line of sight and the position of your fingers when they first come into view will not be far off a right angle. Doubling this angle gives the angle for the other half of your right eye, and the resulting angle, the visual angle, is close to 180°. (Repeating the experiment for the right eye with the left arm outstretched does not work because the nose gets in the way.)

The eyes of many animals are directed forward so that their two fields of view overlap, giving the animal well-developed binocular vision.[3] An overlapping view permits depth perception, of particular importance to hunting animals and others that have to judge distances. The greater the degree of overlap of the two fields, the more complete is the binocular vision. Mammalian predators tend to have eyes that face directly forward, giving the maximum amount of overlap and the highest level of binocular vision but a correspondingly reduced total field of view. Some birds of prey, including owls and hawks, similarly have forward-facing eyes, but reptilian predators tend to have a more limited degree of binocular vision. The degree of overlap between the two fields of vision is only 24° in the alligator, 32° in the monitor lizard, and 34° in the boa constrictor. These reptiles are, nevertheless, successful predators, perhaps because other senses, especially olfaction, are major factors in their hunting success.

Generally speaking, the eyes of the hunted are placed on the sides of the head. This gives them a total field of vision almost twice that of a single eye, that is, nearly 360°. There is little overlap of visual fields, so their binocular vision is not well developed, but their all-around vision enables them to detect the movement of predators from whatever direction they may approach. The relative placement of the eyes can therefore give some guide to an animal's habits, and we can use what we know about eye placement, together with the relative size of the orbit and sclerotic ring, as clues to the visual potentials of hadrosaurs.

The orbit in hadrosaur skulls is relatively large and the sclerotic ring has been preserved in some specimens, giving a good indication of the size of the eye. In our mounted skeleton of *Lambeosaurus*, the sclerotic ring has an outside diameter of approximately three inches (70 mm), which indicates an eye of about the same size as a cow's. The evidence from endocranial casts, which is unfortunately limited to crestless species, is incomplete in the region of the optic lobes, but the cerebral hemispheres, which were probably concerned with sight among other functions, were fairly large. This evidence, taken with the relative size of the eye, suggests that sight was probably well developed. The skull of *Lambeosaurus* has suffered fairly extensive lateral compression, which is usually the case for skulls, and this exaggerates the degree

The eyes of the hunter face forward, giving the animal binocular vision and hence good depth perception. Those of the hunted face the sides, giving poor depth perception but a good all-around field of view.

of lateral placement of the eyes. Even so, it appears that the eyes were placed on the sides of the head rather than in a forward-facing position. This suggests that hadrosaurs were probably specialized for wide-field rather than for binocular vision, which accords with their herbivorous life-style.

The amount of information we can glean for the sense of hearing in hadrosaurs, or indeed for any other dinosaur, is small, but before assessing the data we need to be familiar with the organs of hearing in living forms. Our own ear, like that of most other mammals, has three parts—an outer, a middle, and an inner ear. The outer ear consists of an earflap, the pinna, which leads into a canal that is closed at the bottom by the eardrum. The pinna and the canal leading to the eardrum are found only in mammals. In amphibians, reptiles, and birds, the eardrum is on the surface of the head, behind the eye.[4] Fishes have no eardrum at all. On the other side of the eardrum is the middle ear cavity, inside of which is an articulating series of three small bones in mammals. The outer bone, the malleus, is attached to the eardrum. The inner bone, the stapes, is attached to a smaller membrane stretched across a small opening, the oval window, that leads to the inner ear. The middle bone of the series is the incus. The three bones serve to transmit vibrations from the eardrum to the inner ear. They also amplify the vibrations, partly because the oval window is so much smaller than the eardrum and partly because the three bones function as a lever system. The amplification system provided by the middle ear has undoubtedly been a major factor in the evolution of hearing in terrestrial animals. This is because air does not conduct sound as well as water and land do; airborne vibrations have lower energies and do not travel so far. (Remember those old western movies where the intrepid scout presses an ear to the rails to listen for the approaching train?)

Reptiles and birds have but a single bone in their middle ear, the stapes, but experiments have shown that the amplification of this system is in no way inferior to that of the mammalian middle ear (Manley, 1971, 1972). Indeed, for at least one lizard that has been tested, the amplification has been found to be greater than that of a cat's ear. Whereas modern reptiles have a light and slender stapes that is specialized for conducting airborne vibrations, many extinct species have a massive stapes. A heavy stapes is suited for conducting terrestrial vibrations to the inner ear, as when an animal listens with its head to the ground.

All vertebrates have an inner ear, but there is much variation in its development. It is essentially a small, fluid-filled, and membranous sac (ours is only about an inch, 30 mm, high) embedded in the bone at the back of the skull. The membranous sac, or labyrinth, is separated from the surrounding bone by a thin layer of liquid that serves to transmit vibrations from the stapes. The upper portion of the membranous labyrinth, which is concerned with balance, has three semicircular canals at right angles to one another. Whenever the head is moved the fluid inside the canals is swirled, causing excitation of the sensory cells within the canals. Depending on which way the head is moved, the fluid in one of the three canals will be swirled more violently than in the other two and the impulses that reach the brain indicate position. If we spin around, the fluid in the three canals swirls around too and the sensory cells get so confused that we become dizzy and are no longer capable of balancing.

The hearing function of the membranous labyrinth is conducted in the lowermost portion, in a small tube called the cochlear duct. The anatomy of this region is fairly complex because the roof and floor of the cochlear duct lie in contact with two bony tubes. The cochlear duct therefore forms the inner portion of a triple tube, but we need not be concerned with the details. The floor of the cochlear duct, called the basilar membrane, contains cells that are sensitive to sound waves. It seems that the longer the cochlear duct, and therefore the longer the basilar membrane, the more sensitive the hearing is to high-frequency sounds. The relative lengths of the basilar membrane in different vertebrates are seen to increase from reptiles to birds to mammals. There is therefore an increasing sensitivity to higher-pitched sounds from reptiles to mammals. Mammals, which are able to hear the highest-frequency sounds of all animals, are unique among the vertebrates in having the cochlear duct coiled into a snail-like structure called the cochlea. The cochlea accommodates a longer duct in a relatively small space. The uncoiled cochlear duct of reptiles and birds is called the lagena.

Birds and mammals are about equally sensitive to sound, meaning that their ability to detect very quiet sounds is about the same. Reptiles, on average, are less sensitive. Each animal species seems to have a narrow range of frequencies within their normal hearing range at which they are most sensitive. It is of particular interest to us here that reptiles are most sensitive to low frequencies. Also of considerable interest is the fact that crocodiles, the closest reptilian relatives of dinosaurs, have a particularly well developed inner ear and a basilar membrane that is as long as that of the pigeon. They are most sensitive to sounds within a range of about 100–2,500 Hz (Wever, 1971). Significantly, vocalization plays an important part in their social behavior.

Since the membranous labyrinth is encased in bone, its shape can be reconstructed simply by examining the relevant parts of the skull—the otic capsule. Unfortunately, though, the task is usually not this simple. In fossils the otic capsule may have been lost or only partially preserved. Parts of the original otic capsule may have been cartilaginous and therefore not preserved at all, which is the case in the ichthyosaurs. And even if there is a well-preserved otic capsule, those parts bearing impressions are often inaccessible. Information is therefore limited, but the membranous labyrinth has been reconstructed for two hadrosaurs, both flat-headed ones: *Anatosaurus* (see Ostrom, 1961) and *Lophorhothon* (Langston, 1960). As Weishampel (1981b) has pointed out, the hadrosaur's membranous labyrinth was very similar to that of crocodiles, and it had a well-developed lagena, suggesting that hadrosaurs probably had a sense of hearing comparable to that of crocodiles. So much for the inner ear, but what of the middle ear?

The stapes has been found in at least two hadrosaur specimens, belonging to the genera *Anatosaurus* and *Corythosaurus* (Colbert and Ostrom, 1958). The bone is long and slender, indicating that the middle ear was specialized for transmitting airborne sound waves. The stapes appears to have been preserved in its natural position in the specimen of *Corythosaurus;* it suggests that the specimen had a relatively large eardrum located immediately behind the quadrate. There is also evidence that the oval window was relatively small, indicating that the middle ear provided for a significant amplification of the vibrations received at the eardrum.

The evidence from the middle and inner ear structures strongly suggests that hadrosaurs were able to hear and that their sense of hearing was probably on a par with that of modern crocodiles. Crocodiles, like other reptiles, are sensitive to sounds at the lower end of the frequency spectrum—the same frequencies that hadrosaurs are believed to have been able to generate. The evidence from the organs of hearing therefore reinforces the suggestion that hadrosaurs used their crests to vocalize. We can allow our imaginations to conjure up sounds that might have been heard on a Late Cretaceous dawn.

How much of their time hadrosaurs may have spent in water we do not know. Most hadrosaur skeletons have been found in freshwater deposits, so it is clear that they did not live very far away from water (Dodson, 1971). Perhaps some species, like the giants from Mexico with their deep tails, spent more time in the water than did other species. Some paleontologists have suggested that hadrosaurs, lacking horns or claws or any other organs of defense, may have retreated into the water to escape from predators. But it is doubtful that water would have been much of a deterrent to some of the formidable predators that were abroad during those far-off times.

A brief look at predators

Space does not permit as detailed a treatment of predatory dinosaurs as was given for the hadrosaurs, but it would be useful to make some comparisons between the two types and contrast the adaptations of meat eaters and plant eaters. Several species of carnivorous dinosaurs were contemporaneous with hadrosaurs— from the small *Dromaeosaurus* (six feet long, or 1.8 m) to the massive *Tyrannosaurus*, the "tyrant lizard" (forty-six feet, or 14 m). As there is an almost complete skeleton of *Albertosaurus* in the Royal Ontario Museum, one of the best known of the large theropods, I have chosen this predator to serve as example.

Albertosaurus libratus is a close relative of *Tyrannosaurus rex* but does not appear to have been quite as large, and probably did not exceed thirty feet (9 m) in length. Both species are found in western Canada, and the differences between them are minor. Both belong to the family Tyrannosauridae, often simply referred to as tyrannosaurids or tyrannosaurs. Our specimen of *Albertosaurus*, which was not fully grown, is twenty feet (6 m) long, which is about 20 percent smaller than the hadrosaurs upon which it may have preyed.

Albertosaurus ("Alberta lizard") looks far more slender, racier, than the hadrosaurs, and its individual bones are generally relatively thinner. The ilium, for example, the broad blade that forms the upper part of the pelvis, is less than half an inch thick (less than 1 cm), whereas it is about an inch thick (more than 2 cm) in *Lambeosaurus*. Our skeleton of *Albertosaurus*, mounted in the 1920s, was given an upright posture similar to that of the hadrosaur *Corythosaurus*. As noted above, this posture was unlikely for the hadrosaur, but it probably was appropriate for *Albertosaurus* because its hip joint was quite different from that of a hadrosaur. The head of the femur, instead of being braced against a weak point, abuts against a bony buttress of the ilium. This is because the ilium makes a larger contribution to the anterior rim of the acetabulum (the hip socket) in *Albertosaurus* than it does in hadrosaurs, and the weak point is consequently farther down. The pelvis, and therefore the vertebral column to which it is attached, is able to tilt up to about 60° from the horizontal before the head of the femur abuts against the weak point, whereas in the hadrosaurs this angle is only about 20°. On the evidence of the pelvis, then, it appears that *Albertosaurus* may have been able to stand in

Although this skeleton of the tyrannosaur ALBERTOSAURUS *is mounted in a semi-upright position, close inspection reveals that the head of the femur does not abut against the joint between the pubis and ilium (as it does in the upright hadrosaur skeleton). The ilium is far deeper here than it is in hadrosaurs, indicating that tyrannosaurs were probably capable of a more upright posture. Carnivorous features of* ALBERTOSAURUS *include the sharp, serrated teeth, the short neck, and the clawed hands and feet.*

a semi-upright posture, perhaps to look around for prey. But the vertebral column was probably held horizontally during locomotion, the long tail counterbalancing the weight of the front part of the body about the pelvis, as in the hadrosaurs.

The carnivorous habit of *Albertosaurus* and its relatives are immediately betrayed by the structure of the skull. The sharp teeth

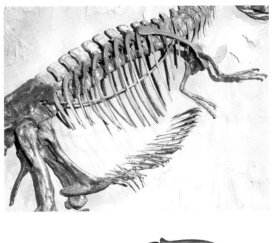

have the appearance of daggers, and closer inspection reveals small serrations on their leading and trailing edges, similar to those of a steak knife. The entire tooth was invested in enamel, including the cutting edges, in contrast with the teeth of herbivores, whose cutting edges display the differential wearing rates of enamel and dentine. The teeth of *Albertosaurus* were no doubt efficient for puncturing and slicing through flesh; they have retained their cutting edges to the present day. The teeth are set in deep sockets, with about as much of the tooth below the jaw line as there is above it. Consequently both jaws are deep. Tyrannosaurs had to have their teeth firmly rooted to prevent them from being ripped out in struggles with captured prey. Teeth were lost from time to time and were continually being replaced, as in the hadrosaurs and other reptiles. Tooth replacement was alternate, a strategy for preventing adjacent teeth from being replaced at the same time, which would have resulted in gaps (Edmund, 1960). The sequential nature of tooth replacement may be seen by marking alternate teeth along the length of a jaw margin. A line drawn through the tips of the marked teeth reveals a regular size gradation, showing the relative ages of the teeth. Each tooth started as a small bud below the jaw line, and as it grew it pushed its way into the pulp cavity of the old tooth. Eventually the old tooth was expelled and the new one took its place.

The front teeth in the upper jaw are smaller than the others, and instead of being flattened from side to side they are flattened from front to back, as are the chisel-shaped incisors of mammals. The front teeth may have been used for tearing meat from bone, in much the same way that a dog uses its incisors. In hadrosaurs, remember the jaw joint permitted a degree of freedom of movement of the mandible, but in *Albertosaurus* and its allies the joint is so constructed that only a hinge-like up-and-down movement is allowed. The movement is determined by the shape of the articular surfaces. The quadrate has a cylindrical articular surface, set at right angles to the length of the skull, and this surface fits into a corresponding groove in the mandible. Carnivores use their teeth primarily for cutting and slicing; any lateral movement in the mandible would be equivalent to trying to cut something with a pair of scissors with a loose joint. The same is true of modern mammalian carnivores. If we watch a cat or a dog feeding we are unable to detect any lateral movement of the mandible, whereas a cow or a sheep rolls its jaw better than any politician.

We saw in the hadrosaurs that the teeth were located posteriorly rather than anteriorly—there were no teeth at the front of the jaws at all. The reverse is true for the tyrannosaurs, which had a wide gape for seizing and killing and for slicing large pieces of flesh. The skull of *Albertosaurus* is far less robust than that of the hadrosaur, partly because of a number of additional openings that perforate the skull (besides the usual orbit, nares, and temporal openings for the jaw muscles). The individual bones are remarkably thin and thus lighter than the bones of the hadrosaur. The bones of the mandible of our specimen of *Albertosaurus,* for example, are only about one-quarter of an inch thick (0.5 cm), compared with a width of about an inch (more than 2.0 cm) for the mandible of *Lambeosaurus.* The mandible of *Albertosaurus,* being restricted to up-and-down movements, was probably not exposed to large lateral forces (unless it were seizing a struggling prey, a point to which we will return shortly) and had only to resist the large vertical forces generated by the powerful bite. These forces were best resisted by having a deep rather than a wide jaw. The same mechanical principle is used in house construction, where planks of wood are used, edgewise on, as supporting beams for the floor above. The hadrosaur mandible, in contrast, appears to have been exposed to large lateral forces during the powerful chewing movements, so it is correspondingly wide as well as deep.

Size for size, then, tyrannosaur skulls are lighter than those of hadrosaurs, but what could be the advantage of this? Animals are notoriously uncooperative at confirming the hypotheses we construct for them, so we should be circumspect in attempting to answer this question. First, it seems reasonable to suppose that hadrosaurs did not need to move their heads very rapidly while going about the peaceful business of cropping plants, whereas tyrannosaurs, as active carnivores, probably did. The cut and thrust of a successful kill would no doubt have involved many fast moves of the head, and any significant reduction in its weight would have increased its speed by virtue of its reduced inertia.

Conversely it could be argued that seizing a struggling prey would have required a robust skull to withstand the large and presumably multidirectional forces involved. Perhaps *Albertosaurus* selected prey that were much smaller than itself, to reduce its chances of personal injury. African wild dogs apparently have relatively weak jaws (see Chapter 5), which might help explain their group killing strategy of wearing their prey down by inflicting small wounds rather than engaging in all-out attacks.

The head of *Albertosaurus* was supported on a short and seemingly powerful neck. We assume the neck was powerful, that is muscular, because the relatively long neural spines and long ribs associated with the cervical (neck) vertebrae provided large attachment areas for the neck musculature. This picture contrasts with the cervical vertebrae of hadrosaurs, whose neural spines were barely developed and whose ribs were relatively small. Modern big cats similarly have well-developed neck muscles, allowing for the powerful movements of the head when capturing and killing prey.

We have seen that modern mammalian predators possess good binocular vision but that the degree of overlap of left and right fields is generally less well developed in their reptilian counterparts. We also noted the difficulties of trying to establish the relative positions of the orbits in fossil skulls because of compression distortion. There is good evidence, however, that there was a good degree of overlap between left and right visual fields in the advanced tyrannosaurs, namely in *Albertosaurus, Tyrannosaurus* and its very close Mongolian relative *Tarbosaurus,* and especially in the "small tyrant," *Nanotyrannus* (Bakker, Williams, and Currie, 1988). The orbits in the diminutive tyrannosaur face directly forward, showing that it possessed well-developed binocular vision. The non-tyrannosaurid theropod *Troodon* ("wounding tooth"; formerly called *Stenonychosaurus*) similarly has forwardly directed orbits, showing that it too probably enjoyed good depth perception (Russell, 1969). This small (six feet, or 2 m), lightly built dinosaur, which was contemporaneous with *Albertosaurus,* had very large orbits, indicating that it had relatively enormous eyes. Evidence from endocranial casts indicates that it had a very large brain, by dinosaurian standards, and it has been suggested that this was an intelligent, keen-sighted, and extremely active little dinosaur. It had well-developed arms, long-fingered hands, and sharp claws, which it used, we assume, for hunting active small prey, like lizards and mammals. The fingers and toes of the tyrannosaurs and their relatives terminated in sharp claws, in contrast to the hooves of their herbivorous contemporaries. It is not hard to imagine them being used as weapons of offense.

It has been suggested that tyrannosaurs, rather than being active predators, were scavengers, feeding only on carrion, but this is difficult to reconcile with some of their skeletal features. Why would a scavenger need such a large and powerful head and neck for feeding on putrefying meat? What is the significance of having a relatively gracile skeleton unless it is for the pursuit of prey?

And if tyrannosaurs did have some measure of depth perception, which seems likely, would this ability have been important to them if they fed on dead meat? Good peripheral vision to detect the approach of competitors would seem more appropriate for scavengers.

One of the major differences between the tyrannosaurs and their Jurassic relatives, as exemplified by *Allosaurus* ("strange lizard"), is the extreme reduction in the forelimb. Not only is the length of the limb smaller in *Albertosaurus* than in the earlier *Allosaurus,* but the hand has only two fingers instead of three. The forelimbs were so small that they barely projected beyond the body and it has been suggested that their main function may have been to act as grapples in assisting a resting individual back onto its hindlegs. By digging the front claws into the ground and applying a backward force, the forelimbs could have anchored the front of the body and prevented forward slippage when the hindlegs were raised beneath the body. Perhaps this was just one of their functions; they may also have been used as grapples to steady a leaping aggressor as it inflicted wounds with its hind claws. Cats, both large and small, often seize a victim with the front paws while kicking in unison with the back ones. The forelimbs may also have been used for manipulating food in the mouth—they could just have reached that far with the head bowed.

In marked contrast to the extreme forelimb reduction seen in the tyrannosaurids are the gigantic forelimbs of an Upper Cretaceous dinosaur named *Deinocheirus* ("terrible hand"), discovered in Mongolia in 1965 (Osmolska and Roniewicz, 1970). As is so often the case, we catch only a tantalizing glimpse of this unusual animal because only its forelimbs and parts of the shoulders were preserved. But just imagine a dinosaur with arms that were eight feet long (2.5 m) and with foot-long (0.3 m), sharply pointed claws. *Deinocheirus,* which may have been related to the ostrich-like ornithomimid dinosaurs, was likely a formidable predator.

Albertosaurus has a relatively longer lower leg than *Lambeosaurus,* because the metatarsus is relatively longer. This suggests that *Albertosaurus* could run faster than *Lambeosaurus.* The clawed toes might have provided greater traction than hooves gave *Lambeosaurus,* traction being of significance for making rapid changes in direction in the pursuit of fleeing prey. *Albertosaurus* had no mechanism for retracting its claws as a cat does, but there was a group of small carnivorous dinosaurs, called dromaeosaurs ("running lizards"), in which this was possible. The best-known example of

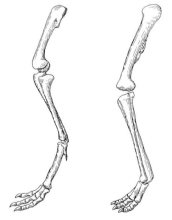

ALBERTOSAURUS (left) *has a relatively longer lower leg than* LAMBEOSAURUS (right), *suggesting it was a faster runner.*

these remarkable dinosaurs is *Deinonychus,* which is known from the Lower Cretaceous of Montana (Ostrom, 1969, 1976a).

Deinonychus was a fairly small dinosaur, about ten feet (3 m) long, but it was robustly built. The forelimbs are long, so too are the three fingers, each armed with an especially sharp claw. The four toes were similarly armed, three pointing forward and one (the "big toe") pointing backward, just as in *Albertosaurus* and most other theropods. What is most unusual about *Deinonychus* is that the innermost of the three forwardly-directed toes had a massive sickle-shaped claw, which is jointed, like the claw of a cat, with an extensive wraparound articular surface. This toe was capable of being swung right back, clear of the ground, the way that cats retract their claws when they walk. Retraction helps to keep claws sharp. Dogs, in contrast, are unable to retract their claws, which is why their claws are blunt. The pelvis of *Deinonychus* is also modified in that the pubis has swung back and lies parallel with the ischium. Although not unprecedented among theropods, this is nevertheless an unusual pelvic condition because the pubis is usually inclined forward. The femur was also modified, having an extra process on its trailing edge that has been interpreted as an attachment area for a large leg-retractor muscle. Both features, it seems, were adaptations for a powerful downward slashing motion of the foot. One can easily imagine the terrible damage that could have been inflicted by the sickle claw when it was bared in anger. An equally unusual feature of the skeleton of *Deinonychus* is that the tail, save for a short section at its root, is ramrod straight, this being achieved by a series of overlapping bony rods. Close examination reveals that these rods are actually bony extensions of the processes of the vertebrae. The bony rods in the dorsal (upper) part of the tail are elongated anterior zygapophyses, and the ventral ones are chevron bones that have been drawn out anteriorly. The ornithomimid dinosaurs similarly have long zygapophyses, but they are nowhere near as long as they are in *Deinonychus.* The stiff tail, like the unusual pelvis and femur, was probably a modification correlated with the kicking action of the hindlegs.

Dromaeosaurus, from the Upper Cretaceous of western Canada, was a scaled-down version of *Deinonychus,* reaching lengths of only about six feet (2 m). These seemingly agile little dinosaurs may have hunted in packs, like African wild dogs. Imagine the chilling scene of a swarm of determined "running lizards" closing in on some hapless hadrosaur, and slashing it to death.

A Matter of Scale

T H E Victorians had a penchant for dressing their children as adults. Sons wore suits and daughters crinolines, and both were expected to look and act like perfect miniatures of Papa and Mama. The Victorians were missing the point, of course—there are greater differences between children and adults than a mere disparity in size. And the same holds true for the differences between small and large animals.

One of the obvious differences between small and large things is their difference in weight, or, more strictly, mass,[1] but the discrepancy is far greater than most people imagine. Think of two objects of the same general shape, one of which is twice the size of the other. How much greater is the mass of the larger object? Provided the two objects are both solid, doubling the size increases the mass by a factor of eight. This is because volume, and therefore mass, increases according to the *cube* of the size increase. There is a very easy way we can demonstrate this, just by looking at things about the home. Many consumables come in small, regular, and large sizes, often in containers that are essentially the same shape—barbecue sauce, scotch, and sugar come immediately to mind. I have a regular and a king-sized bottle of barbecue sauce in front of me; the smaller one is seven-and-a-half inches tall (190 mm), the other nine-and-a-half inches tall (245 mm). The increase in size is therefore 9.5/7.5 = 1.27, which is not very much. But the cube of this size difference, the increase in mass, turns out to be 2.05. So the larger bottle should have twice the mass of the smaller one, as is confirmed by the information on the labels— the smaller bottle contains sixteen ounces (455 ml), the larger one thirty-five ounces (1,000 ml). These observations may be surprising if you have not given such things much thought before. Think what this means as far as people's weights are concerned. A person who is, say, six feet four inches tall (1,930 mm) would weigh twice as much as someone who is only five feet tall (1,524 mm). People who guess weights at fairs are well aware of this principle, even though they may be unfamiliar with the theory behind it.

Another important difference accompanies size increase, namely the change in relative areas, and this, like the change in mass, has important biological consequences. A simple experiment with sugar cubes, real or imagined, illustrates this point. A single cube has a volume of 1, a mass of 1, an area of base of 1, and a total surface area of 6. The pressure, or stress (mass/area) acting on the base is therefore $1/1 = 1$, and the total area-to-volume ratio is $6/1 = 6$. What happens when we double the size of the cube, that is, double the linear dimensions? To make a cube whose edge is two sugar cubes long requires a total of eight cubes. Each of the six faces of this larger cube comprises four sugar cubes. The volume of the cube is now 8, the mass is 8, the area of the base is $2 \times 2 = 4$, and the total surface area is $6 \times 4 = 24$. The stress acting on the base is now $8/4 = 2$, and the area-to-volume-ratio is $24/8 = 3$. Doubling the size of the cube has therefore doubled the stress acting on the base and halved the area-to-volume ratio.

Let us make one more cube, this time one with an edge that is three sugar cubes long, that is, a cube that is three times larger than the original one. We need 27 sugar cubes this time because each of the six faces comprises nine cubes. The volume of the cube is now 27, the mass is 27, the area of the base is $3 \times 3 = 9$ and the total surface area is $6 \times 9 = 54$. The stress acting on the base is now $27/9 = 3$, and the area-to-volume ratio is $54/27 = 2$. Tripling the size of the cube therefore triples the stress acting on the base and diminishes the area-to-volume ratio to one-third the original. A pattern emerges: the stress on the base increases according to the increase in linear dimension, while the area-to-volume ratio decreases according to the reciprocal of the linear dimension. Simply stated, as things get bigger the stresses acting on them increase while their relative surface areas get smaller— provided they retain the same shape. And all of this comes about because mass and volume increase with the cube of the linear dimension while areas increase with the square. Both of these relationships have major biological consequences, which will be referred to in the chapters that follow. Here, however, we will focus on stress and its immediate implications for weight bearing.

We have seen that the vertebral column is the primary supporting structure for the whole body and that (for four-legged animals) the fore and hind limbs, through their connections with the pectoral and pelvic girdles, have to support this entire mass. The total area of bone available to support the body is the combined area of cross section of the bones of the four legs. Supposing the limb bones of large animals had exactly the same shape as those of smaller ones, that is, all the linear dimensions were scaled to the same degree. This would mean that if photographs of the bones were taken and all printed to the same size they would be identical. The bones would be said to have geometric similarity (as in the case of the sugar cube example). If this were the case the stresses acting on the bones would increase with increasing body size. The limb bones of the larger animals would therefore experience greater stresses than those of smaller ones. Since the compressive strength of bone has fairly fixed limits, a point would be reached when the burgeoning giant would no longer be able to support its own weight. What to do?

The most obvious solution to the problem is to increase the diameters of the limb bones more rapidly than the increase in length. This would mean that progressively larger animals would have progressively more robust limb bones. This strategy is actually used in building construction. The concrete supporting columns at the base of a high-rise building have to support the weight of the entire structure, but those that are half-way up only have to support half of that weight. Consequently the thickness of the columns is greatest at the base and decreases toward the top. Do larger animals have relatively thicker bones? Well, it all depends on which animals are examined. Two separate studies of ungulates (hoofed animals) showed that larger animals do have relatively thicker leg bones (McMahon, 1975; Alexander, 1977). We could certainly go through a collection of modern skeletons and pick out the femora (thighbones) of progressively larger animals—say a dog, a deer, and a hippo—that would show a nice gradation of increasing robustness. But the series would be wrecked if we tossed in an elephant femur because it is relatively slender, even though the elephant is the heaviest of all land animals. When a

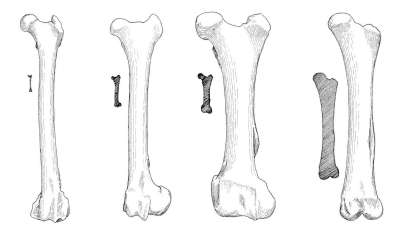

Larger animals are often thought to have relatively thicker bones than smaller ones. While this may hold true for selected groups of animals, as depicted by the first three femora (thigh-bones) shown here, it does not hold for a wide range of groups. From left to right, the femur of the dog, red deer, hippopotamus, and African elephant. (The relative sizes of the bones are shown in silhouette.)

large enough variety of animals is assessed, the overall picture is more nearly one of geometric similarity (similar shape) than of increased robustness (Alexander et al., 1979). How, then, do large animals cope with the increasingly large stresses on their leg bones? By analyzing movie footage of a running elephant and a running buffalo, Alexander and his colleagues concluded that the larger animals keep their feet on the ground for relatively longer segments of each stride, which reduces the peak stresses acting on the limbs (Alexander, Langman, and Jayes, 1977; Alexander et al., 1979). They found that the fraction of the time a given foot is on the ground, the *duty factor,* was 0.49 for the elephant and 0.35 for the buffalo, which was about one-fifth the elephant's weight. This means that none of the elephant's feet were off the ground for more than about half of the time, for the running speeds recorded. Now this all makes perfectly good sense but it may not be correct.

Measuring duty factors accurately is a difficult task and requires a high-speed movie camera. Andrew Biewener, a functional anatomist at the University of Chicago, has pointed out that since such equipment was not used by Alexander and his colleagues, their errors could be as high as 30 percent. Biewener (1983) measured the duty factors of five different species ranging in size from mice to horses. These were measured at increasing speeds as the animals ran on treadmills. He also measured the angles that their limb segments made to the vertical. Furthermore, by reference to a collection of skeletons, he estimated the relative curvature of the limb bones of these five species, together with thirty-two other species besides.

When duty factors were compared at the same equivalent speeds (that is, at the trot-gallop transition; see Chapter 3), Biewener found that they were about the same for the large animals as they were for small ones. Admittedly, he did not measure the duty factor of the elephant, and it might be that the elephant does have a longer duty factor. But as far as the other large animals studied were concerned, stresses on their limb bones were obviously not reduced the way that Alexander and his colleagues had suggested. So how were they reduced? Significantly, the angles of the limb bones decreased with increased body size; the limb bones of larger animals were held more nearly vertically, therefore reducing the bending stresses acting upon them. Larger animals also tended to have straighter bones, and this again appears to be a strategy for reducing stress.

Biewener (1989) also found that the changes in posture that accompany increases in body mass reduce the forces that the muscles have to exert to support the animal's weight. The forces generated by the leg muscles were found to scale in proportion to (body mass)$^{0.74}$, which means that as animals get heavier the stresses in their muscles become relatively smaller. The stresses in the leg muscles of an elephant would therefore be similar to those of a mouse.

Anatomists have long recognized that heavy-bodied animals hold their limb bones straight so that their legs function as vertical supporting columns, and such animals are described as being graviportal. The largest living land animal, the elephant, is graviportal. A study of its skeletal adaptations, together with the physiological and behavioral modifications that accompany the attainment of large body size, will provide valuable insight into the largest of all dinosaurs, the sauropods. A study of the elephant also provides insights into some of the problems of collecting data and why the data for living animals, far less for extinct ones, are so often less reliable than we should like them to be. Because of the large sizes involved, the weights will usually be given in tons. An English ton is 2,240 pounds (called the "long ton" in the United States) and a metric ton (spelled *tonne* to distinguish it from the English unit) is 1,000 kilograms, which is 2,200 pounds. The measures used here are approximate, so I will not distinguish between them.

Elephants: The largest living land animals

There are only two living species of elephant, the African and Indian, though there were many more species in the past. The African elephant is the larger of the two, weighing up to five or six tons. Can we not be a little more specific in stating its maximum weight? Actually we cannot, and for two good reasons. First there is the problem of weighing an animal as large as an elephant.

Just think about it. You are in the middle of a game park in Kenya and have just seen an enormous bull elephant—the biggest you have ever seen—and you want to record its weight. How are you going to do it? Even if there were a weigh-station for trucks nearby, which there is not, it is unlikely you could persuade the elephant to cooperate and check in. One method which has been used, but which is satisfactory neither from the elephant's point of view nor from the standpoint of accuracy, is to kill the elephant, cut it up into manageable pieces, then weigh these separately, allowing a percentage (usually 3 percent), for the loss of blood and other body fluids.

Because elephants have been slaughtered in such large numbers over the years, a large number of weight estimates are available, many of these accompanied by length measurements. It is therefore possible to use these compiled data to obtain an estimate of the weight of an elephant from a length measurement, such as the height at the shoulder. Shoulder height can be estimated by eye, but this is not very accurate, and it is best to obtain a direct measurement. You have to tranquilize the elephant first, and this is probably not a trivial matter in itself. Because of all these problems body weight estimates for elephants are inherently inaccurate.

The second problem in obtaining the maximum weight of the elephant is that, unlike most mammals, elephants continue growing long after reaching sexual maturity, and perhaps never completely stop growing throughout life (Eltringham, 1982; Laws and Parker, 1968, fig. 3). Since they live to such great ages, probably into their late fifties or sixties, there is a wide range of body weights among adults. This is especially true for mature males because they grow to larger sizes than do females. Contrast this with the situation in, say, a herd of impala. Here all the mature individuals would be close to the same size. These various problems in estimating body weights should be borne in mind when we consider estimating body weights for dinosaurs.

By a remarkable coincidence—unfortunate for the Toronto Zoo but fortunate for me—Tantor, the zoo's magnificent African bull elephant, died at the time I was writing this chapter. The zoo generously donated the body to the Royal Ontario Museum, so I was able to check out several features for myself, including the inaccuracy of weight estimates. Tantor was a big elephant, apparently the largest in North America, with a shoulder height of eleven feet two inches (3.4 m). The usual estimating procedure gave him a weight of about five-and-a-half tons, but he actually weighed six-and-a-half tons (14,300 lb; 6,500 kg).

There are several graviportal features of the elephant skeleton that are readily understandable in terms of weight bearing. The long bones (the femur, tibia and fibula in the hindlimb and the humerus, radius, and ulna in the forelimb) are kept essentially vertical so that the legs function primarily as vertical columns. Since the bones are loaded as columns rather than as beams there is less advantage in their being tubular. Indeed it is now desirable under these conditions to maximize the area of cross section, to provide the largest area of bone material, so the bones no longer have large marrow cavities. If the bones remained hollow they would need to have larger diameters to achieve the same cross-sectional areas.

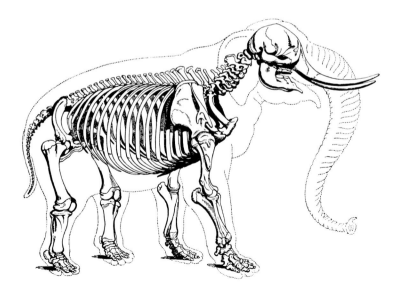

High stresses in the limb bones of the African elephant, attributable to the large body mass, are minimized by the vertical orientation of the limbs. Compare the straight limb bones of the elephant with the angled limb bones of the camel on page 42.

The pectoral and pelvic girdles, like the limbs, have a vertical orientation, which reduces the bending stresses. The feet are seemingly digitigrade; in other words, the metacarpus (palm) and metatarsus (sole) are raised off the ground. In life, however, the foot is cushioned by an extensive elastic pad so that the foot is functionally plantigrade. The pad is remarkably resilient. When I pushed against one of Tantor's pads it felt like dense foam rubber. Having such a low Young's modulus enables the pad to absorb much of the stress of the footfall—like an overweight jogger wearing well-cushioned jogging shoes. The front feet are much larger than the hind ones, showing that they carry more of the weight (the weight of the head is carried over the front legs).

Elephants have large heads. This is especially true for the African species, and particularly for the male, which has larger tusks than the female. Tantor had abnormally small tusks, but his head still weighed a staggering three-quarters of a ton (750 kg). More than two-thirds of this was attributable to the weight of the soft parts—the skin, muscles, ears, and especially the trunk. When all of these had been removed the head weighed only 484 pounds (220 kg), most of which was attributable to the skull. One of the reasons the skull is relatively light is the occurrence of numerous internal spaces—sinuses—in the bones; a somewhat similar strategy is seen in the tyrannosaurs. There are also skeletal modifications to facilitate the support of the head. The neck is relatively short and robust, providing a large attachment area for the muscles—once again, a strategy similar to that seen in the tyrannosaurs. (It is of interest to note that the elephant, like other mammals, has seven cervical vertebrae; the neck appears to be short because each one of these is compressed, the centrum being a relatively thin disk of bone.) The neural spines of the vertebrae in the shoulder region are both tall and robust, providing a large attachment area for the muscles and for the ligament that extends to the back of the skull (called the nuchal ligament—to be discussed later). There is therefore an extensive musculature for moving the head. Since the neck is so short, a standing elephant is unable to reach the ground with its head, and even if it could, its tusks would prevent it from being able to pick up food. The elephant has to rely on its trunk for food gathering; in the wild, individuals that suffer severe injuries to the trunk starve to death. Although we can learn a great deal about an animal from its skeleton, there is considerably more to be learned from watching it in the wild and from studying its soft anatomy.

Elephants resolve much of the problem of supporting such a large body weight through their behavior. They live unhurried lives and seldom do anything quickly. They walk rather ponderously, keeping their legs relatively straight throughout, and although they can run at a fair pace, easily outrunning a man, their speed is achieved by virtue of a long stride, not by rapid movement. Their run is a relatively straight-legged amble and they are unable to gallop. Nor do they go in for jumping, bounding, or similar athletic activities. And although wild elephants sometimes rear up on their hindlegs to reach up higher into trees, and captive ones can perform some unusual feats at the circus, they do all these things rather slowly and deliberately. In this way they avoid placing undue stresses on their skeletons, especially bending stresses, and thus avoid injuring themselves. Injuries nevertheless do occur. Falling down may be fatal. Elephants are able to lie down to sleep, but they avoid lying down for more than about an hour at a time because of the compression damage they can do to the muscles and nerves underneath their great weight. They therefore spend most of their time on their feet, which must be made more comfortable for them by virtue of their extensive foot pads.

Elephants are sensitive to the heat and avoid the midday sun beneath the shade of trees. If they are exposed to the sun for long periods of time they suffer heat stroke, probably because of their having a relatively small surface-area-to-volume ratio and hence a relatively small area for the dissipation of excess heat. They appear to lack sweat glands, or if they do have any they are too few to make a significant contribution to evaporative heat loss. Elephants use their trunks to spray water on their hides and to daub themselves with mud when the opportunities arise, probably to help them shed excess heat. The relatively large ears of the African elephant serve primarily to dissipate heat. The thin skin covering the ears is richly supplied with blood; the blood then loses some of its heat to the cooler environment. The elephant may facilitate the process by "flapping" its ears, and it has even been suggested that the elephant's tendency to walk into the prevailing wind, especially during hot weather, may aid heat dissipation (Eltringham, 1982).

The metabolic rate, the rate at which chemical processes occur within the body, is higher in small animals than in large ones, as will be discussed in detail in the next chapter. Indeed all processes are speeded up in smaller animals, from the rate at which the heart beats to the rate at which offspring are produced; smaller animals also have shorter life spans (Hill, 1950; Lindstedt and Calder, 1981). The life of a small animal is therefore a bustle of activity which produces large numbers of offspring but which ends after only a few short years. Large animals, in contrast, amble through life at a gentler pace, leaving far fewer offspring behind but living for many more years. A one-ounce (25 gm) mouse, for example, has a resting heart rate of about 700 beats per minute, takes about 160 breaths a minute, has a gestation period of about twenty days and a life expectancy of only two or three years. A three-ton elephant, in contrast, has a resting heart rate of about twenty-five, takes about ten breaths every minute, has a gestation period of twenty-two months, and may live into its sixtieth year.[2]

One of the implications of the elephant's relatively low metabolic rate is that it requires less food than a smaller mammal, relative to its body weight. In absolute terms, of course, a five-ton elephant consumes more food in a day than a one-ounce mouse, but five tons of mice would eat many times the quantity of food a single elephant would eat. In spite of its relatively lower metabolic rate, an elephant has to eat such a prodigious amount of food that it spends about 75–80 percent of its time eating, both day and night, and this has much to do with the fact that elephants are herbivorous and their food is relatively low in nutrients.

Plant food is made up of two main parts: the contents of the plant cells (the cell sap) and the cell walls. The cell sap is the nutritious part: the sugars, proteins, and starches that are readily digestible by the enzymes secreted by vertebrates. (Enzymes are proteins that facilitate chemical reactions in living organisms, in this case the digestion of large food molecules into smaller ones.) The tough cellulose walls, in contrast, cannot be digested by vertebrate enzymes. Herbivores have to rely on the bacteria in the gut to break down the cellulose.

Except for rare food items like fruits, root tubers (potatoes, for example), and fresh young grass shoots, cell contents constitute but a small part of plant material. Herbivores therefore have to take large quantities of the relatively indigestible fibrous material in their diets, and this has to be slowly broken down in the gut by the action of bacteria. Gut capacity increases in step with body weight, and so does the time that the food remains in the gut (Demment and Van Soest, 1985). This means that large herbivores can consume relatively larger quantities of fibrous materials than smaller animals can and that the food remains in the digestive tract for longer periods of time. They are therefore able to exploit a wider range of plant foods, and since there is always a higher percentage of lower-quality plant foods in the environment (those with higher percentages of fiber) giant animals, like the elephant, have the competitive edge over smaller ones. This is especially true when higher-quality food is less abundant, as in times of drought.

Being large often means being tall too, and before leaving the living world for the world of the Mesozoic, we need to consider some physiological problems associated with being tall. Naturally our example will be the tallest of living animals—the giraffe.

Giraffes and blood pressure

An adult giraffe is about sixteen to nineteen feet tall (5–6 m), and its head is about ten feet (3 m) above the level of its heart. In order for the blood just to reach the level of the head, the heart has to discharge blood at a pressure about one-third of that of the atmosphere, which is about twice the pressure of the blood leaving our own hearts.

Pressure is usually expressed in terms of the height of a column of liquid that the pressure can support. One could, for example, measure the water pressure available for watering the lawn simply by turning on the garden hose and hoisting the end of it up in the air until it was sufficiently high that the municipal pressure could no longer force the water out of the nozzle. Atmospheric pressure[3] is equivalent to the pressure exerted by a thirty-two foot water column (10 m); since the municipal water pressure is likely to be higher than this, it is more convenient to use a liquid denser than water to measure pressure, because a liquid with more mass per unit of volume will exert the same force in smaller quantities than lighter liquids do. Mercury has been the liquid of choice for centuries, and atmospheric pressure is equivalent to a thirty-inch (760 mm) column of mercury. This unit of measure should be familiar to anyone whose blood pressure has been taken: the average pressure of the blood leaving a healthy human heart during its contraction is 120 mm of mercury, abbreviated to 120 mm Hg.[4]

The blood leaving the giraffe's heart has to do more than just reach the level of the head, it has to be at a high enough pressure to pass through all of the fine vessels, the capillaries, that supply the brain and other organs. To achieve this the blood leaves the heart at a pressure of 200–300 mm Hg, which is probably the highest blood pressure of any living animal (Warren, 1974; Hargens et al., 1987). A giraffe's blood pressure is so high that it would probably rupture the blood vessels of any other animal, but two mechanisms appear to prevent this. First, the arterial walls are much thicker than in other mammals. Second, the fluid that bathes the cells of the body is maintained at a high pressure; this is largely achieved by the thick skin, which is tightly stretched over the body and which functions like the anti-gravity suits worn by pilots of fast aircraft.

Although heart rates usually decrease with increasing body weight, the giraffe, which weighs a little over a ton, has a heart rate of about sixty-six, which is comparable to our own. Presumably this relatively high rate in such a large animal is imposed by the need to generate high blood pressures.

The unpleasant dizzy sensation we sometimes experience when standing quickly from a lying or sitting position (especially if we are not in good shape) is caused by a temporary fall in the pressure of the blood reaching the brain. Giraffes sometimes lie down to rest, and when they regain their feet they have to do so in stages, first squatting and waiting for a few moments before standing fully erect. This behavioral modification presumably allows the vascular system to stabilize. If a giraffe suddenly raised its head to its full height, it would probably become faint. Conversely, when a giraffe is drinking, its head is lower than its heart, which increases the blood pressure to the brain. The giraffe reduces the extent of this pressure increase by splaying its front legs while drinking, which reduces the difference in height between the head and the heart.

We are not very tall, but the difference in height between our lower legs and our heart is sufficiently great to cause problems in returning blood to the heart. This is primarily because the blood in the veins—the vessels that return blood to the heart—is at such low pressures. The blood that left the heart at such high pressure has to be distributed to all parts of the body. The blood is forced through vessels of decreasing size, from arteries to arterioles, and then through the extensive networks of capillaries that supply the various tissues with blood. The capillaries, the smallest of all vessels, may be little wider than the diameter of a red blood cell. It therefore follows that the blood leaves the capillaries at a very low pressure. From the capillaries the blood passes into venules and finally into the veins. The blood from the legs has to make its way, against gravity, back to the heart, and its return is greatly assisted by the pumping action of the leg muscles. This works because veins have valves that allow blood to flow in only one direction, toward the heart; when the veins are massaged by the leg muscles the blood is squeezed toward the heart. The extreme discomfort that soldiers experience when forced to stand still on a parade ground is largely due to the absence of this pumping action. So much blood may pool in the legs that the reduced blood return to the heart may lower the blood pressure to the head sufficiently to cause fainting. The problem of returning blood to the heart is even more acute in the giraffe because its legs are so much longer. The tightly fitting skin covering the legs appears to reduce the extent of blood pooling, and the pumping action of the muscles, which results in extreme fluctuations in blood pressure in the legs when the giraffe is moving, also helps.

Another problem posed by the possession of a long neck is the large volume of air in the trachea, the tube that connects the back of the throat with the lungs. This air is unavailable for respiration and the space it occupies is consequently referred to as the dead space. The dead space has a volume of about five pints (2.5 l) in the giraffe. Since this air has to be moved each time the animal breathes, the rate of ventilation has to be increased to compensate for the reduced air flow. A resting giraffe takes about twenty breaths per minute, compared with our twelve and an elephant's ten; this is a very high respiratory rate for such a large animal. Although the elephant is extremely large, and the giraffe exceptionally tall, both would be dwarfed by some of the dinosaurian giants.

The implications of being a giant

Finding adequate words to convey the enormous size of a sauropod dinosaur is not easy. It may help to compare the weight of *Brachiosaurus,* one of the heaviest sauropods, with that of a Boeing 727 airliner.[5] But nothing can compare with standing beside the mounted skeleton of the beast in the museum in East Berlin. An adult elephant could probably walk between those legs without brushing against the rib cage, and if a six-foot man stood beside the front foot, the top of his head would only just reach up to the dinosaur's elbow. The immensity of the sauropods so impressed some paleontologists that they refused to believe they could have supported their weight on land, which would have obliged them to live in the water. This aquatic interpretation, which was held by most of the early paleontologists, may have been inspired by Richard Owen, the first anatomist to examine the skeletal remains of sauropods. Owen's studies led him to conclude that they were aquatic, but this was probably because he had mistakenly identified the rather incomplete material he had to work with as being crocodilian rather than dinosaurian. (Coombs, 1975, gives a good account of the early history of researches.)

Sauropods, like many other dinosaurs, were obviously heavy animals, but how do we arrive at estimates of their weight? The method is simplicity itself. A model of the appropriate dinosaur is made or purchased; there are now many good models available, such as those produced for the British Museum (Natural History). Its volume is then measured by the Archimedean principle of weighing the model in air then in water. The difference between the two weights is due to the volume of water displaced by the model. To convert this weight difference into the volume of water displaced, that is, into the volume of the model, simply requires dividing it by the density of water. The volume of the model is then multiplied by the cube of the magnification factor to estimate the volume of the full-sized dinosaur. For example, if the model was 1/40 full size, its volume would be multiplied by $40^3 = 64,000$. This estimate of the volume would then be multiplied by an estimate of the density of the dinosaur to obtain an estimate of its weight. Needless to say, the method is subject to many sources of error, not the least of which is the accuracy of the model, and error is exacerbated by the fact that model volumes, which are relatively small, are multiplied by such large numbers.

Reptiles, unlike most mammals and birds (but like the elephant), appear to continue growing throughout life, though the rate of growth slows down with increasing maturity. Consequently there is a wide size range among adult dinosaur skeletons (similarly for other reptiles) and a wide range of body weights within a given species. Weight estimates for dinosaurs should therefore be taken only as a very rough guide, just as modern animal weights are. I will use estimates from Edwin Colbert (1962), a dinosaur specialist who has done so much to popularize his subject, together with some more recent estimates given by Alexander (1985).

The controversy over whether sauropods were aquatic or terrestrial is over a century old. One of the pieces of evidence that

Estimated body masses of dinosaurs

Genus	Estimated weight (tons, in rounded numbers)	Reference
Theropods		
Allosaurus	2	Colbert, 1962
Tyrannosaurus	7.5	Alexander, 1985
Sauropods		
Apatosaurus	28.0	Colbert, 1962
Diplodocus	18.0	Alexander, 1985
Brachiosaurus	78.0	Colbert, 1962
Ornithopods		
Camptosaurus	0.5	Colbert, 1962
Iguanodon	5.5	Alexander, 1985
Corythosaurus	4.0	Colbert, 1962
Anatosaurus	3.0	Colbert, 1962
Stegosaurs		
Stegosaurus	3.0	Alexander, 1985
Ankylosaurs		
Palaeoscincus	3.5	Colbert, 1962
Ceratopsians		
Protoceratops	390 pounds (177 kg)	Colbert, 1962
Styracosaurus	4	Colbert, 1962
Triceratops	6	Alexander, 1985

has been used to support the aquatic interpretation is the dorsal position of the nares. In both *Camarasaurus* ("chambered lizard") and *Brachiosaurus* ("arm lizard") the nares are placed high on the skull, while in *Diplodocus* ("double-beam lizard") they are placed right on the very top. Edward Drinker Cope and Othniel Charles Marsh, pioneers of American paleontology and early students of sauropods, were both impressed by this feature. They drew the obvious parallels with the high nares of living aquatic animals, like crocodiles and whales. However, there are some terrestrial animals, such as the elephant and the tapir, that have their nares placed high on the head, and there are some aquatic animals that have their nares placed low down on the skull. The latter group includes most marine turtles, otters, and seals, together with extinct marine reptiles, the ichthyosaurs and plesiosaurs. The narial evidence is therefore equivocal.

Regardless of the narial evidence, an aquatic interpretation is difficult to justify on the grounds of the difficulties the animals would have had inflating the lungs while submerged. As noted earlier, water pressure increases by one atmosphere for every thirty-two feet (9.8 m) of depth. In *Brachiosaurus* the lungs were about twenty-two feet (6.5 m) below the level of the head; therefore they would have experienced a pressure of about two-thirds of an atmosphere (522 mm Hg) when the animal was submerged. The lungs would have needed to generate a negative pressure well in excess of 522 mm Hg in order to inspire air, and this seems very unlikely.

Sauropods, in contrast to many other herbivorous dinosaurs like the hadrosaurs and ceratopsians (horned dinosaurs), have relatively few teeth, and these are not arranged to form extensive grinding surfaces. This suggested to some paleontologists that the teeth were suitable only for cropping soft aquatic plants. Aside from the fact that not all aquatic plants are soft, the type of vegetation eaten does not necessarily indicate the mode of life of the consumer (Coombs, 1975). The hippopotamus, for example, which spends most of its time in the water, eats considerable quantities of terrestrial plants, while the moose, which spends much of its time out of the water, consumes large quantities of aquatic plants. Again the evidence is equivocal.

The most significant clue to the life-styles of sauropods was stumbled upon quite by accident in Texas, one day late in November 1938. Roland Bird, a paleontologist with the American Museum of Natural History, was nearing the end of a disappointing field season when he heard about some unusual dinosaur footprints in the bed of the Paluxy River, near the town of Glen Rose, Texas. Being especially interested in trackways, he decided to make the necessary detour and spend a short time talking with the locals and prospecting in the riverbed. He did not find anything particularly unusual at first and set about collecting two rather fine sets of tracks that had been made by a pair of large carnivorous dinosaurs walking close together. That these tracks had been made by carnivores was clear from their distinctive three-toed print and the impressions of sharp claws.

With the intention of making absolutely sure that there were no other tracks, he cleared an area of about a yard wide all around the tracks he had collected. As he shoveled away he came upon a round pothole that was about a yard (1 m) in diameter and partly filled with silt.

"When I dug into it and threw back a few shovelfuls for a looksee, my heart nearly jumped out of my mouth. There, right at my

Sauropod dinosaur tracks uncovered in the bed of the Paluxy River, Texas.

very feet, was a depression totally unlike any I had ever seen before, but one I instantly surmised must be a sauropod footprint" (Bird, 1939, p. 260).

Bird had indeed discovered a sauropod footprint, and he correctly identified it as having been made by a right hindfoot. By following the direction in which the toes were pointing he tried to estimate the position of the next footfall of the same foot. With mounting excitement he cleared away the river debris, but he found nothing. He ran his shovel along farther and farther, and eventually his shovel hit the rim of the next depression. He was amazed to find that it was thirteen feet (4 m) in front of the first, for he had not anticipated such a long stride length.

The following season Bird returned with a field crew, who excavated some of the trackways and sent them back to New York. A second trackway was subsequently discovered in Bandera County, Texas, and the information gathered at the two localities has added considerably to our knowledge of these gigantic dinosaurs. With one exception the trackways were made by four-footed animals, and they were fairly deep, which strongly suggests that the sauropods were walking on land, on fairly soft mud, rather than moving along in water. The prints made by the forefeet are smaller and shallower than those made by the hindfeet, showing that most of the weight was borne by the back legs. This conclusion accords with skeletal evidence; most sauropods have hindlegs that are considerably larger than the front ones. What is more, the footprints accord with some recent modeling procedures that show that the center of mass (or center of gravity) of a sauropod was at about the level of the pelvic girdle (Alexander, 1985).[6] Since these two separate lines of evidence agree with the footprint data in showing that most of the weight was carried by the hindlegs, we can be confident that the trackmaker was moving on land rather than being buoyed up in the water. Furthermore, scrolls of mud were preserved that had been squeezed out at the sides of the feet as the weight of the animal pressed down on the ground. If the animal had been wading in water it seems likely that these scrolls would have been swirled away by the movements of the water.

The footprints were so numerous in one area that they overlapped one another—they had apparently been made on the same occasion. Twenty-three individual sets of prints, all pointing in the same direction, bear testimony to the passage of a group of sauropods all those millions of years ago. What seems likely is that this area was once an access route, probably running beside a lake or a river. In one or two of the trackways a groove had been channeled between the footprints; this has been interpreted as the mark made by the tail dragging along the ground. And along with the sauropod tracks were those of large carnivores. Some of these had been imprinted by the sauropod tracks, showing that the carnivores had passed by before the herd of sauropods. Were they waiting in ambush for the sauropods further along the way? Other carnivore tracks crossed over the sauropod prints and had clearly been made by predators that had travelled the route after the passage of the sauropods. In one case the track of a carnivore was imprinted alongside that of a sauropod. When the sauropod trackway swung to the left that of the carnivore followed. Was the predator following the sauropod? Perhaps the herd was being stalked, and perhaps other predators were waiting in ambush for them further along the way. But it is equally likely that the carnivores were merely travelling the same route.

The Texas trackways are compelling evidence that sauropods could support their own weight on land. But this does not mean that they never went into the water, nor that they may not have spent much of their time in water, as do present-day hippos and elephants when given the chance. Indeed, one of the trackways provided direct evidence that sauropods were accustomed to taking to the water. This short trackway, discovered some little distance from the rest, was unusual in that it was comprised almost entirely of prints made by the forefeet. The stride length was about six feet (just under 2 m) and the prints were about two feet (60 cm) across (Bird, 1944). The explanation that was given for this trackway is that the animal was floating in water, and using its front legs to push itself along. At one point the animal turned to its right, and there was a partial impression of a hindfoot there, suggesting that the sauropod had used a back leg to change course. The sauropods, as shown by the estimates of their body weights, were considerably heavier than the largest living land animal, the African elephant. It is of considerable interest to see how their skeletons were engineered to cope with such a burden.

Apatosaurus (meaning "deceptive lizard," formerly referred to as *Brontosaurus,* "thunder lizard"; see Riggs, 1903) was not the heaviest of sauropods, but with a length of about fifty-seven feet (17 m) and an estimated weight of some twenty-eight tons, it was about five times heavier than an African elephant. How did its skeleton support such a weight? The vertebral column appears to have been arched between the two limb girdles, and this shape would have been maintained partly by the shape of the individual vertebrae, partly by ligaments, and partly by the action of the hypaxial muscles (the ones that lie below the level of the transverse processes of the vertebrae and in the abdominal region). The neural spines are tall, especially in the dorsal vertebrae (the ones lying between the two girdles) and in the anterior caudals (the first tail vertebrae), which also have long chevrons. The long processes provided both for a long leverage and for a large attachment area for the ligaments and muscles that moved and

maintained the shape of the vertebral column. The vertebrae in the pelvic region are especially robust, and five of these are fused together to form the sacrum, to which the pelvic girdle attached.

The individual vertebrae are massive, especially the dorsal ones, which have centra that are about one foot in diameter (35 cm). In life these probably were loaded as beams, with compressive stresses dorsally, tensile stresses ventrally, and no bending stresses in the middle. There was therefore no need for the vertebral centra to be heavily ossified in the middle, and indeed there are deep excavations on either side of the centrum, the pleurocoels. The overall shape of the centrum resembles an engineer's I-beam; this strategy is analogous to the tubular design of limb bones, which are also loaded in bending. A similar economy of material, where bone is concentrated in areas subjected to the highest stresses, is seen in the bony struts and flanges, as opposed to solid bone in the vertebrae. The presence of pleurocoels in these and other

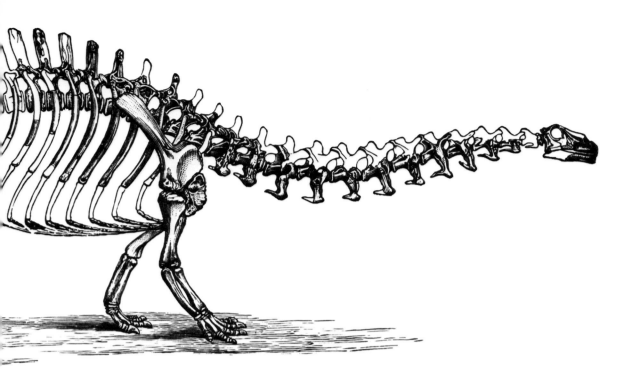

Marsh's reconstruction of the sauropod dinosaur APATOSAURUS, *Upper Jurassic, Western U.S.A., about sixty feet (18 m) long. Marsh used the skull of* CAMARASAURUS *in this reconstruction.*

saurischian dinosaurs has been interpreted as evidence that they may have had an air-sac system like that of birds.

Most of the dorsal vertebrae articulate together by a well-developed ball-and-socket joint between their centra. This arrangement added to the stability of the joint and hence to its load-bearing capacity. Furthermore, there is an additional set of articulating surfaces below the level of the anterior and posterior zygapophyses that strengthened the vertebral column in the dorso-ventral (up and down) plane.

The sauropod presumably went to a considerable effort just to hold its long neck and tail clear of the ground, primarily because of the great leverages involved. Dragging the tail may have alleviated the problem of support somewhat, though there are reasons to believe that the tail may usually have been kept clear of the ground, as will be discussed later. Regardless of how the tail was carried, the head was certainly kept elevated, and it would have

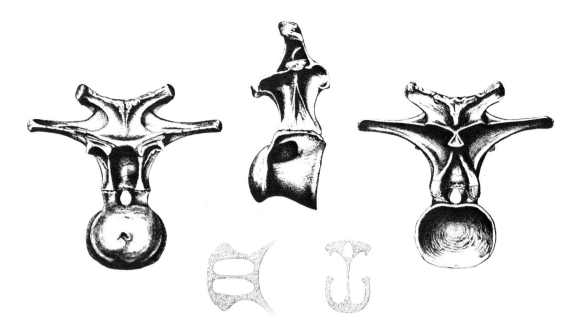

A dorsal vertebra of the sauropod dinosaur CAMARASAURUS. *The anterior, left-side, and posterior views are shown at top. Below are cross-sections of the centrum: horizontal on the left and transverse on the right (note the I-beam structure). The lateral surfaces of the centrum are deeply excavated, which reduces the weight of the bone without sacrificing strength.*

been capable of movements in both the vertical and lateral planes. But the neck may not have been as flexible as we have previously supposed. Relatively little attention has been given to determining how much mobility there was in the sauropod vertebral column, but the recent mounting of a partial skeleton of *Cetiosaurus* ("whale-like lizard"), an early sauropod from the Middle Jurassic of England, presented the opportunity of examining a well-preserved neck in detail (Martin, 1987). It was discovered that neck movements were considerably restricted because of the way that the vertebrae articulated with one another. The head was believed capable of reaching the ground but it was estimated that the neck could not have been raised beyond an angle of about 30 degrees above the horizontal. There may have been a physiological reason for this restricted range of motion: a smaller range would have limited the fluctuations in blood pressure experienced by the brain, a point which will be returned to later. Even though head movements may have been limited in some or all sauropods, the effort involved in holding it clear of the ground and in moving it must have been considerable.

Head support in living quadrupeds is assisted by the nuchal ligament, an extensive structure running from the vertebrae of the shoulder region to the back of the skull (Dimery, Alexander, and Deyst, 1985). Unlike most other ligaments, which are made of collagen and are white, this one is primarily composed of elastin fibers and is a buff yellow. Elastin is an elastic protein, which, like rubber, has a very low Young's modulus. Tension in the ligament holds the head up, relieving the epaxial neck muscles of much of the burden. Furthermore, when the hypaxial muscles contract, depressing the head, the ligament is stretched. Because of its low Young's modulus the nuchal ligament stores a large amount of strain energy, and since it can be extended fairly easily the ligament allows for a fairly large range of motion. When the muscles relax the strain energy is released, assisting the epaxial muscles in elevating the head.

Many sauropods have neural spines that are prominently forked in the neck and anterior dorsal regions, and these probably acted as a guide for a massive nuchal ligament. The possibility that this bony groove may have housed a large muscle instead of a nuchal ligament was investigated by Alexander (1985). His deductions showed that although such a muscle would have been large, it would probably not have been capable of generating large enough forces to elevate the head and neck.

In contrast to the hadrosaur skull, a heavy structure because of its extensive dental batteries and deep jaws, the skull of a sauropod is small and lightly built. Not only are the individual bones rela-

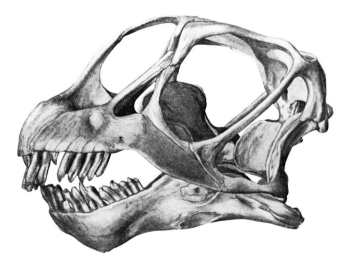

The heads of many large dinosaurs were not as heavy as one might suppose. Large openings, like the ones in this CAMARASAURUS *skull, reduced the weight.*

tively thin but there are large perforations in the skull. The teeth, which are relatively few, are fairly long, and while some are rod-like, as in *Diplodocus,* others are more spatulate, as in *Brachiosaurus.* The teeth extend well back in the jaws in some sauropods, like *Brachiosaurus,* but in many others, including *Diplodocus,* they are confined to the front of the skull. The strategy of the sauropod was therefore to have a small but mobile head which served for food gathering but not for food processing. The latter function was almost certainly performed by the gut, probably in a muscular gizzard containing stones for grinding up food, like that of a bird.

So much for the head and neck—what of the tail? Was it kept clear of the ground? Like so many other paleontological questions, this one will probably go unanswered because of the lack of data. However it seems reasonable to suppose that there would have been much variation in postures from one species to another, possibly even among individuals of the same species too. There are also reasons to suspect that the tail may have been kept clear of the ground, at least in some sauropods. David Norman, in his outstanding book on dinosaurs (1985), suggested that the nuchal ligament may have been continued caudally (backward) along the tops of the neural spines of the dorsal vertebrae and along those of the caudal vertebrae. A ligament of this size would have functionally connected the neck with the tail so that the two would have counterbalanced each other, which makes good mechanical sense. Norman thought that this may account for the rarity of tail-drag marks in sauropod trackways. Negative evidence is unsatisfactory though, and there is some positive evidence that the tail was kept clear of the ground. This evidence is provided by the occurrence of the osteological condition called DISH (Chapter 2) in which articulating bones become fused together by bony outgrowths spanning the joint between them. This condition appears to be nonpathological (not caused by disease or injury); it seems to occur in response to localized regions of high stress. DISH has been detected in the caudal regions of half of the specimens of *Apatosaurus* and *Diplodocus* that have been examined (Rothschild, 1987). The area of fusion, involving from two to four contiguous vertebrae, seems to occur at about the same level in all skeletons, between caudal vertebrae 20–23. It is extremely unlikely that the fusion was a result of injury, not only because it occurs at the same location in different individuals but also because it does not disrupt the articular surfaces of the zygapophyses. The fused vertebrae occur at about the point where a gently curving tail would come close to the ground, and it has been suggested that fusion at this point would have kept a significant portion of the tail clear of the ground. The evidence supports the interpretation that at least some sauropods kept most if not all of their tail off the ground.

The weight supported by the vertebral column was transmitted to the legs by the pectoral and pelvic girdles, both of which are characteristically robust, even in the relatively lightly built sauropod *Diplodocus*. In most sauropods (*Brachiosaurus* and its allies excepted) the hindlegs are longer than the front ones and the pelvic girdle is correspondingly larger than the pectoral girdle. The scapula, a huge bone in sauropods, provided a large attachment area for the muscles and ligaments that strapped it to the rib cage. The coracoid, which is often fused with the scapula, is similarly large. The three-part pelvic girdle is a massive structure, with the ilium forming a great thick blade of bone that was rigidly attached to the sacrum. The bones of both legs are straight and fairly robust, much like those of the elephant, and the terminal positions of their articular surfaces show that they were kept essentially vertical. They were therefore loaded as columns, as would be expected, and the cross-sectional supporting area was maximized by being of solid bone, with no marrow cavity. Sauropod feet are very similar to those of the elephant. Although they appear to have been digitigrade, we can be quite certain that there was an extensive pad of connective tissue beneath the palm and sole, as in elephants, because the pothole shape of their footprints resembles an elephant's print. A low Young's modulus for the material of this pad would be a reasonable assumption, for one of its functions would have been to absorb some of the impact energy of the footfall.

The forces required of the skeletal muscles to move such a heavy body is difficult to imagine, and muscle energy ultimately translates into food requirements. How much food would a mature sauropod eat in a day? If sauropods were warm-blooded, like elephants, we could make some predictions about their food requirements by extrapolating from the elephant on the basis of the general relationship between metabolic rate and body weight (page 99). This estimate would be quite large, and we would then have to ask whether such large animals could manage to gather enough food in a day to satisfy a warm-blooded appetite, especially with their relatively small heads. I do not want to get into the matter of energetics now because that is the subject of

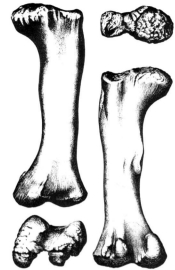

The left femur of the sauropod dinosaur CAMARASAURUS. *The shaft is solid and the terminal placement of the articular surfaces shows that it was held in a vertical position.*

the next chapter, but it is convenient now to test whether our knowledge of sauropods supports a view that they had high metabolic rates, like those of birds and mammals, or low rates, like those of modern reptiles.

Robert Bakker, whose radical ideas on dinosaurian energetics and life-styles have probably stimulated more research on dinosaurs than those of any other paleontologist, considers that all dinosaurs had high metabolic rates. (As will become evident in Chapter 6, this is not the view that I hold.) From this starting point emerged a picture of sauropods as active, warm-blooded animals that were capable not only of running but also of rearing up on their hindlegs to browse on tall trees and to defend themselves against predators (Bakker, 1978, 1986). Not all paleontologists share this view, but Bakker's ideas have had such far-reaching effects that they have been the subject of serious biomechanical discussions and have influenced artists' restorations of sauropods in the most respected places.[7] How likely is this picture of sauropods?

The African elephant reaches a maximum body weight of between five and six tons, which is about one-third that of *Diplodocus,* one-fifth that of *Apatosaurus,* and about one-thirteenth that of *Brachiosaurus.* If sauropods had mammalian levels of metabolism their food requirements would be like those of scaled-up elephants. We have seen that the food requirements of mammalian herbivores increases at a slower rate than their increase in body weight. Actually the increase is in accordance with the body weight raised to the power of $\frac{2}{3}$ (Brown and Maurer, 1986), which is less than the increase in metabolic rate (which scales with the body weight raised to the power $\frac{3}{4}$). We want to estimate the food requirements of a sauropod-sized elephant, but we will be conservative, and err on the side of underestimating its appetite. Let us take a value of $\frac{1}{2}$ instead of $\frac{2}{3}$ for the power function. Raising a number to the power of $\frac{1}{2}$ is the same as taking its square root. That is, the food requirements of a scaled-up elephant are proportional to the square root of the increase in weight. Accordingly, *Diplodocus* would need to consume $3^{1/2} = 1.7$ times as much as an elephant, *Apatosaurus* a little more than twice as much, and *Brachiosaurus* almost four times that of an elephant, and these estimates, remember, are very conservative.

African elephants spend about 75–80 percent of every twenty-four hours feeding, so in terms of their food requirements they would appear to be close to the maximum size permissible for a mammalian herbivore living in Africa today. Observations on (smaller) Indian elephants in Sri Lanka (Eltringham, 1982) show that they too spend a considerable part of their time feeding, so the preoccupation with food is not something that is peculiar to the African environment. Although the elephant is the largest land mammal alive today, it is not the largest one that has ever lived. *Baluchitherium,* an extinct rhinoceros which lived about 35 million years ago during the Oligocene Period, reached a shoulder height of about eighteen feet (5.5 m). This mammalian giant would have attained an estimated weight of about twenty-six tons—four times that of the elephant.[8] But how could such a large animal have possibly obtained enough food when the African elephant has to spend eighteen hours out of every twenty-four feeding? There would simply not have been enough hours in the day. This is just the sort of conundrum that animals seem to delight in setting for biologists!

Although we cannot guarantee the correct solution, it seems likely that the answer probably lies in the question of gut efficiency. The elephant's digestive tract has a high processing rate; its droppings contain large amounts of undigested food. Other mammalian herbivores, including the ruminants (animals that chew the cud), have more efficient systems and can obtain more nutrients from their food. Although *Baluchitherium,* being a rhinoceros, was not a ruminant, it seems that it must have had a more efficient gut than that of the elephant. Even so, it is difficult to imagine that *Baluchitherium* would have had much time for anything else besides eating.

Given that a twenty-six ton *Baluchitherium* was able to obtain enough food to satisfy a mammalian level of metabolism, it is conceivable that an eighteen ton *Diplodocus* or a twenty-eight ton *Apatosaurus* could have done the same—though not a seventy-eight ton *Brachiosaurus.* But these conclusions hold only if sauropods had a gut efficiency as high as that of *Baluchitherium,* and we have no way of knowing whether they did. If sauropods had a lower digestive efficiency, like that of the elephant, then it is very unlikely that even *Diplodocus* could have obtained enough food to support a mammalian level of metabolism—not unless sauropods enjoyed a richer diet. The African elephant, we know, has to include some fairly low-quality food in its daily diet, especially during the dry season, and if sauropods were able to obtain higher-quality food they would have been able to survive on less food than elephants need. Given their enormous sizes and their apparently gregarious behavior, though, it seems likely that they would have made tremendous demands upon the local flora. Therefore they may not have been able to consume significantly higher quantities of high-quality food than do elephants—just consider how much high-quality food an individual *Brachiosaurus* would have needed to equal the daily food intake of four elephants. And *Brachiosaurus* was not the largest of sauropods. Jim Jensen's (1985) *Ultrasaurus* ("beyond lizard"), found in Colorado in 1979, had an estimated length of more than ninety-eight feet (30 m), compared with seventy-four feet (22.5 m) for the Berlin specimen of *Brachiosaurus*. This is a length increase of 1.32, so the weight increase would be $(1.32)^3 = 2.3$, giving an estimated weight of almost 180 tons. Such an enormous animal would have had an appetite equivalent to that of at least six elephants, assuming a similar digestive efficiency.

To summarize the evidence for food requirements, it is conceivable that the smaller sauropods *could* have had high metabolic rates like those of birds and mammals, but only if their digestive systems had been more efficient than that of the elephant. And even if this had been the case, it is inconceivable that the giants among them could have obtained sufficient food to have maintained such a high metabolic level. Furthermore, there are other physiological consequences of the sauropods' large sizes that would have severely restricted the scope of their activities.

Because of its long neck the giraffe's heart has to discharge blood at pressures of between 200 and 300 mm Hg to supply blood to the brain. Sauropods had even longer necks. Estimates from mounted skeletons give values for the vertical distance between the heart and head of almost seven feet (201 cm) in *Apatosaurus* and twenty-two feet (655 cm) in *Brachiosaurus;* their hearts would have needed to discharge blood at pressures of 216 mm Hg and 568 mm Hg, respectively (Hohnke, 1973). If sauropods had stood on their hindlegs to reach up even higher into the trees, as Bakker proposed (though not for *Brachiosaurus*), the change in posture would have resulted in a greater vertical distance between heart and head. This would have required even greater blood pressures. Bakker discounted the problem of blood pressure by suggesting that sauropods could have used contractions of the neck muscles to assist the pumping action of the heart. Such muscle contractions, he noted, are important in living animals for returning venous blood to the heart. But this argument is invalid because the pressures involved in returning blood to the heart are so very low. For example, the pressure of the blood in our own superior vena cava, the main vein returning blood to the heart, fluctuates between 5 and −2 mm Hg (the negative pressure is because of the negative air pressure within the thorax). The pressure of the blood in the veins of our legs, which are massaged by the leg muscles, is not likely to be much higher, so it is obvious that the pumping action of skeletal muscles has a very limited effect on blood pressure. Short of speculating that sauropods had evolved an ancillary heart, which seems very fanciful to me, we have to accept that they, like all other vertebrates, had to rely solely on the heart for generating blood pressure. And a very large heart it must have been.

Whales, the largest of all living animals, have the largest of hearts. A fifty-ton whale has a heart of about 440 pounds (200 kg). Since the whale is aquatic its vascular system is not subject to the forces of gravity, and it therefore does not have to generate very high blood pressures. Vertebrate hearts appear to respond to increasing pressure demands by having relatively thicker walls. The heart of a fifty-ton sauropod, it has been estimated, would have had to have been about eight times heavier than the heart of a fifty-ton whale in order to withstand pressures in excess of 500 mm Hg (Seymour, 1976). Now a heart that weighs 1.6 tons is difficult to imagine, but regardless of whether these estimates and assumptions are valid, this line of reasoning serves to underscore the magnitude of the circulatory problems of large land animals that have long necks.

When a sauropod had its head fully raised, the pressure of the blood reaching the brain would have been considerably lower than that leaving the heart. When the head was lowered to the ground, however, the blood reaching the brain would have been at higher pressure. The additional pressure due to this effect has been estimated for *Brachiosaurus* to have been 508 mm Hg (Hohnke, 1973). The blood pressure in the lowered head, then, would reach a staggering 1076 mm Hg, about twice that which the giraffe would encounter under similar circumstances. The giraffe appears to be able to reduce the flow of blood to its capillaries by precapillary vasoconstriction (Hargens et al., 1987). Perhaps sauropods possessed a similar mechanism, but they may have alleviated the problem behaviorally, by reducing the extent of the vertical movements of the head. As noted above, there is skeletal evidence that they may have done so.

Yet another implication of the long neck is the large volume of the dead space. The giraffe, with its ten-foot neck (3 m), has a dead space of about five pints (2.5 l). That of *Brachiosaurus*, with a thirty-foot neck (9.2 m), would be three times as much, about fifteen pints (7.5 l). This large dead space would have obliged the brachiosaur to breathe at a disproportionately high rate for its body size, as the giraffe does.

What does all this mean?

The idea that sauropods reared up on their hindlegs, raising their heads even higher than their necks allowed, can be dismissed on the grounds of blood pressure problems alone. Even holding their heads as high as depicted in most mounted skeletons would have presented serious problems. Perhaps, on the other hand, they did not raise the head much higher than the rest of the body. Keeping the head low would have considerably reduced the demands upon the heart. In any event sauropods, like giraffes, would have avoided raising and lowering their heads rapidly, thereby minimizing changes in the pressure of blood going to the brain. We can therefore dismiss the idea, popular in some restorations, that sauropods threw themselves into combat with predators. Gone too is the notion that sauropods could trot and gambol along as they are seen to do in illustrations by Bakker and others (Bakker, 1986; Czerkas and Olson, 1987). There are other reasons for arriving at these conclusions, and these will be discussed in the next chapter, but there are still more lessons to be learned from the largest living land animal, the elephant.

Elephants in the wild lead unhurried lives. They walk at a leisurely pace in search of food and doze in the shade of trees at the height of the day. They do run, but not very fast, and their straight-legged gait is more like a speeded-up walk than a trot. They are unable to gallop and do not jump over obstacles. But in spite of their large size they are quite agile. They are able to climb quite steep slopes and will sometimes rear up on their hindlegs to reach up into trees. But they do these things fairly slowly and deliberately, not only because of the inertia of their great bulk but also, one presumes, to prevent injuries. Minor falls, like a tumble down a slope, can result in serious injury. Elephants avoid placing themselves at risk by making deliberate movements. Given the sedate habits of a five-ton elephant, what are the possibilities that an eighteen-ton *Diplodocus*, a twenty-eight-ton *Apatosaurus*, or a seventy-eight-ton *Brachiosaurus* would be relatively more active? Rather remote I believe.

The advantages of gigantism

Gigantism is not the prerogative of dinosaurs. Indeed the trend of evolving toward larger size is so common in the fossil record that the phenomenon has been described as a rule, named Cope's Rule in honor of Edward Drinker Cope, who first discussed it. Like most generalizations there are many exceptions; some of the earliest ichthyosaurs, for example, are the largest ones, and the last of the group are only of modest size. It is generally true, however, that most groups of animals have evolved from smaller ancestors, and many lineages of animals did evolve toward large size (Stanley, 1973). The trend must have some selective advantages. We have already seen one advantage in terms of relative energy requirements. Larger animals need less food per unit body weight than smaller ones, so, unlike large cars, they are more fuel efficient, on a mass-specific basis. But since large animals produce far fewer offspring than smaller ones and have longer generation times, their population densities are always much smaller. There are always more mice in a field than there are deer. This has led to the assumption that the advantages that a large species enjoys because of fuel economy is offset by smaller population densities, but there is evidence, and for a wide variety of both animals and plants, that larger-bodied species gain a disproportionately large share of the resources within a given local ecosystem (Brown and Maurer, 1986; du Toit and Owen-Smith, 1989). (Think of competition between species as a competition for resources, each species trying to gain the largest share of food to convert into their own biomass.) There is, therefore, a strong selection pressure for the evolution of large size just in terms of utilizing the resources of the environment.

Another relationship that varies with body size is the cost of locomotion. So far this relationship has mainly been investigated for warm-blooded animals, and although it might equally well apply to other terrestrial vertebrates too, we have to be cautious in applying the results to dinosaurs. Investigations have shown that the costs of locomotion decrease with increasing body weight and that this holds true across a wide range of birds and mammals (Taylor, Heglund, and Maloiy, 1982). For example, on a weight-for-weight basis, a one-ounce (30 gm) mouse uses six times the amount of energy per second when running at the trot-gallop transition speed than does a 660-pound (300 kg) horse. Why should this be so? Well it is not because small animals are mechanically less efficient than large ones. In fact, it has been found that the locomotory costs of animals, in terms of the energy required to move a given mass of animal through one stride, is remarkably constant, again across a wide range of species (Heglund and Taylor, 1988).[9] There are two reasons for the high locomotory costs of small animals: small animals take more strides per second than large ones do and, as mentioned earlier, smaller animals generate larger forces in their leg muscles relative to their body mass. For example, it is estimated that a 660 pound (300 kg) horse would exert only about 17 percent of the force in a given mass of its leg muscles that a 0.66-pound (0.3 kg) squirrel would exert when both were running at the same equivalent speed (Heglund, Taylor, and McMahon, 1974). It is tentatively concluded from evidence like this that another advantage of the sauropod's large size may have been the decrease in the relative costs of locomotion.

Large animals run faster than smaller ones, as we saw in Chapter 3, but the largest ones are not necessarily the fastest. The African elephant, for example, has a top speed of only about 22 mph (35 kph) compared with a zebra's 43 mph (70 kph—data given by Garland, 1983). There appears to be an optimal size for fast running. Garland (1983) suggested an optimal body weight of about 260 pounds (120 kg), but Coombs (1978) gave a smaller estimate of 110 pounds (50 kg). Once again this is a relationship which has been investigated for only one group of animals (mammals), but it seems reasonable to expect that the same principle should apply to other tetrapods too. It therefore seems likely that sauropods were unable to run as fast as elephants.

One of the most obvious advantages of being large is that of discouraging predators. As there are no really large reptiles living today that can be used for assessing the deterrent value of large body size, we will look at some mammals. Attention will then be turned to the only living land reptile of any size—the Komodo dragon, a large predatory monitor lizard that lives in Indonesia.

Mammalian predators The hunting behavior of the African lion, which has been extensively studied in recent years, provides some useful comparative data.[10] Adult lions weigh between 240 and 400 pounds (110–180 kg) and hunt alone, in pairs, or in small groups. A solitary adult can kill animals that are as much as twice its own weight, but larger prey, such as buffalo, which average

925 pounds (420 kg) and which may reach 1,870 pounds (850 kg), are usually attacked by several individuals working together. Lions prey largely upon wildebeest (550 lb; 250 kg), zebra (660 lb; 300 kg), and buffalo; most of their other prey, like Thompson's gazelle (44 lb; 20 kg), are smaller than themselves. Lions usually make no attempt to attack elephant (13,200 lb; 6,000 kg), hippopotamus (4,000 lb; 1,800 kg), rhinoceros (3,000 lb; 1,400 kg) or giraffe (2,600 lb; 1,200 kg). Animals weighing more than about 2,000 pounds (1,000 kg) are therefore relatively safe from lions, and from other predators too.

Lions can run fast, up to about 37 mph (60 kph), but only for short distances. Their prey, however, can usually run just as fast (buffalo) or faster (wildebeest and gazelles, 50 mph; 80 kph) and can sustain their speed over longer distances. The only advantage the lion has is its greater acceleration, and it is therefore essential for the lion to approach its prey as closely as possible, usually to within 100–160 feet (30–50 m) before making its charge. Most charges end in failure. The average hunting success of a solitary lion is about 20 percent, and this is approximately doubled when two or more hunt together. Of cardinal importance to the success of an attack is the lion's ability to stop the prey as soon as contact has been made. The fleeing prey is usually seized by the hindquarters and knocked to the ground. Success is largely determined by the difference in weight between hunter and hunted. A lion has far greater difficulty with an animal that is much larger than itself, like an adult buffalo, than with an animal that is nearer its own weight. And it is important that a victim be killed rapidly because a prolonged struggle may lead to serious injury to the lion, especially if the weight difference between them is large. Flaying hooves and stabbing horns are additional hazards, but the lion is very efficient at killing and usually dispatches its victim within minutes with a deep bite to the neck or throat.

African wild dogs, which hunt in packs, weigh between 37 and 44 pounds (17–20 kg) and usually select victims that do not exceed 140 pounds (65 kg). Larger prey are sometimes taken, but usually only with difficulty because, although they hunt in packs, their jaws are relatively weak, which makes it difficult for them to subdue their victims. One pack seen attacking a large wildebeest calf took eight minutes to drag it down, and another pack took five minutes just to chew through the abdominal wall of a female wildebeest. Wild dogs have been known to kill adult zebra, which are ten times their own weight, but this is unusual.

Adult cheetahs weigh between 75 and 120 pounds (35–55 kg) and prey upon Thompson's gazelles, which seldom exceed 44 pounds (20 kg). Larger prey are taken (even wildebeest), but the preference appears to be for animals that do not exceed 130 pounds (60 kg) because of the difficulties cheetahs have in subduing large prey. Leopards weigh about the same as cheetahs, but they do not experience such difficulties with larger animals. Their preferred prey size is in the range 45–155 pounds (20–70 kg) but they will take animals of up to three times their own weight.

Some generalizations may be made from this survey of mammalian predators. Solitary hunters tend to select prey that are about their own weight but will attack animals up to three times as large as themselves. When hunting in groups predators will take larger game, but it is unusual for carnivores to attack prey that exceed about five times their own weight. The large herbivores, those exceeding about a ton (1000 kg), appear to be immune from predators, although their offspring frequently fall victim to them. It is important to note that the data for living mammalian predators can serve only as an approximation to the possible predator-prey relationships that may have existed among dinosaurs. Dinosaurs were not mammals, and although they may have been warm-blooded, and some of them may have been as active as mammals, we cannot assume that they behaved like mammals. Nor was the Mesozoic environment the same as that of present-day Africa.

The Komodo dragon This large lizard has been veiled in as much mystery as the Indonesian islands from which it comes. There have been reports of its exceeding twenty feet (6 m) in length and of its attacking villagers and their livestock, but it seems unlikely that the ora, as it is known locally, exceeds lengths of about ten feet (3 m). The average length is probably around six feet (1.8 m—Auffenberg, 1981). Body weights are very variable and depend upon the amount of food eaten. The ora has a prodigious appetite and can increase its weight by as much as 80 percent over its fasting weight. The average weight of large individuals (7–8 ft; 2.2–2.5 m) is about 100 pounds (45 kg). Although oras do hunt small animals like rats and mice and birds, they prefer larger prey. The large ones have a preference for deer, with wild boar as a second favorite. They are voracious predators and will occasionally enter villages to attack goats and domestic cattle.

There have even been reports of their having killed and eaten villagers, though such incidents are rare.

The average weight of deer on Komodo Island is 110 pounds (50 kg), and mature stags can weigh as much as 440 pounds (200 kg). The ora selectively attacks smaller individuals, however, and the average weight of deer eaten is 30 pounds (13.6 kg). Sometimes mature stags are killed and even mature buffalo weighing as much as 1,300 pounds (590 kg)—about ten times the weight of the dragon—but these larger animals are relatively immune from attack (Auffenberg, 1981, p. 286).

In marked contrast to the mammalian predators discussed above, which pursue their prey at high speeds over distances of 100–160 feet (30–50 m), the ora is unable to chase down its victims. As a consequence it has to position itself within about a yard (1 m) of a potential victim to stand any chance of a successful attack. It relies on two basic hunting strategies. One tactic is to wait in ambush along the game trails used by the deer. If a deer approaches too close it is seized, usually by a leg, where the ora's sharp teeth and powerful jaws inflict severe wounds, severing tendons and lacerating muscles. The prey is usually thrown to the ground and when the opportunity arises the leg grip is exchanged for a damaging bite to the neck or to the abdomen. The hapless victim is often dispatched by evisceration. The second tactic is for the ora to approach its prey by stealth, usually using the cover of tall grasses for concealment. As the ora is so low to the ground it has to stop every so often to stretch up its neck and peer at its prey over the top of the vegetation.

The sense of smell plays an important part in hunting. Oras are able to follow the scent trials of their prey, and they even seem able to distinguish pregnant females from other individuals. One of their strategies is to stay close by so that they can seize the offspring as it is born. Sometimes their harassment of a pregnant cow causes a miscarriage, ensuring them, or another ora, of a meal. Although several individuals may sometimes be seen devouring a carcass, oras appear to be solitary hunters and there is no substantive evidence of group hunting.

To summarize, Komodo dragons sometimes kill animals that are more than ten times their own weight, but they tend to attack individuals weighing less than themselves. They are unable to pursue their prey and therefore either use a sit-and-wait tactic or stalk their prey by stealth.

Allosaurus, a common Jurassic carnivore, weighed about two tons (Colbert, 1962). Some considerably larger individuals have been found, so the two-ton estimate should probably be considered an average adult weight. Herbivorous dinosaurs that were contemporaneous with *Allosaurus* include *Diplodocus, Apatosaurus, Brachiosaurus,* and *Ultrasaurus* (weighing about 18, 28, 78, and 180 tons, respectively). But there were also much smaller herbivores, including *Camptosaurus* (1/2 ton) and *Stegosaurus* (3 tons). If carnivorous dinosaurs exercised the same caution as modern mammalian ones, we would expect that solitary individuals of *Allosaurus* may have preferentially attacked *Camptosaurus.* They may also have tackled *Stegosaurus,* though some care would have been needed because of the latter's formidable tail spikes. It seems less likely that *Diplodocus* would have been attacked, and the larger sauropods would probably have been left well alone. But the carnivorous dinosaurs may have hunted in packs, a suggestion that has been made by other paleontologists (Ostrom, 1969; Farlow, 1976). If *Allosaurus* did hunt in packs it may have attacked *Diplodocus,* though the largest sauropods were probably still safe. Juveniles of all species may have been attacked though, just as young elephants and rhinos are sometimes attacked by modern carnivores.

Sauropods appear to have been relatively less abundant during the Cretaceous than the Jurassic, especially toward the end of the period when there was a wide variety of hadrosaur-sized herbivores for the predators to prey upon. The Late Cretaceous carnivore *Tyrannosaurus* was heavier than *Allosaurus* (7.5 tons compared with 2 tons) and had the potential to attack herbivores that were about half its size, like *Corythosaurus,* for example, which had an estimated weight of four tons. This suggested to Farlow (1976) that group hunting may have been more important for allosaurs than for tyrannosaurs.

Body size and longevity

Body weight, as we have seen, has a major influence on most aspects of an animal's life, from its relative food requirements to how fast it can run. There is also a strong correlation between body weight and longevity. Large animals live longer than small ones, and this has some fascinating implications for sauropods. The life span for a wide variety of mammals and birds (Lindstedt and Calder, 1981)[11] has been found to vary approximately according to (body weight)$^{1/4}$. This means that a sixteen-pound coyote would be expected to live approximately twice as long as a one-pound guinea pig because it is 16 times larger and $(16)^{1/4}$ = 2. Like any generalizations pertaining to living things, this rule has many exceptions, and, since the arithmetic involves a power function, small errors in data are magnified, so predictions are likely to be very approximate. Furthermore, it must be borne in mind that life-span data are likely to be even less reliable than body-weight data. Think of the problems of trying to establish how long animals live. There are few opportunities for collecting such data in the wild—that would require observations of known individuals, possibly over long periods of time—so most information has to come from zoo records, and zoo animals cannot be considered exemplary of their kind. There are also data from the recollections of people who have kept various animals, but these tend to be apocryphal in nature. I have heard, for example, of a tortoise that is supposed to have survived on some Pacific island since the time of a visit from Captain Cook, over two hundred years ago. For all of these reasons any estimates of life spans that are deduced from body weights are going to be very approximate. As far as our example of the coyote living twice as long as a guinea pig, it is approximately correct because coyotes are said to live to a maximum age of fourteen years, guinea pigs six years (Spector, 1956).

Most of the data that have been collected for the relationship between body weight and life span pertain to mammals and birds, but it appears that the same relationship applies to cold-blooded animals too (Peters, 1983; Calder, 1984).[12] We can therefore make some extrapolations from elephants to sauropods, but it must be emphasized that these calculations are entirely speculative; they serve only to give us an idea of the ages these great dinosaurs may have attained.

Our weight estimates for *Diplodocus, Apatosaurus, Brachiosaurus,* and *Ultrasaurus* are approximately 3, 4.7, 13, and 30 times the six tons we've assigned to the elephant. Their respective life spans would therefore be expected to scale according to these numbers raised to the power 1/4, that is, to 1.32, 1.47, 1.90, and 2.34. If the elephant lives sixty years,[13] we would estimate 60 × 1.32 = 79 years for *Diplodocus,* and 88, 114, and 140 years for the others.

Modern reptiles grow much more slowly than mammals and birds—about one-tenth as fast—and they continue growing throughout life, though at slower rates with increasing years. The daily increase in weight of a growing reptile is directly related to the adult body weight attainable by the species. For example, a young lizard of a species that attains a body weight of only a few ounces will have a daily weight increase that is the smallest fraction of an ounce, whereas a young alligator might put on a quarter of an ounce a day. If it is assumed that dinosaurs were like modern reptiles in their growth patterns, rather than like mammals or birds, an estimate of the daily growth rate for a particular dinosaurian species can be obtained by extrapolation from that of modern reptiles. If an estimate can be made of the body weight of the hatchling dinosaur, therefore, it is at least theoretically possible to estimate how long the hatchling would take to reach adult body weight. This method was used by Ted Case (1978) to obtain age estimates for *Hypselosaurus,* a sauropod for which eggs have been found. He gave a predicted age of between 104 and 208 years, depending on whether the maximum growth rate was maintained throughout life or not. These predictions were based on an adult body weight estimate of 5.3 tons. Sauropods, then, may have lived to great ages. It is even conceivable that their unhurried lives spanned from one century to the next.

If I were an artist I would paint a tranquil picture of sauropod life. In the foreground would be a motionless giant, head barely raised above its body as it stared blankly into space. Another would be shown with its neck reaching down to browse on leaves, its movements gentle and unhurried. Most, if not all of the tail, would probably be held clear of the ground, but I might introduce some dense foliage here so that a firm decision could be postponed. And off in the distance would be a third individual, dashing along at its top speed of an animated walk. But sauropods have long legs, and they take long strides, and if I could enter that majestic presence I might have to run my hardest just to try and keep up.

What's Hot and What's Not

THE traditional view of dinosaurs was that they were cold-blooded, like modern reptiles—in other words, that they were relatively inactive, slow-moving, and generally unresponsive creatures. Birds and mammals, on the other hand, being warm-blooded, were all the things that the reptiles were not. It was not difficult to understand why birds and mammals eventually succeeded where the dinosaurs and their reptilian kin had failed. Simply stated, birds and mammals were considered superior to the reptiles, the inheritors of the Earth when the great Age of Reptiles was eclipsed by the Age of Mammals. But if reptiles are in some ways inferior to mammals, why are there more reptilian than mammalian species alive today? And how do we rationalize the fact that many reptiles, including snakes, crocodiles, and monitor lizards, feed largely on mammals and birds? Nor are reptiles necessarily inactive and slow-moving either, as anyone who has tried to catch a lizard will attest, and the complex behavioral patterns of some species disqualifies them from being considered unresponsive. In this chapter we will take a close look at the activity levels of modern reptiles to see how these compare with those of birds and mammals. We shall also be concerned with another challenge to traditional views, namely, the claim that dinosaurs were warm-blooded.

Robert Bakker of the University of Colorado was not the first to propose that dinosaurs were warm-blooded,[1] but it is probably fair to say that he has been the most active and controversial advocate of the thesis, and the one who can be credited with its recent revival. His evidence and conclusions, published in the early 1970s, were vigorously challenged by some (Bennett and Dalzell, 1973) but generally accepted by others (Dodson, 1974; Ostrom, 1974a). Indeed so much attention was focused on the issue that it was made the subject of a symposium at the annual meeting of the American Association for the Advancement of Science in 1978 (reported in Thomas and Olson, 1980). Although no consensus was reached, one conclusion did emerge: the thermal strategies among living vertebrates, far less extinct ones, are more complex than a simple dichotomy between warm- and cold-bloodedness.

The starting point in this debate must be a discussion of basic biochemistry and exercise physiology, but since I am neither a biochemist nor a physiologist it will, most assuredly, be kept simple. We shall see that there are fundamental differences in the rates of metabolism between birds and mammals on the one hand and all the other vertebrates on the other. We shall also see that metabolic rates affect not only body temperatures and their constancy (or lack of it), but also the ability to sustain muscular activity—in a word, stamina.

Metabolism: the furnace within

Like other mammals, and like birds, we are warm-blooded. Our body temperature is maintained at a fairly high and constant level independent of ambient (external) temperatures. It is maintained within such narrow limits, in fact, that fluctuations are interpreted as signs of illness. Human temperatures are usually measured by placing the thermometer in the mouth (less often in the rectum), because we wish to know the temperature of the central region of the body—the core temperature. Peripheral temperatures, that is, temperatures measured at the extremities, such as arms and legs and ears, may vary according to ambient temperatures, but core temperatures remain constant.

Constant core temperatures of around 98° F (37° C) are maintained by many mammals. Other mammals, including sloths, monotremes (such as the duck-billed platypus), and marsupials (opossums, koala bears, and their Australian relatives), have lower metabolic rates and correspondingly lower body temperatures, by a few degrees; they may also have less precise regulatory mechanisms. Brian McNab, who has made a major contribution to our knowledge of thermal strategies, has shown that diet is a major factor (1986). Mammals that feed of fruit and woody plants, for example, tend to have lower metabolic rates than those that feed on herbaceous plants. Birds generally have higher body temperatures than mammals, averaging about 104° F (40° C), and they tend to have higher metabolic rates as well (McNab, 1983).

The high body temperatures of birds and mammals are maintained whether they live in the tropics or the frozen tundra. The body temperature of penguins and polar bears are no lower than those of egrets and elephants. The heat required to maintain these high temperatures is generated within the body by the cells. All living cells generate small amounts of heat as a by-product of the chemical processes occurring within them. This chemical activity is called metabolism. Because different organisms have different rates of metabolism, the rates have to be measured under the same conditions before they can be compared. The standard method is to measure the metabolic rate when the animal is completely at rest and when it has not just eaten a meal. Animals have to be at rest because active cells, especially muscle cells, generate much more heat than resting cells—about 80 percent of the energy output of a contracting muscle cell appears as heat. And the animal has to be fasting because the process of digestion generates additional heat. There is one more condition—the animal has to be kept resting within a particular temperature range. This is especially important for birds and mammals because if the temperature is too low they increase their metabolic rate—they may also start shivering to generate more heat. If the temperature is too high they decrease their metabolic rate. Mammals can also shed excess heat by sweating and panting—birds can pant too.

The temperature range within which these changes do not occur is called the *thermal neutral zone;* we would describe such a temperature range as being comfortable. Although reptiles and the other animals that are not warm-blooded do not have a thermal neutral zone, most have a preferred temperature range and it is usual to keep them within this range when measuring their metabolic rates. Furthermore, since their metabolic rates increase with temperature, it is necessary to state the temperature at which measurements were taken. The metabolic rate of an animal when measured under these conditions, the *basal metabolic rate,* may be thought of as the minimum level of chemical activity required to sustain life.

Mammalian and avian cells appear to be physiologically different from those of all other animals in that they have the highest rates of metabolism. The basal metabolic rates of small birds and

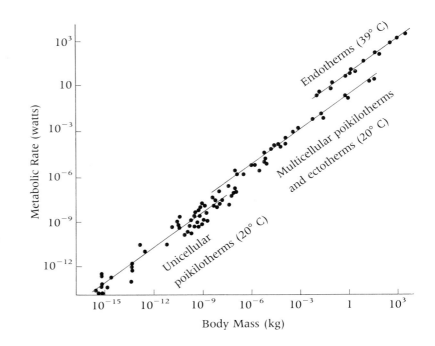

mammals are at least ten times higher than those of equally sized reptiles, amphibians, and fishes (and multicellular invertebrates), and these, in turn, are about ten times higher than those of unicellular organisms (Peters, 1983). A resting frog, for example, generates about 30 joules per kilogram per hour compared with about 250 J/kg/hr for a bird of similar size.[2] (Metabolic rate is often expressed in terms of watts, a watt being one joule per second. Therefore 250 J/kg/hr is 0.07 watts/kg.) As noted in Chapter 5, large animals have relatively lower metabolic rates than smaller ones, and this rule applies to each of the three metabolic groups considered here (mammals and birds; reptiles, amphibians, fishes, and multicellular invertebrates; and unicellular organisms). Consequently, when (logarithmic) graphs are plotted of metabolic rates against body mass, three straight lines are obtained, one below the level of the next, each with a same slope of about $\frac{3}{4}$.

Most of the metabolic power of an animal appears as heat. Since smaller animals have higher metabolic rates than larger ones, pound for pound, it would make more sense to snuggle up to a lot of small animals on a cold night than to one large one. The resting metabolic rate of a 1,100-pound (500 kg) moose, for example, is equivalent to the power output of seven sixty-watt light bulbs, but 1,100 pounds of mice would generate metabolic power equivalent to about nine hundred light bulbs (Peters, 1983). Considerably less comfort would be found in snuggling up to an equivalent weight of lizards or toads, mainly because of their much lower metabolic rates.

Thermal strategies

It is by virtue of their high metabolic rates—aided by the insulation provided by their fur and feathers—that mammals and birds are able to maintain high and constant body temperatures. They remain at much the same temperature regardless of ambient temperatures and regardless of their activity levels—even in their sleep. This strategy is described as *endothermy* ("heat from within") or *tachymetabolism* ("fast change"—for formal definitions see Bligh and Johnson, 1973; Cabanac, 1987). The term *homeothermy* ("equal heat") is also used, emphasis being placed on the constancy of the body temperature rather than on the source of heat, but we will use the term *endothermy* here. Some mammals and birds, especially the smaller ones, are unable to maintain their high body temperatures in the cool of the night and their temperatures accordingly fall by a few degrees. Roosting bats and hummingbirds and hibernating bears allow their temperatures to approach ambient temperature. This particular strategy, which economizes on energy requirements, is described as *heterothermy* (*heteros* means "different"). Why endotherms should maintain such high body temperatures (around 98° F or 37° C) has been the subject of much discussion. One of the suggestions that has been made is that enhanced muscle performance may have been a primary factor (McGowan, 1979b).

The metabolic rates of most animals are too low to enable them to be endothermic, but there are benefits to be being warm and reptiles warm themselves by absorbing heat from the sun. Most reptiles are inactive in the cold light of dawn and have to bask in the sun before they can become very active. But once they have reached their working temperature they can keep their bodies at a fairly constant temperature: when they are too hot they seek the shade, when too cold they bask in the sun. Active reptiles have average body temperatures of about 95° F (35° C), which is only a little lower than that of most mammals, but there is prob-

A Galápagos land iguana. Reptiles spend much of their day sunning themselves. By alternately basking and seeking shade they are able to maintain fairly constant body temperatures.

ably much variation from species to species. As their major source of heat is external to the body, they are said to be *ectothermic* (*ektos,* "outside").

Aside from seeking a more agreeable environment, most animals, including amphibians, fishes, and invertebrates, have no control over body temperature, which accordingly fluctuates with ambient temperatures. Such a thermal strategy, or rather lack of it, is described as *poikilothermy* (*poikilos,* "various"). Being poikilothermic does not necessarily mean having a widely fluctuating body temperature. Fishes living in Antarctic waters, for example, experience an almost constant temperature ($-1.9°$ C) that varies by only one-tenth of a degree throughout the year (Somero and De Vries, 1967). Tropical fishes similarly live in a stable temperature regime, but most poikilotherms are exposed to widely fluctuating temperatures.

When animals that are not endothermic are exposed to a temperature gradient they usually select a preferred temperature range, which is called the *normal activity range*. This range is often quite narrow, while in other cases it is quite broad. It is a universal property of matter that chemical reactions are speeded up by raising the temperature, and it has been found that speeds are approximately doubled with each 10 C° increment,[3] a phenomenon referred to as the Q_{10} effect. Consequently, within the normal activity range of an animal, an increase in temperature brings about an increase in its activity rates. If a particular lizard had a normal activity range of 20–40° C, raising the temperature from 25° C to 35° C would theoretically cause a doubling in the rate of all its body functions, from the speed of locomotion to the rate at which food is digested. In practice, though, the Q_{10} effect often brings about less than a doubling of rate (Harlow, Hillman, and Hoffman, 1976; Huey and Bennett, 1987; Whitehead et al., 1989).

The rationale for an animal selecting a particular normal activity range probably has to do with the way that enzymes function. Enzymes are proteins that catalyze the chemical processes occurring within living cells. Most enzymes work best within a particular, and usually narrow, temperature range. If an enzyme is exposed to temperatures outside this range, its catalytic powers are reduced or eliminated. Each species has its own system of enzymes, adapted to its own normal range of temperatures. Antarctic fishes are adapted to life at −1.9° C while tropical fishes are adapted to life at about 25° C. The Antarctic fishes are operating at their maximum efficiency at −1.9° C just as the tropical fishes are operating at their maximum efficiency at 25° C. Because these fishes have such narrow normal activity ranges, any temperature change is likely to be detrimental. Raising the water temperature to just 6° C, which is still rather cold, killed half of a sample of the Antarctic fishes within a week (Somero and De Vries, 1967).

The obvious advantage an endotherm has over other animals is that it is always in a state of readiness to flee, fight, or go about its business, regardless of ambient temperatures. This was graphically illustrated for me while prospecting for fossils in an underground cave. I came upon the carcass of a sheep that had fallen to its death, through a crevice, some months before. The flesh was in an advanced state of decay and well-fed flies scurried all over it as they went about their unsavory business. They made no attempt to fly away, even when rudely harassed with the end of my trowel, and I soon realized that they were unable to do so. Although it was a hot day the inside of the cave was quite cold, and their flight muscles were presumably below their working temperature. The flies were simply unable to generate sufficient power for takeoff. Had they been endotherms they would have been able to fly instantly. But endothermy is a very expensive strategy.

Being endothermic requires the intake of large quantities of food. It has been estimated that a nine-ounce (250 gm) mammal or bird would use about seventeen times more energy during the course of a mild day than a lizard of similar size (Nagy, 1987); they therefore have to eat that much more food. A 100-pound (45 kg) cougar would eat about five times as much as a crocodile of similar weight (reported in Farlow, 1980). Notice that the disparity between the food requirements of endotherms and ectotherms decreases with increasing body size. This is because the slope of the metabolic rate–body mass graph is slightly steeper for ectotherms than it is for endotherms, so the two graphs tend to converge toward the large end of the scale. Birds and mammals, particularly the smaller ones, have to spend most of their waking hours foraging for food, but the frog sits idly on his lily pad and watches the world go by.

The difference in food requirements between endotherms and ectotherms is partly due to the endotherms' higher capacity for generating energy at the cellular level. This is reflected in the higher density of mitochondria in their cells (mitochondria are minute cellular bodies that are concerned with energy exchanges). But the primary reason that endotherms require more food is that they maintain their high body temperatures during the night as well as during the day (Nagy, 1987).

A compromise strategy to being endothermic is to elevate one's temperature just before periods of activity. This minimizes the costs of being alive during times of inactivity and permits activities to be conducted over a wide range of ambient temperatures. Many flying insects have evolved this strategy, which is properly referred to as *behavioral temperature regulation*.[4] (Many authorities refer to it, incorrectly, as endothermy because the source of heat is internal.) The core temperature of a grounded moth or a bee varies with the ambient temperature, but in flight the wing muscles generate enough heat to keep the thorax at a fairly constant and high temperature, in the region of 86–95° F (30–35° C). Prior to takeoff these insects undergo a preflight warm-up procedure that lasts for five to ten minutes (Heinrich, 1974, 1987). They simultaneously contract both sets of wing muscles—the ones that depress the wings and those that elevate them. This generates heat without any wing movements, aside from a slight shivering. As the muscles warm up they contract faster and with a greater force. When the working temperature is reached they have sufficient power to sustain flight, and the insect is able to take off. The ability of these insects to maintain high thoracic temperatures is assisted by the thermal insulation afforded by the fine bristles that cover the thorax.

The increased power that insects obtain from warming up their flight muscles derives from the Q_{10} effect. But not all animals can take advantage of this effect. Fishes, for example, being surrounded by water, quickly lose the heat generated by their actively contracting muscles. This is because the heat that leaves the muscles is lost when the blood passes through the gills. Some fishes, like the tuna, are able to maintain some of their swimming muscles at a relatively high and constant temperature, through modifications in their blood vascular system. The tuna has a heat interchanger in the form of a dense concentration of capillaries. Warm blood that is on its way to the gills to be oxygenated passes through capillaries in the interchanger. Here it comes into close contact with capillaries carrying cold, oxygenated blood from the gills. Heat passes from the warm to the cold blood and is therefore conserved. The ability of the tuna to elevate its body temperature by muscular activity was observed as long ago as 1835 by a British physician named John Davy. Davy, who was voyaging in the tropics, was amazed to find that the blood of the tuna was relatively warm, about 10 centigrade degrees (18 F°) higher than the surrounding water. Tunas are very active fishes that swim continuously and are capable of reaching high speeds. The larger ones, which have higher body temperatures than the smaller ones, can achieve a temperature of about 95° F (35° C). The warmest part of the body is a strip of muscle that runs down each side of the fish. It is brown, partly because of its rich blood supply. Those of us who enjoy our fish served on the bone will have noticed this strip of muscle in many other fishes besides the tuna, but relatively few species are capable of keeping this muscle warm. The tuna frequently travels between the warm upper layers of the sea and the deeper cold layers. By maintaining its body at an elevated and fairly constant temperature, it is able to maintain the same powerful swimming movements regardless of ambient temperatures. The tuna is also able to elevate the temperature of its stomach, and this, again by virtue of the Q_{10} effect, increases the rate of digestion of its food (Carey, Kanwisher, and Stevens, 1984).

The swordfish, a close relative of the tuna, does not keep its muscles warm, but it does have a brain heater. This specialized tissue, which is rich in mitochondria, generates enough heat to keep the brain and eyes between 18 and 25 F° (10–14 C°) above ambient temperatures (Carey, 1982). Like the tuna, the swordfish makes daily excursions between the warm surface layers and the cold ocean depths, and the brain heater reduces the extent of the

Very large reptiles, like this Galápagos tortoise, maintain fairly constant body temperatures by virtue of their thermal inertia.

temperature fluctuations on the central nervous system. It is thought that this heater enables the swordfish, an active predator, to maintain its vigilance at all times, regardless of fluctuations in ambient temperatures.

Modern reptiles, as we have seen, are ectothermic. As they can only absorb heat through their skin, the smaller ones heat up (and cool down) more rapidly than the larger ones, because of their relatively large area-to-volume ratio. Spiny lizards weighing between one-half and three ounces (12–85 gm) can raise their body temperatures from 81 to 109° F (27–43° C) after fifteen minutes of basking in the Arizona sun (Bogert, 1959). Alligators weighing two ounces (50 gm) take a little over a minute to warm up by one centigrade degree, whereas one that weighs twenty-eight pounds (13 kg) can take more than seven minutes for the same increase in temperature (Colbert, Cowles, and Bogert, 1946). Large reptiles, then, by virtue of their relatively small area-to-volume ratios, tend to warm and cool relatively slowly. This phenomenon is described as thermal inertia.

The Komodo dragon, discussed in Chapter 5, is a large reptile with an average weight of about 100 pounds (45 kg). They have a certain degree of thermal inertia, and large individuals are able to keep their body temperatures several degrees above ambient temperatures during the cool of the night. This ability is enhanced by their tendency to retreat into burrows. An important consequence of this habit is a reduction—by as much as one or two hours—in the time they have to spend basking in the early morning sun to raise their temperatures to active levels (Auffenberg, 1981).

The Galápagos tortoise, which weighs about 450 pounds (200 kg), has a thermal inertia so large that although ground temperatures fall by about 36 F° (20 C°) during the night, its body temperature falls only by about 5 F° (3 C°—Mackay, 1964). Their body temperatures therefore remain fairly constant throughout the twenty-four-hour period; this thermal strategy is referred to as *inertial homeothermy*. I had the opportunity of seeing the effectiveness of this thermal strategy for myself, during a trip to the Galápagos Islands. Our party had climbed to the top of a volcano, some three thousand feet (1,000 m) above sea level, and were camped overnight on its rim. Although it had been quite hot during the day, the temperatures plummeted at night and we all felt cold, even in our sleeping bags. As I emerged from my tent the following morning, cold and stiff, I encountered a tortoise who had spent the night around our camp. He was wide awake and active, and when I touched his body with my cold fingertips he felt quite warm.

The leatherback turtle, which is found in oceans throughout most of the world, is even larger than the Galápagos tortoise, and its body mass, which exceeds 1,300 pounds (600 kg), confers considerable thermal inertia. As a consequence of this inertia the heat generated by the muscles during swimming enables the turtle to maintain a body temperature some 32 F° (18 C°) higher than the surrounding water (Frair, Ackman, and Mrosovsky, 1972; Mrosovsky, 1987). This is remarkable when one considers how rapidly water, especially cold water, conducts heat away from the body. (The high thermal conductance of water explains why a frozen chicken will thaw more quickly in a bowl of water, even cold water, than in air.)

So far we have seen that all cells generate heat as a by-product of their (resting) metabolism and that the basal metabolic rate of birds and mammals is the highest of all animals. This rate is at least ten times higher than that of other multicellular animals, which, in turn, is about ten times higher than that of unicellular organisms. Birds and mammals, by virtue of their high metabolic rates, are able to maintain high and constant body temperatures. The obvious advantage of being endothermic is always being ready to perform at the maximum effort, regardless of ambient temperatures, but the costs of endothermy are very high. Other animals are able to take advantage of the increased activity rates that elevated temperatures bring, through the Q_{10} effect, without paying the high price of endothermy. These alternative thermal strategies include heterothermy (bats, many birds, hibernating bears), behavioral temperature regulation (some flying insects and some fishes), ectothermy (most living reptiles), and inertial homeothermy (Galápagos turtle, leatherback turtle). Having considered the thermal consequences that arise from differences in metabolic rates, we turn now to the other major consequence of metabolic rate—an animal's ability to sustain muscular activity.

Biochemistry and stamina

Basal metabolic rates are usually measured not by assessing the heat output of a resting animal, but rather by measuring the amount of oxygen that an animal uses in a given time. This is largely a matter of convenience because it is easier to measure the amount of oxygen an animal removes from its air supply than it is to measure its heat output. Oxygen consumption is measured because oxygen is used during metabolism to release energy from the substrate (food).

The first step in the metabolism of carbohydrates (compounds of carbon and, in a 2:1 ratio, hydrogen and oxygen) is that glucose molecules are broken down into pyruvate molecules. If oxygen is supplied to the cells these pyruvate molecules pass through a cycle of chemical reactions called the Krebs cycle. This chemical cycle releases energy, together with the by-products of heat, water, and carbon dioxide. This metabolic process is described as being *aerobic* because it requires the supply of oxygen to the cells. The energy that is produced during metabolism is stored in the form of chemical energy, locked up inside a chemical compound called ATP (adenosine triphosphate). ATP is formed when ADP (adenosine diphosphate) is linked up with a phosphate molecule, and the energy required to do this is the energy that was produced by the metabolic process. This chemical energy can subsequently be released by breaking down the ATP into ADP and phosphate. ATP functions as a temporary storage device, like a spring-loaded mousetrap waiting to go off. And once the energy has been released, the ADP and phosphate can be re-used to form ATP again, if more energy is supplied by metabolism. ATP is therefore the energy currency used by cells. Having considered what metabolic rate is in theory, we are ready to see how it is measured in practice.

There you are, lying on a rather uncomfortable couch in a very clinical lab with your face covered by a plastic hood that is hooked up to an oxygen analyzer. We are going to measure your basal metabolic rate first, so we have made sure that you have not eaten for several hours. And we want to keep you as quiet as possible so that you will be using oxygen at the lowest possible rate and the reading we get will be your basal metabolic rate. But you do not like wearing the mask, and you are feeling a bit uneasy about the whole thing, so your metabolic rate is already a little above its minimum level. We now ask you to move your arms. Moving your arms requires additional supplies of ATP to the muscles of your arms, so your metabolic rate has to be increased; your oxygen consumption increases accordingly. This is not very strenuous exercise and when you are asked to return to resting again you do not have to lie there puffing and panting to regain your breath. Your oxygen consumption begins dropping, indicating a decline in metabolic rate to something approaching the basal level. But we have not finished with you yet . . .

Pretty soon we have you on a treadmill, walking along at a steady pace. Your metabolic rate is now several times higher than your basal rate was. We then increase the speed of the treadmill, hence your walking speed, and your oxygen consumption increases in the same proportion. If we doubled your speed your oxygen consumption, hence metabolic rate, would be doubled. You are in quite good shape—you do a bit of jogging and like a game of tennis—so when we ask you to tell us when the treadmill has reached a comfortable running speed, you choose a fairly brisk pace. The treadmill operator never exercises—she is not the athletic type—and would not be able to run as fast as you are running. You are happy to keep this speed up for the next ten minutes, and when we stop the machine you are not puffing and panting, you are just breathing normally. Your heart and lungs and vascular system were obviously supplying enough oxygen to your muscles to sustain the Krebs cycle. The muscle cells were therefore receiving an adequate supply of ATP for their continued contractions.

You are now back on the treadmill and we are going to go for broke! We increase the speed, slowly, and your oxygen consumption rises linearly. So far so good. You are now accelerating up to your comfortable running pace and your oxygen consumption continues to increase. But as we increase the speed there comes a point when your oxygen consumption reaches a maximum—it is about six times higher than it was when you were lying down. If you were in really good shape it would be about eleven times higher—the highest documented increase is 27.5, reported for an exceptionally fit Scandinavian skier.[5] We keep on increasing your running speed but your oxygen consumption does not go any higher.

You are now running at a fair pace and breathing quite heavily, but I am sure you can run much faster! We want to get you up to your maximum speed as quickly as possible so we crank the machine up a lot faster.

You are approaching your top speed and although you are breathing very heavily, and going quite red in the face, your rate of oxygen consumption has not risen above the maximum level you reached a few minutes ago. We will explain what all this means in a minute, but we want to finish the experiment first.

I bet you never thought you could run this fast! Well, it probably seems fast to you but your speed is actually only about twice as high as it was when you reached your maximum oxygen consumption. You are looking a bit tense at this point and the other people in the lab are beginning to moralize. But before I can slow down the machine and let you get off you rip off the face mask and make a leap for terra firma. And there you stand, hanging on to the end of a bench and gulping down air for the next two minutes. I don't think you want to talk about exercise physiology just yet.

When you were running at the slower speeds you were performing aerobically. Your muscles (and all other cells) were being provided with oxygen at a rate high enough to break down glucose into pyruvate and then into carbon dioxide and water in the Krebs cycle. Sufficient ATP was therefore formed to satisfy all of the energy demands. As your speed increased you still performed aerobically, as was shown by your rising oxygen consumption, but then you reached your maximum rate of oxygen consumption, which was about six times higher than it was at your basal metabolic rate. And no matter how hard you breathed after that, your lungs and heart and vascular system were unable to increase the supply of oxygen to your muscles. But you still continued to make increased demands upon your muscles because you continued to run even faster, and the energy for this extra demand had to be met by *anaerobic* metabolism, metabolism without oxygen.

At the point immediately before aerobic metabolism has to be supplemented by anaerobic metabolism, oxygen is being supplied at a rate that is just sufficient to oxidize all of the pyruvate that is being formed from the breakdown of glucose. But as more pyruvate is produced to meet the increased energy demand there is no longer enough oxygen to oxidize it all. This additional pyruvate cannot be allowed to enter into the Krebs cycle along with the rest. Instead, it is converted into lactate, and this accumulates in the muscles as lactic acid. This anaerobic process, the breakdown of glucose to pyruvate and then to lactate, is called *glycolysis.* Glycolysis, which could be thought of as the incomplete combustion of glucose, produces only about *one-thirteenth* as much ATP as aerobic metabolism, so it is less efficient.[6] Anaerobic metabolism also has the disadvantage that it results in the accumulation of lactic acid in the muscles, and since lactic acid eventually inhibits muscle contractions anaerobic sprinting exercise cannot be continued for very long. When the anaerobic activity ceases, the accumulated lactic acid has to be broken down, which requires oxygen. This explains why we continue breathing heavily even after we have finished our strenuous exercise—we need the extra oxygen to complete the breakdown of the lactic acid. We are, in effect, paying off an *oxygen debt,* and although this term is widely used in this context, it is actually an oversimplification of what is actually happening (Shephard, 1984).

Middle- and long-distance runners run aerobically for most of a race, and if they had to stop part-way through they would not find themselves puffing and panting to repay an oxygen debt. But when the finishing line comes into sight they make an all-out effort and their exercise becomes anaerobic. Sprinters perform their entire event anaerobically and accumulate so much lactic acid in their muscles that it can become a limiting factor to their performance. Aside from interfering with the contractions of the muscles, high levels of lactic acid can be painful. Winners are often those who can withstand the pain and keep up their pace. But they soon recover after the event, probably within a minute or two, and could then run again if they had to.

The amount of glucose available for metabolism is fairly limited—there is only about an ounce (25 gm) circulating in our blood. The main carbohydrate reserve is glycogen (Shephard, 1984). Glycogen is a starch (sometimes referred to as animal starch) and is stored mainly in the muscles, where it is readily available, and also in the liver. We have a total glycogen store of about one pound (0.5 kg). Because of its importance as a substrate for muscle metabolism athletes, especially those of endurance events like marathons, frequently engage in "carbohydrate loading." By eating lots of starchy foods like pasta for two or three days before their event, they can double their glycogen store. And because glycogen molecules bind with water, they also gain an additional store of water—about three pounds for each pound of glycogen. Incidentally, this bound water accounts for most of the weight we lose when we go on starch-reduced diets.

Because aerobic metabolism is much more efficient than anaerobic metabolism, it is a more desirable way of powering an animal's activities. The major limitation is the rate at which oxygen can be supplied to the tissues. A measure of this rate is the amount by which an animal is able to increase its oxygen consumption above that of the resting condition. This is referred to as the *aerobic scope,* or the aerobic metabolic scope. The aerobic scope gives a measure of the range of speeds that an animal is able to sustain. Birds and mammals have aerobic scopes of between five and ten times, meaning that their maximum rate of oxygen consumption is between five and ten times higher than their resting rate. Aerobic scope can be improved by training, certainly in mammals. I do not know what my aerobic scope is at present, but it is definitely higher now than it used to be because I go out jogging almost every day. I can therefore sustain a faster running speed, without incurring an oxygen debt, than I could before I started jogging regularly. Training increases oxygen consumption mainly because it changes the vascular system and lungs. These changes include an increase in the size of the heart, which increases the volume of blood pumped at each beat; an increase in the blood supply to the muscles, by the development of additional capillaries; an increase in the airflow to the lungs; and an increase in the rate of diffusion of oxygen and carbon dioxide across the respiratory surface of the lungs.[7] Comparative data for reptiles is meager, but it seems that they are unable to improve their aerobic scope by training (Bennett, 1982); we will shortly see why this is probably so. But before making any other reptilian comparisons we should take some time out to consider skeletal muscles.

Skeletal muscles and fiber types

A skeletal muscle, say the biceps, looks simple enough from the outside, but a slice through the middle reveals a series of structures within structures, like those Russian dolls that fit inside one another. The muscle cells themselves have diameters of only about 10–100 micrometers (a micrometer, also called a micron, is one-thousandth of a millimeter, or one-millionth of a meter), but they are very long, sometimes as long as the entire muscle, which is why they are usually called muscle fibers. Within each fiber is a tight bundle of fibrils, and within each fibril are filaments of the proteins actin and myosin. These very fine filaments interdigitate with one another, and a muscle contraction occurs when they move past one another, increasing their degree of overlap. The muscle fibers are bundled together into fascicles and the fascicles are often organized into separate divisions of the whole muscle, called muscle slips. The entire muscle is supplied with a nerve, which itself comprises a bundle of individual nerve fibers, called axons. Each axon is branched at its distal end (nearest the muscle) and each branch passes to a single muscle fiber. A given axon therefore supplies several muscle fibers. The whole unit—axon plus the muscle fibers it supplies—is called a motor unit. When a muscle fiber is stimulated by its nerve branch it contracts, and it contracts to its maximum extent. There is no graded response; a muscle fiber either contracts fully or it does not contract at all (often called the all-or-none principle). A nerve fiber works the same way—it either conducts an electrical impulse or it does not. When a nerve fiber conducts an impulse, it passes it on to each of the muscle fibers within its motor unit. Therefore the motor unit functions as a single entity.

If I wanted to scratch the end of my nose I would contract my biceps muscle quite gently. But if I wanted to swat a fly I would use a much greater force. The reason we can obtain a graded force of contractions is that we can increase the number of motor units that fire off. If we are jogging along at a gentle pace only a percentage of our motor units are in use, but this percentage increases as we increase our effort and approaches 100 percent at full speed.

Although muscle fibers are essentially all the same, they are not identical because some are specialized for aerobic, and others for anaerobic metabolism. At first only two major types of fiber were identified, the slow-twitch and the fast-twitch fibers, according to their speed of contraction. However, it soon became apparent that there were two types of fast-twitch fibers, so three fiber types are now recognized (Peters, 1989). The slow-twitch fibers have a high capacity for aerobic metabolism and are very resistant to fatigue, but they generate only small forces. One of the fast-twitch types has a fairly high capacity for aerobic metabolism, therefore a high resistance to fatigue, and generates moderate forces. The other fast-twitch fiber has a low aerobic and a high anaerobic capacity and is therefore readily fatigued, but it generates large forces.

All of the fibers in a given motor unit belong to the same type. There are accordingly three types of motor units: slow ones that can contract for long periods without tiring but that generate only small forces; fast ones that can carry on for a reasonable time and that generate moderate forces; and fast ones that are readily fatigued but that generate large forces. What is particularly interesting from a functional standpoint is that as an animal changes gait from a walk to a trot to an all-out gallop the three progressively more powerful motor units are brought into action. Some muscles tend to predominate in one fiber type over another. Since the slow-oxidative fibers are rich in myoglobin, a pigment similar to hemoglobin, muscles that predominate in slow fibers are red. Muscles that predominate in fast fibers, in contrast, appear white. We have probably all seen examples of these differences at the dinner table. Game birds, which belong to the same family (Galliformes) as the chicken and turkey, are very good at sprint-starts, but they are unable to fly fast for very long. Ducks, however, are endurance fliers; they can stay in the air for hours at a time during their long migratory flights. This difference in performance is reflected in their flight muscles—the breast meat of a chicken is white, but that of a duck is red.

The tuna, we have seen, maintains its body temperature above that of the seawater by virtue of the heat generated by the actively contracting muscles. Tunas are endurance swimmers and can maintain their aerobic cruising for hours on end. Examination of their body muscles reveals that red fibers are distributed among the white ones. These probably supply most of the sustained swimming power, but it seems that some of the white fibers are also called upon for this purpose, especially in larger individuals, which tend to have relatively fewer red fibers than smaller individuals (Graham, Koehrn, and Dickson, 1983). The mackerel, which belongs to the same family as the tuna (Scombridae), also has a large number of red fibers among the white, which gives the meat a dark color. There is also a strip of brown muscle running along the length of the body, just beneath the skin, which comprises red fibers. This thin block of muscle, which is also found in many other fishes, appears to be used for continuous cruising—look out for it next time you eat mackerel or herring.

More on stamina: why have reptiles got so little?

Reptiles, and most other multicellular animals, have basal metabolic rates that are an order of magnitude lower than those of birds and mammals (that is, lower by a factor of ten). They have a very limited capacity for aerobic exercise and are therefore unable to sustain any fast activities. This is not because they have a limited aerobic scope—reptiles are able to increase their oxygen consumption by a factor of between five and ten, the same as birds and mammals—it is just that they have such small rates of consumption to start with. A two-pound (1 kg) iguana lizard, for example, has basal and maximum levels of oxygen consumption of two and nine millimeters per minute, whereas the values for a mammal of similar weight are nine and fifty-four millimeters (reported in Bennett and Ruben, 1979). The relative increase is approximately the same in each case, but the meager energy supply available to the lizard allows it to sustain a maximum walking speed of only 0.3 mph (0.5 kph), compared with 2.5 mph (4.1 kph) for the mammal.

Birds and mammals are well known for their extensive wanderings, and their ability to keep up their flying or their running for such long periods is attributable to their extensive capacity for aerobic metabolism. Reptiles, in contrast, are confined to living their lives at a much gentler pace, interspersed by only brief periods of intense activity. But they can be very fast during these periods—just consider how quickly lizards dart from place to place, or the speed with which a venomous snake can strike its victim. Young crocodiles (less than 6 ft or 2 m long) have even been observed galloping at speeds of almost 40 mph (65 kph), which is faster than many mammals can run (Zug, 1974). But they can only sustain these speeds over short distances because they have to rely upon anaerobic metabolism.

Since a reptile's aerobic performance is so modest compared with what it can achieve anaerobically, their maximum speeds may be ten to thirty times their maximum sustainable speeds. Mammals, in contrast, have much higher levels of aerobic performance and their anaerobic performance does not give them such a wide margin of improved speed. As a consequence, a mammal's maximum running speed is only about twice as high as its maximum sustainable speed (Bennett and Ruben, 1979). Anaerobic metabolism therefore provides a greater margin of scope for activity in ectotherms than it does in endotherms.

Most reptiles are predatory (there are few herbivorous ones—Zimmerman and Tracy, 1989), and they usually adopt a sit-and-wait hunting strategy. They spend much of their time resting, or moving aerobically and therefore relatively slowly. In both instances their energy requirements are an order of magnitude below those of endotherms, and they are therefore very economical in their food requirements. Their major limitations are that they are unable to sustain high levels of activity for long periods. Whereas endotherms can sprint anaerobically and recover within minutes, reptiles take considerably longer to recover. They seem to lack the ability to rid their muscles of the accumulated lactic acid very quickly. This was well illustrated by a study conducted on the salt-water crocodile of Australia (Bennett et al., 1985). Individuals of different sizes were captured and the time taken for them to exhaust themselves with their struggles to escape was recorded, together with their recovery time. Animals with body weights of less than two pounds (1 kg) struggled for only about five minutes, after which time they became limp and lifeless. Larger animals took longer to become exhausted, between ten and twenty minutes for 20–220 pounders (10–100 kg) and more than thirty minutes for larger ones. Blood samples taken from the exhausted animals showed very high levels of lactic acid, and several of the larger individuals showed levels higher than any that have ever been reported for animals following strenuous anaerobic activity. Most of the animals had partially recovered within two hours. Large reptiles therefore seem capable of longer periods of strenuous anaerobic activity before becoming exhausted.

Reptiles, then, in contrast to mammals and bird, have a limited capacity for aerobic activity and this severely limits their sustainable activity levels. They lack the stamina for anything more than a relatively slow jog, and when they need to run fast, or participate in other intense activities, they rely on glycolysis. We have attributed this lack of stamina to a metabolic rate that is an order of magnitude below that of birds and mammals, but there are other contributing factors. These include the way they ventilate their lungs, their posture, and the anatomy of their heart. The relationship between these various parts is best considered from an historical perspective, as set out in a most interesting paper by David Carrier of the University of Michigan (1987).

From water to land: the evolution of stamina

Many living fishes are capable of feats of stamina comparable to those of birds and mammals. Sharks, forever on the move, reach impressive speeds, and the numerous accounts of swordfishes ramming boats and whales, sometimes driving their swords in up to the hilt, attest to the high speeds they attain. And many fishes are endurance swimmers. One school of tuna cited by Carrier covered a distance of 266 miles (428 km) in a single day, achieving an average speed of 11 mph (18 kph). But fishes have low metabolic rates, like those of reptiles, so how can they do these energetic things? First, the costs of aquatic locomotion are about one-tenth those of locomotion on land (Tucker, 1975), one important factor being that fishes do not have to spend energy in supporting their body weight. Another consideration is that as swimming speed increases, the flow of water over the gills increases, facilitating the uptake of oxygen. A third factor that needs consideration is the metabolic rate of high-performance fishes like the tuna. Trying to measure the basal metabolic rate of an animal that never stops moving is obviously an impossibility. The next best thing is to measure the *least observed metabolic rate* (Cabanac, 1987), which is the metabolic rate of an animal that an investigator is trying his level best to keep as quiet and inactive as he possibly can! Some researchers in Hawaii measured the least observed metabolic rate of some small tuna that were kept quiet and undisturbed in a circular tank (Gooding, Neill, and Dizon, 1981).

So that their results could be compared with those of other animals, they calculated what the tunas' metabolic rates would have been at zero swimming speed—that is, they estimated their basal metabolic rate. They found that this was two to five times higher than that of a typical fish of similar size. Given their elevated metabolic rate, high body temperature, and high activity level, the tuna should probably be regarded as being as endodermic as a bird or mammal.

When fishes left the water for the land they took their relatively low metabolic rates with them. They took their undulatory ways of moving, too. In the water they moved by bending the body from side to side, but on land they now had finny feet at their sides for extra leverage. Most modern reptiles, and amphibians, have retained the same locomotory pattern. But because the costs of moving on land are at least ten times higher than those in water, the early tetrapods were unable to sustain the same levels of activity as their ancestors had enjoyed in the water. Their lives, we can imagine, were played out in a minor key, periods of slow locomotion punctuated by periods of rest. Breathing was accomplished by lungs, and these, it is safe to conclude, would have been ventilated by raising and lowering the ribs. We use a similar mechanism, but, like other mammals, we also have a diaphragm. This muscular membrane seals the floor of the thoracic cavity and its movements assist those of the ribs in lowering and then raising the pressure of the thoracic cavity. The thoracic cavity behaves like a bellows, drawing air into the lungs then expelling it again.

There must have been strong selection pressures for the evolution of bursts of rapid activity, both to capture prey and to avoid being captured. But the primal tetrapods were faced with a problem here because the side-to-side bending for locomotion interrupted their breathing movements. The same is also true for modern lizards (Carrier, 1987). Under these conditions rapid locomotion *has* to be anaerobic. The reliance on glycolysis for high-performance exercise in modern reptiles is therefore an inescapable consequence of their having retained the primitive sprawling gait. This explains why the aerobic scope of reptiles does not appear to be improved by training—no matter how many hours a lizard puts in on the jogging track, he cannot increase his body's capacity to deliver oxygen to his muscles. Fishes, as mentioned earlier, do not have the same respiratory problem as reptiles because of the increased flow of water over their gills. Very active swimmers, such as swordfishes, apparently have to keep swimming all the time in order to force sufficient water—ram-jet fashion—through their gills. This is probably why these fishes become moribund and eventually die after they have been hooked on a line.

The lizard's heart is three-chambered, with left and right auricles (upper chambers) and a single ventricle (lower chamber). Venous blood (deoxygenated) from the body flows into the right auricle and then into the ventricle. Oxygenated blood from the lungs passes into the left auricle and then into the ventricle, where it mixes with the deoxygenated blood. This is a rather inefficient system because partially oxygenated blood passes to the lungs and partially deoxygenated blood passes to the rest of the body. But it does make sense from the lizard's point of view because its lungs are essentially inactive during rapid locomotion and do not need to be supplied with blood. And the blood that was destined for the lungs is shunted to the rest of the body, where the need for blood, even blood that is not fully oxygenated, is greatest.

Walking tall: the possible benefits of the erect posture

Not all descendants of those early tetrapods retained the primitive sprawling gait. In two independent lineages an erect posture was evolved. One of these gave rise to the mammals and the other to the dinosaurs and birds. The erect posture, in placing the legs beneath the body instead of out at the sides, changed the locomotory movements from side-to-side to up-and-down. This movement, rather than interfering with breathing, now actually assists it and allows for breathing to occur at the same time as running: instead of alternate lungs being compressed then expanded as the body moves from side to side, which only shunts air from one lung into the other, both lungs now move in unison, which augments the bellows action of the thorax. Running mammals capitalize on this action by matching their breathing rhythm to their stride frequency, and the to-and-fro movements of their visceral mass (gut, liver, and other abdominal organs) acts in concert with the diaphragm like a piston (Bramble, 1989).

Several other anatomical changes accompanied changes in posture and limb ventilation, namely: the development of long transverse processes on the vertebrae for the attachment of axial muscles involved in respiratory movements of the ribs; the development of a diaphragm; and the development of a four-chambered heart, with separate left and right ventricles. This heart is actually two pumps in one; the right side pumps blood to the lungs at a relatively low pressure (it must not be too high, otherwise fluid would be forced out of the blood vessels and into the lungs—Ostrom, 1980), while the left side pumps blood at high pressure to the rest of the body.

The crocodile and its extant relatives appear to contradict this scenario for the evolution of locomotory stamina in tetrapods. This is because crocodiles have all of the anatomical attributes of an endotherm: a four-chambered heart (though it operates at relatively low pressures); a diaphragm; well-developed transverse processes; and some capacity for erect locomotion, suggesting some ability to breath during locomotion. But for all that they are still ectothermic. However, Late Triassic forms appear to have been fully erect and may therefore have been endothermic. Carrier (1987) suggested that the ectothermy of modern crocodiles may represent a secondary reduction in metabolic rate, an evolutionary change perhaps correlated with their aquatic life-style.

A last word on stamina

Here we are, halfway through the chapter, and hardly anything about dinosaurs has been mentioned. Before bringing them into the picture I want to tie up a few loose ends by saying something about invertebrate animals. If lack of stamina is the hallmark of modern reptiles and amphibians, the same certainly cannot be said of insects. Mosquitoes and blackflies, and ants and wasps, and all those other summertime pests seem to have more stamina than we do, and they are not even endothermic like us. The reason they are able to maintain very high levels of aerobic activity probably lies in their very efficient respiratory apparatus, which delivers air directly to the cells by a network of exceedingly fine tubules. This tracheal system works efficiently only at relatively small body sizes, for which the surface-to-volume ratios are high. This is most fortunate for us because we are spared the prospect of being attacked by four-foot-long mosquitoes.

Many other invertebrates are small and aquatic, like the members of the plankton. Their small size gives them a relatively enormous surface area so they can obtain sufficient oxygen from the water to sustain their indefatigable swimming activities as they drift along with the currents. But enough of the minutia of life, let us get on with something really big.

Dinosaurs: unravelling the evidence

According to Bakker all dinosaurs were endothermic, with high metabolic rates and high body temperatures, like birds and mammals (Bakker, 1971, 1972, 1975a, 1975b, and 1986). This conclusion was reached on several lines of evidence, including the possession of Haversian bone, limb posture, and predator-prey ratios. How substantive is this evidence?

Haversian bone, as we saw in Chapter 1, is highly vascular. It is found in mammals and birds, but does not usually occur in other vertebrates. This type of bone is capable of rapid growth and also permits the calcium (from the mineral portion) to be absorbed into the bloodstream rapidly. Both features, Bakker reminds us, are associated with endothermy, and so he took this as evidence that dinosaurs were similarly endothermic. Aside from the fact that this "guilt by association" argument is logically unsound, not all mammals have highly vascular bone and some living reptiles that are known to be ectothermic have bone with a mammalian appearance (Bouvier, 1977). Indeed, the Haversian bone sample taken from a tortoise that was illustrated in Reid's (1987) paper dealing with bone histology (the microscopic structure of bone) is indistinguishable from that of a mammal. Reid stressed the point that there is no type of bone that can be used to identify endotherms with certainty. However, we must not be too hasty in rejecting the evidence of bone histology because the picture is not quite as simple as it at first appears.

Most textbooks give the impression that Haversian bone is typical of mammals, undoubtedly because most of these books are about medicine and Haversian bone happens to be characteristic of our own species (Currey, 1960). Haversian bone is only one of several bone types found in mammals. One of these other types, called *fibro-lamellar,* is typically found in large mammals (Currey, 1984). We need not be concerned with its appearance; suffice it to say that it looks much like Haversian bone, for which it is sometimes mistaken, but the layers of bone that surround the vascular spaces tend to be elongate rather than rounded. Fibro-lamellar bone is associated with rapid growth, as in cattle, which reach adult body size within a year or so of birth.

Armand de Ricqlès, a specialist in bone histology and an advocate of dinosaurian endothermy, drew attention to the fact that fibro-lamellar bone is found in dinosaurs (1976). He concluded that dinosaurs similarly had rapid rates of growth, hence high metabolic rates. Since the appearance of his earlier work, however, he has had the opportunity of examining more material, including some bone from a half-grown sauropod (Ricqlès, 1983). While the bone was of the fibro-lamellar type, as seen in other dinosaurs, there were areas that showed distinct zones of cyclical growth, like the growth rings of a tree. This provided fairly convincing evidence that, at least for part of the individual's life, its growth was seasonal. Such growth is seen in living reptiles but not in endotherms, and Ricqlès concluded that sauropods were probably inertial homeotherms. Similar observations have been made for other sauropod material (Reid, 1981).

Evidence from nest sites of the hadrosaur *Maiasaura peeblesorum,* which Jack Horner of the Museum of the Rockies, in Montana, has been working on since the late 1970s, suggests that growth rates may have been high, at least in this species (Horner and Makela, 1979). This tends to corroborate Ricqlès's correlation between bone histology and rapid growth. But, as pointed out elsewhere (Reid, 1987), rapid growth does not necessarily mean that metabolic rates were also high. Suppose that the chain of causality were not fibro-lamellar bone → rapid growth → high metabolic rate, but fibro-lamellar bone → rapid growth → *high body temperature.* After all, one of the manifestations of high metabolic rate in mammals is high body temperature and temperature, because of the Q_{10} effect, is probably the most important factor as far as growth is concerned. Bakker tells us in his book (Bakker, 1986) that some commercial crocodile farmers keep their animals at high temperatures to accelerate their growth rate, a point that is documented elsewhere (Joanen and McNease, 1989). Bakker's point here is that dinosaurs achieved high body temperatures by having high metabolic rates, but there is an alternative strategy for achieving this end (I will return to this strategy a little later). I do not mean to say that *some* dinosaurs did not have a high metabolic rate, but I do take issue with Bakker's premise that *all* dinosaurs had the same metabolic strategy. Why should dinosaurs have had similar physiologies when they had such diverse anatomies? This same question has been asked by other paleontologists too (Hopson, 1976; Ostrom, 1980).

A second "guilt by association" line of reasoning used by Bakker pertains to erect posture (Chapter 3). Birds and mammals are essentially the only vertebrates[8] that share this feature with dinosaurs, and since they are endothermic Bakker reasoned that dinosaurs must have been endothermic too. However, no causative linkage between the two phenomena was established. Furthermore, we have seen that a good case has since been made correlating erect posture with improved lung ventilation during locomotion.

Without wishing to belabor the point, I must repeat that endothermy is expensive. Lions need to consume their own body weight in prey about every nine days and wild dogs every seven days (Schaller, 1972), whereas the carnivorous Komodo dragon eats its own weight only about every ninety days (Auffenberg, 1981). Endothermic carnivores therefore require relatively larger numbers of animals upon which to prey than do ectothermic carnivores. Bakker (1980; also see 1972) found, for several modern African mammalian communities, that the ratio of the biomass (total mass of all animals) of predators to the biomass of prey animals was about 1:100. (That is, for every ton of lions there are 100 tons of potential prey animals.) Somewhat higher, but still relatively low values of about 3:100 were obtained for some fossil mammal communities from the Cenozoic Era (post-Cretaceous). However, the ratio for some Permian fossil reptiles, which were almost certainly not endothermic, was about 45:100. These results suggested that predator-prey ratios could be used to indicate the presence of endothermic carnivores in fossil communities. Bakker calculated the ratio for a number of different Cretaceous dinosaur assemblages and found that the ratios were about 2:100—not significantly different from the Cenozoic mammal communities—and this was taken as evidence that the predatory dinosaurs were endothermic.

Great caution is required when inferring the relative abundances of animals from their buried remains because the samples are not truly representative of the animals that once lived together (Chapter 2). And even if the fossil samples were an accurate reflection of the actual numbers of carnivores and herbivores, it cannot be assumed that the carnivores were actually responsible for limiting the numbers of herbivores—their numbers may have been limited by the availability of vegetation. By the same token, it cannot be assumed that the number of carnivores was being limited by the numbers of herbivores—the carnivores may not have been feeding exclusively on the herbivores; we do not even know for sure whether they were feeding on any of them, though the possibility that they were not does seem unlikely. Both of these points have been made by Ostrom (1980) and elaborated upon by other paleontologists (Beland and Russell, 1980). Furthermore, since the disparity between the food requirements of endotherms and ectotherms appears to decrease with increasing body size (p. 136) the dividing line between them becomes indistinct at large body sizes anyway (Farlow, 1980).

Is there *any* substantive evidence that can be used to indicate the possible thermal strategies of dinosaurs? Before attempting to answer this question, we should remind ourselves of a few important points. First, dinosaurs represent a diverse assemblage of animals and we would not expect them to have shared the same thermal strategies. Second, dinosaurs were unique and since none survive today we should not expect to be able to match their thermal strategies with those of extant animals. The third point is that endothermy is a very expensive strategy that has been adopted by relatively few animals. The impression is often given that endothermy is superior to other thermal strategies, but endotherms do not enjoy superiority over other organisms. That is not to say that there are not considerable benefits to being endothermic, only that these tend to be offset by the high costs involved. As a reminder, these benefits are: (1) constant body temperature that allows the animal always to be in a state of preparedness to function at optimum levels, regardless of ambient temperatures; (2) relatively high body temperatures (often in excess of 95° F or 35° C), possibly to enhance muscle performance; and (3) high sustained activity levels.

Could dinosaurs have attained any of these benefits without paying the high costs of endothermy? Most dinosaurs were considerably larger than the 450-pound (200 kg) Galápagos tortoise, whose thermal inertia confers relatively stable body temperatures. Most dinosaurs probably had fairly constant body temperatures as an inescapable consequence of their large size. To suggest that these temperatures might have been relatively high would seem entirely speculative, but a mathematical modeling procedure developed by James Spotila suggests that a hypothetical dinosaur-sized animal (body diameter of three feet or 1 m) with a reptilian level of metabolism could have attained temperatures of up to 100° F (38° C) in an equable climate (Spotila, 1980). Some indication of the confidence that can be attached to this modeling procedure is provided by the close correspondence that was attained between predicted and actual body temperatures for known animals. Thus when the relevant data for the modern elephant was entered into the model, the predicted body temperature given was 40.5° C, which is within a few degrees of the actual temperature (36.0–38.2° C). It therefore seems reasonable that large dinosaurs (hadrosaur-sized and larger) probably did maintain a relatively high body temperature, by virtue of their thermal inertia.

A similar modeling procedure has recently been conducted for the hadrosaur *Maiasaura* and its hatchlings (Dunham et al., 1989). The assumptions were made that *Maiasaura* had a metabolic level equivalent to that of a scaled-up modern lizard and that it lived in a climate similar to that of present-day Louisiana. The results suggested that while adult body temperatures may have remained fairly constant during a twenty-four-hour period—fluctuating by only one or two degrees—body temperatures may have ranged from about 10° C in winter to about 33° C in summer. Because of their smaller thermal inertia, hatchlings were considered to have experienced much wider daily temperature fluctuations but to have reached higher body temperatures than the adults. The overall conclusion was that hadrosaurs of the size of *Maiasaura* (an elephant-sized dinosaur) could not have remained active throughout the winter—primarily because of digestive problems (most ectotherms require a body temperature of about 20° C for digestion to occur). The researchers did concede, however, that their climatic parameters may have been too severe—the Late Cretaceous climate may have been more equable than they supposed. They also pointed out that they had taken no account of the heat that would have been generated by the fermentation of food in the gut.

A more significant point is that a metabolic rate of modern lizards was assumed. While this was perfectly reasonable in the absence of more suitable data, it must be remembered that dinosaurs were quite unlike any animals alive today. Dinosaurs cannot be treated as scaled-up lizards; the extensive development of fibrolamellar bone in dinosaurs, for example, suggests significant physiological differences between the two groups. According to Reid (1987), the histological evidence suggests some type of physiology intermediate between modern ectotherms and endotherms. I suspect that if modeling procedures were conducted for a hypothetical hadrosaur with a metabolic rate somewhat higher than that of modern reptiles, the problem of overwintering in the Late Cretaceous of North America would cease to exist.

The discovery of Late Cretaceous dinosaur remains from the high latitudes of Alaska is good evidence that dinosaurs could tolerate relatively low temperatures (Brouwers et al., 1987). The

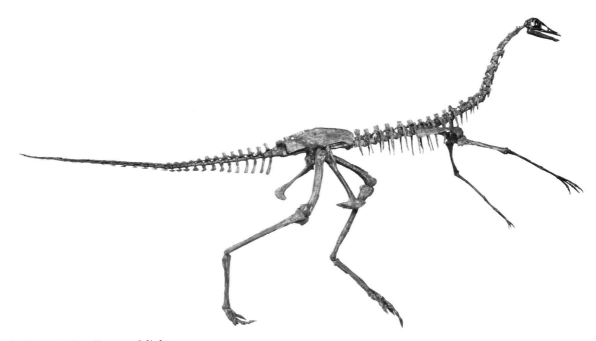

The long hindlegs and light build of the ostrich-sized dinosaur ORNITHOMIMUS *suggest fleetness of foot. (The ribs in this specimen are incomplete.)*

dinosaurs represented include hadrosaurs and a tyrannosaur. It seems unlikely that they would have migrated south for the winter because of the great distances involved, and the occurrence of juvenile hadrosaurs lends support to the idea that the site was a year-round habitation.

One of Bakker's most persuasive arguments that dinosaurs, or to be more precise *some* dinosaurs, were very active animals is the fact that some possessed cursorial adaptions. Who could doubt that *Ornithomimus*, with its long slender hindlegs and lightly built skeleton, was not a fast runner? The same is true for many more dinosaurs, including *Deinonychus* and its allies (Chapter 4). The tyrannosaurs, though larger and less slenderly constructed, also possess features that suggest the ability to run fast, and the same holds for the hadrosaurs, though to a lesser extent. Must these dinosaurs have had high metabolic rates to support their running abilities?

This reconstruction fancifully shows an ORNITHOMIMUS *stealing eggs.*

We know that modern reptiles, with their low metabolic rates, are lacking in stamina and can keep up bursts of speed, anaerobically, only for short distances. The fact that their lungs are essentially inoperable during such activity must have acted as an evolutionary straitjacket to any significant improvements in their performance, but dinosaurs did not suffer from this constraint.

Their erect gait would have allowed them to breathe while running, conferring a considerably wider aerobic scope than is seen in present-day reptiles. We may also predict, with some confidence, that dinosaurs, like their archosaurian relative the crocodile, had a four-chambered heart. How else could long-necked species like the sauropods have achieved a pressure sufficiently high to deliver blood to the head?

These improvements over modern reptiles in the rates at which oxygen could be delivered to the muscles may not have permitted dinosaurs to sustain running speeds as high as those of modern mammals, but anaerobic metabolism would still have been available for short bursts of intense activity, such as making hunting attacks or fleeing from predators. Large crocodiles can keep up their anaerobic struggles to escape capture for over half an hour, so it is reasonable to expect that dinosaurs, being considerably larger, could have maintained a similar level of activity for even longer. Crocodiles accumulate large quantities of lactic acid in their muscles during intense activity, causing exhaustion and a recovery period of up to two hours. If it is true that dinosaurs had greater ability to supply oxygen to their tissues, the time taken to repay an oxygen debt should have been considerably shorter than in the crocodile. They should therefore have been able to recover from anaerobic exercise more rapidly, and if anaerobic metabolism were restricted to bursts of short duration they might have avoided complete exhaustion and recovered within minutes rather than hours. Cursorial dinosaurs could therefore have been as active as their skeletal anatomy suggests that they were, without their having had the high metabolic rates and associated high costs of modern endotherms.

High metabolic rates would probably have been detrimental to most dinosaurs because of the problem of dissipating excess heat from their relatively small surface areas. Spotila's mathematical model predicted that even a modest-sized dinosaur would have suffered severe heat stress if it had possessed a high metabolic rate. Elephants are apparently prone to heat stoke, so the problem would have become insurmountable for the considerably larger sauropods. And even with a low metabolic rate it is likely that the larger sauropods would have guarded against too much exertion to minimize the additional heat load generated by their active muscles. The need to avoid overheating, of course, is concordant with the picture of sauropods that was painted in Chapter

5. Their long necks and long tails, possessing relatively larger area-to-volume ratios than their globular bodies, were probably important sites for shedding excess heat. Alexander (1989) reminds us that skin is slightly permeable to water and that crocodiles lose much water through their skin. He concluded from this that much heat could have been shed from a sauropod's body through evaporative loss of water. So much for the largest and the middle-sized dinosaurs, but what about the smallest ones?

Small dinosaurs, like *Compsognathus* ("elegant jaw"; 2–4 ft; 0.7–1.4 m), *Segisaurus* ("found in Segi Canyon"; 3 ft; 1 m), *Avimimus* ("bird mimic"; 5 ft; 1.5 m), *Dromaeosaurus* (6 ft; 1.8 m), and *Ornitholestes* ("bird robber"; 6.5 ft; 2 m), which are all theropods, were probably not heavy enough to have been inertial homeotherms. They were all lightly built and had well-developed cursorial features suggesting that they were capable of running fast. Were these little dinosaurs ectothermic, becoming active only after sunning themselves in the warming rays of a new day, or had they evolved endothermy? We shall probably never know the answer to this question, but there are some scraps of circumstantial evidence that suggest the latter might have been the case, evidence pertaining to body coverings.

The skeleton of *Avimimus*, from the Late Cretaceous of the USSR, has a very avian appearance. S. M. Kurzanov, the Russian paleontologist who described the specimen in 1981, pointed out a most interesting feature. The trailing edge of the ulna has a raised bony ridge that Kurzanov interpreted as an insertion area for feathers. Birds do not have a ridge like this, having instead a series of small tubercles to which feathers attach. No evidence of feathers themselves was found, and whether this dinosaur really did have feathers is not known. However, feather impressions are known for *Archaeopteryx* ("ancient wing"), the earliest bird from the geologically older Upper Jurassic of Germany. This animal is anatomically so similar to small theropod dinosaurs that it would have been described as such but for traces of feathers (Ostrom, 1975; see also the Epilogue, below). We naturally associate feathers with birds and with flight, but several authors believe that feathers probably first evolved to serve as insulation and became modified for flight only secondarily (Evart, 1921; Bock, 1969; Ostrom, 1974b). Only evidence of wing and tail feathers has been found, and these serve primarily for flight, but it seems extremely likely that insulatory contour feathers would also have been present, as in modern birds. The presence of an external insulating layer seems to rule out the possibility that *Archaeopteryx* was an ectotherm. However, P. J. Regal, who believes that feathers initially evolved as solar shields to keep out the sun, has argued that the presence of feathers would not prevent an animal from absorbing solar radiation (1975). This sounds contradictory but Regal proposed that the feathers would have been capable of being raised and lowered. He points out that many modern birds, like the roadrunner, raise their feathers while basking in the sun. Presumably this allows them to absorb radiant heat. The fact that *Archaeopteryx* was a flier implies that it did have a high metabolic rate, for it is difficult to visualize sustained flight without a sustained and high level of energy input. This conclusion tends to corroborate that feathers, at least in *Archaeopteryx,* served to insulate the body against heat losses.

Small theropod dinosaurs may have been feathered, and perhaps the absence of feather impressions should be viewed as absence of data rather than as evidence for the absence of feathers. Feathers are rarely preserved as fossils and easily overlooked even when they are, as in the case of two specimens of *Archaeopteryx* that had formerly been misidentified as reptiles simply because traces of their feathers were so faint. It must be pointed out, though, that a careful examination of *Compsognathus* has failed to reveal any evidence of feathers, and it was found in the same fine-grained limestones as *Archaeopteryx* (Ostrom, 1978). It is tempting to suggest that small theropods probably had thermal strategies similar to *Archaeopteryx* because of their similar anatomies, but that would be entirely speculative. All we can say is that there is some circumstantial evidence that the smallest theropods *may* have been endothermic.

To summarize our conclusions so far:

1. Most dinosaurs, because of their large size, could not have avoided maintaining fairly constant body temperatures, a strategy described as inertial homeothermy. It may safely be assumed that hadrosaur-sized dinosaurs, and larger ones, were inertial homeotherms.

2. The body temperatures of the inertial homeotherms may have been relatively high, comparable to those of modern birds and mammals. Daily fluctuations may have been only a degree or two, but seasonal variations may have been greater.

3. The metabolic rate of an inertial homeotherm may have been a little higher than that of a modern reptile of similar weight, but it would probably have been more typical of modern reptiles than of birds and mammals.

4. Erect posture would have given dinosaurs a more efficient respiratory system than that of modern lizards. Together with other modifications, including a four-chambered heart, improved respiration would probably have given them a fairly wide aerobic scope, permitting them higher levels of aerobic exercise than modern reptiles are able to attain.

5. The sauropods, because of the relatively small surface-area-to-volume ratio imposed by their immense size, may have had problems shedding excess body heat. This would almost certainly have prevented them from having had a high (avian or mammalian) level of metabolism.

6. The smallest theropods, like *Compsognathus* and *Dromaeosaurus*, were too small to have been inertial homeotherms. They possess skeletal adaptations that indicate high levels of activity and it is possible that some of them may have been endothermic. Possession of feathers in the closely related earliest bird, *Archaeopteryx*, is persuasive evidence for endothermy.

If we keep in mind what a wide range of body temperatures, metabolic rates, and climatic tolerances is found among living mammals, we should not be surprised to find an equal diversity among dinosaurs.

Dinosaurs and their young

If most dinosaurs achieved constant, high, body temperatures by virtue of their large size, what about their offspring? Would they have been large enough to have been inertial homeotherms too? The available evidence, which is scant, indicates that hatchling dinosaurs were moderately large compared with the offspring of most modern birds and mammals. This is probably because most adult dinosaurs were larger than most adult birds and mammals, a point graphically made by Nicholas Hotton (1980) of the Smithsonian Institution.

The earliest dinosaur eggs to be found were ascribed to *Proto-*

ceratops, a small, horned dinosaur from the Upper Cretaceous of Mongolia, which reached a length of about six feet (3 m). The eggs are large compared with those of birds, up to about eight inches (20 cm) long. From the large egg size, and from the remains of young ones that have been found, it has been estimated that the hatchlings were about ten inches long (25 cm—Colbert, 1961). *Psittacosaurus,* also from Mongolia but geologically older than *Protoceratops,* is now generally regarded as being an early member of the horned dinosaurs, though it is lacking horns itself. It is about the same size as *Protoceratops* (adult length 6.5 ft; 2 m), and Walter Coombs, who has made a study of some juveniles of the species (1982), estimated that the hatchlings would have been about nine inches long (23 cm). This is about the size of an adult pigeon. Coombs made the point that the psittacosaur juvenile was one of the smallest dinosaur individuals known. It shares this distinction with an almost complete skeleton of a hatchling prosauropod dinosaur (the group that may have given rise to the sauropods), named *Mussaurus patagonicus,* from the Upper Triassic of Patagonia, Argentina. This individual had an estimated total length of eight inches (20 cm—Bonaparte and Vince, 1979). This species was identified on the basis of juvenile material and the size of the adults is not known. The hadrosaur *Maiasaura,* from Montana, was about twenty-three feet long (7 m), and hatched offspring with an estimated length of twelve–fourteen inches (30–35 cm—Horner, 1984).

On the available evidence, then, hatchling dinosaurs appear to have been considerably larger than the hatchlings of modern birds, but far too small to have been inertial homeotherms. Their size would have obliged them to have been ectothermic, but, like present-day ectotherms, their body temperatures could have been kept high, at least during the day. Jack Horner, who has been doing pioneer work on the hadrosaur nest sites in Montana, makes a good argument that the youngsters were afforded some measure of parental care (Horner and Gorman, 1988). The hatchlings would have been vulnerable to attack from predators and the presence of the adults would have reduced this threat. On the other hand, the hatchlings would have been equally vulnerable to being stepped on by their twenty-three-foot-long (7 m) parents, which is why Horner believes that staying in the nest would have been the safest place for them. The evidence for parental care is not substantial, and some of it has been shown to be equivocal. For example, it was thought that the wear-facets on the teeth of the hatchlings provided evidence that they were feeding while in the nest, which would strongly suggest that they were being fed by their parents. Horner has since found, however, that wear-facets are present on the of teeth unhatched individuals too.[9] Regardless of the paucity of good evidence, the suggestion that the youngsters were given parental care makes a great deal of good sense.

In one of the nests the juveniles were only about eighteen inches long (45 cm), but in another they were much larger, about three feet (1 m), which suggests an extended period of parental care (Horner, 1984). It is conceivable that parents may have brooded the young after sundown, to keep them warm and maintain a high growth rate. The prospects of a four-ton parent lowering itself down close enough to its offspring to warm them without crushing them may sound overly fanciful, but crocodiles, as we shall see in the next chapter, are capable of some remarkably gentle manipulations of their offspring. I do not want to give the impression that parental care is a well-established fact for dinosaurs. In this regard we should heed the warnings of Walter Coombs (1982), who reminds us that the available evidence for parental care in dinosaurs is scant and equivocal, and although some dinosaurs may have provided extended care to their hatchlings, parental care is unlikely to have been universal. It is also possible that some dinosaurs may have borne their young alive, at a relatively advanced stage of development, as do some living reptiles.

Thermal strategies of other Mesozoic reptiles

Pterosaurs have long been considered likely candidates for endothermy. Flying is very demanding, for energy has to be supplied continuously—a flying animal cannot risk having to end its exertions abruptly, due to exhaustion. Wide aerobic scope is therefore a prerequisite for flying, and so too is a relatively high metabolic rate, needed to supply energy at a sufficiently rapid rate; both conditions are met by endothermy. High muscular activity generates large quantities of heat, and as more power can be obtained from warm muscles it is reasonable to expect that pterosaurs maintained high constant body temperatures, at least during flight. But many pterosaurs were small, gull-sized animals, far too small to have been inertial homeotherms. Even the largest of their kind, whose wingspan is comparable to that of a light airplane, had relatively light bodies, and it is difficult to visualize how high and constant body temperatures could have been maintained in pterosaurs of any size in the absence of insulation.

Evidence of insulation was announced as early as 1927, when the German paleontologist F. Broili described impressions of a hair-like body covering in a well-preserved specimen of the pterosaur *Rhamphorhynchus.* His findings were not widely accepted, but supporting evidence was provided in 1971 by a Russian paleontologist named Aleksandr Sharov. Sharov described the impressions of a furry body covering in a new pterosaur which he described as *Sordes pilosus,* derived from the Latin *pilosus,* meaning "hairy."

Evidence for the thermal strategies of other Mesozoic reptiles is somewhat circumstantial, but one of the best cases for resolution can be made for the ichthyosaurs. Some exceptionally well-preserved specimens of these marine reptiles have left their body outlines preserved as a carbonaceous film. These outlines reveal a body form like that of the tuna and its relatives, the scombroid fishes (Chapters 9 and 10), and suggest that some ichthyosaurs, at least, were capable of maintaining high swimming speeds. Perhaps they, like the tuna, may have had elevated body temperatures. It must be emphasized, however, that this did not necessarily apply to all fast-swimming ichthyosaurs, no more than it applies today to all scombroid fishes. The swordfish, for example, is a large scombroid fish that appears to be every bit as active as a tuna, but it is not endothermic, as the tuna appears to be. Nor does an elevated body temperature necessarily imply high metabolic levels comparable to those of birds and mammals.

We have seen that marine turtles, by virtue of their thermal inertia (and also of the insulating properties of their shell), are able to maintain a body temperature that is 32 F° (18 C°) higher than the temperature of the surrounding water. Many ichthyosaurs had body masses far exceeding that of the leatherback turtle; they may also have had a fatty insulating layer beneath the skin, like the blubber of whales (see Chapter 8). We can conclude, therefore, that the larger ichthyosaurs, just by virtue of their size, could probably not have escaped being inertial homeotherms and that they probably maintained body temperatures greater than that of the surrounding sea. But high temperatures by no means implies metabolic rates as high as those of birds and mammals; instead, I suggest that they had a typically reptilian level.

Strategies and vulnerabilities

It is generally believed that the close of the Mesozoic Era was marked by a general cooling trend in world climates (Axelrod and Bailey, 1968). If most dinosaurs were inertial homeotherms, as argued here, and had typically reptilian levels of metabolism, they may have been vulnerable to this climatic deterioration. Modern mammals with low metabolic rates, like the sloth, appear to be vulnerable to sudden falls in temperature, which probably explains why they are restricted to the tropics. Elephants, because of their large size, have relatively low metabolic rates. They are also essentially naked, but they can acclimatize to cold climates quite readily by increasing the thickness of their hair. An extreme expression of this strategy occurred during the Ice Age, when the relatives of the elephant, the mammoths and mastodons, grew thick coats to withstand the cold.

As far as we know dinosaurs were naked, too, and were therefore unable to respond to climatic deterioration by growing a thick coat. Their large size would also have made it difficult for them to find shelter from the elements. These two factors, taken with their (presumed) low metabolic rate, would seemingly have made them vulnerable to climatic deterioration. The possibility that the dinosaurs were the victims of a climatic change has always been a popular explanation for their extinction, a question that will be explored more fully in Chapter 12.

SEVEN

Brains and Intellect

H *OMO SAPIENS,* the wise man, has an unusually large brain for the size of his body, and we are all well aware of the intellectual superiority this gives us over other animals. It is therefore perfectly natural for us to correlate large brains with intellect, and to be so condescending when discussing the intellectual abilities of other species. Dinosaurs, with their disproportionately small brains for their immense bodies, have traditionally been viewed as having been cerebrally incompetent. In fact, a popular explanation for their demise is that they were just too stupid to survive. But how reliable is the correlation between brain size and mental ability?

The average weight of the adult human brain is about 2.8 pounds (1.4 kg), though there is a wide range of variation (2.2–4.4 lb; 1–2 kg). Although there have been claims of intellectual superiority by some people with large heads, there is absolutely no evidence of any correlation between brain size and intelligence. The human brain is about three times larger than that of our closest living relatives, the great apes, but some large animals have bigger brains than ours. An adult African elephant, for example, whose a brain is almost four times larger than ours, has the largest brain of any living land animal. Whales have even bigger brains; that of the sperm whale, which is six times the size of man's, is probably the largest brain that has ever existed. Does this mean that elephants and whales are intellectually superior to humans? There is certainly no question that whales are extraordinarily intelligent animals, but they are probably no brighter that we are (McIntyre, 1974). Absolute brain size is obviously not a reliable

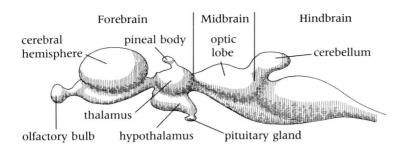

Forebrain Midbrain Hindbrain

cerebral hemisphere · pineal body · optic lobe · cerebellum

thalamus

olfactory bulb · hypothalamus · pituitary gland

Diagrammatic representation of the main features of the vertebrate brain.

guide to intellect. What is needed is a method of assessing relative brain sizes. If the average brain size for a given body size could be calculated, it would be possible to determine whether a particular animal had a large or a small brain for its size. But before considering how this can be done, we need to know something about the anatomy of the brain, for the relative development of different parts of the brain is of considerable importance in interpreting an animal's behavioral potentials.

There is a great deal of variation in the structure of the brain throughout the vertebrates but a basic pattern can be recognized, so it is possible to describe a generalized vertebrate brain. While we have an approximate idea of the function of the brain's different parts, our knowledge is based largely upon studies of our own brain and the brains of certain other mammals. Our account of the form and function of the various regions of the vertebrate brain is therefore overgeneralized and serves only as a broad guide.

The brain is essentially a thick-walled tube in three main parts, the forebrain, midbrain, and hindbrain. Arising from the front of the forebrain are two swellings, the olfactory bulbs, which are associated with the sense of smell. Most birds have a very poor sense of smell and correspondingly small olfactory bulbs, whereas reptiles and mammals generally have a well-developed olfactory sense and correspondingly large bulbs (higher primates, including man, being notable exceptions). A swelling of the roof and sides of the forebrain constitutes the cerebrum, which is usually a paired structure, each half being called a cerebral hemisphere. This region is so large in humans and some other mammals that it overlies most of the rest of the brain. The cerebrum is associated with the integration of the senses and with what is commonly referred to as intelligence.

The word *intelligence* comes from the Latin, *intelligere,* "to understand," but there is little agreement among specialists as to how intelligence should be defined. Definitions include the ability to adapt to new circumstances, the ability to learn (Grech, Crutfield, and Livson, 1969), and the capacity to construct a perceptual world (Jerison, 1973).

Many different tests have been devised to assess intelligence in humans and other animals. Some of these test the ability to solve problems, and others test the readiness with which an individual learns, but it is important to realize the limitations of these procedures. In the first place, the tests are usually conducted in a laboratory, far removed from the real world of the animal. An animal may fail a test through no shortcomings in its mental capacities but because the task it is expected to perform is far removed from its normal behavioral patterns. As a hypothetical example, a test which required an animal to push open a door using its nose would probably be more successfully completed by a dog than a cat merely because dogs frequently use their noses for such purposes, whereas cats do not. Another point is that we cannot always equate intelligence with learning abilities. Chimpanzees are considered to be far more intelligent than dogs but they are apparently far more difficult to house-train (Burghardt, 1977). Lastly, although hundreds of tests have been performed on animals, these have been confined to relatively few species, notably the laboratory rat, the chimpanzee, and the pigeon. Few reptiles have been tested and the study of reptilian behavior is in its infancy. Sweeping generalizations that certain types of animals are less or more intelligent than others are therefore unfounded, but they continue to be made. There is also a great tendency to underestimate the capabilities of animals, but as more and more investigations are conducted upon animals in their natural environment, it becomes apparent that they are capable of far more than we originally anticipated. When we refer to the intellectual abilities of animals, then, we must remember that we are on uncertain ground.

The walls of the cerebral hemispheres are referred to as the cerebral cortex. In the most intelligent animals, which include man, the great apes, whales and dolphins (the Cetacea), the cerebral cortex is very thick and highly convoluted; in the less intelligent mammals it is smooth. The cerebral cortex is thin and unconvoluted in fishes, amphibians, reptiles, and birds and appears to be largely associated with olfaction.

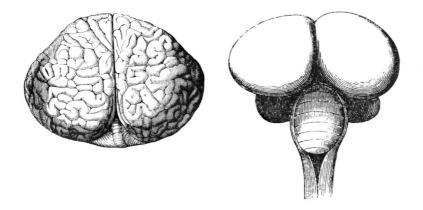

Left: *In the dolphin's brain, the convoluted cerebral hemispheres are so large that they obscure the rest of the brain.* Center: *A bird brain* (LARUS) *has large but unconvoluted cerebral hemispheres.* Right: *In the turtle* (CHELONE), *the cerebral hemispheres are small and unconvoluted.*

The floor of the cerebrum is thickened, forming a pair of swellings, the corpora striata. These are especially prominent in birds, which correspondingly have relatively large cerebral hemispheres. The classical interpretation is that birds are incapable of intelligent behavior because their brain lacks a thick cerebral cortex. Nevertheless, they are capable of some very complex behavioral patterns, including nest building, courtship displays, and navigation, though these are largely instinctive. Instinctive behavior is inherited, and it is characteristically stereotyped and inflexible, unlike learned behavioral patterns, which can be modified according to changing circumstances. The corpora striata have been interpreted as the seat of instinctive behavior, and birds were viewed almost as automatons, largely incapable of modifying their behavior through experience. But this view changed as we became aware that birds are capable of some remarkable feats of learning.

I saw a wonderful example of this for myself a few summers ago. My wife and I were visiting Stratford-upon-Avon, but our most memorable recollections of the trip had nothing to do with Shakespeare's birthplace but with what we saw in the car park as we were leaving. A sparrow was systematically working his way along the rows of parked cars as if he were checking their license plates. We could scarcely believe our eyes at first and had no idea what he was doing. He would hop up to a car, pause in front of it for a few moments while he inspected it, then move on to the next one. This happened several more times before we realized what he was really looking for—insects impaled on the radiator grill and fender! And when he found them he fluttered up and pecked them down before moving on to the next car. We watched in wide-eyed amazement as he completed work on one row of cars, then moved on to the next row. We often wonder how many other sparrows in Stratford-upon-Avon have learned to exploit the tourist industry in this fashion.

In some learning experiments birds have actually outperformed some mammals (Stettner and Matyniak, 1968). The center for their learning behavior is now understood to be the corpora striata and not the cerebral cortex, as it is in mammals.

The hind portion of the forebrain is called the diencephalon. The walls are referred to collectively as the thalamus; the roof is called the epithalamus, and the floor is called the hypothalamus. The thalamus is concerned with the integration of sensory activities, the hypothalamus with controlling the internal functions of the body. A small lobe hanging from the floor of the hypothalamus, the pituitary gland, produces a large number of hormones that coordinate certain physiological processes, including growth. Gigantism is often the result of an enlarged and overactive pituitary gland, and it is noteworthy that large dinosaurs also appear to have had particularly large pituitaries (Edinger, 1942). Arising from the epithalamus is a small, stalked appendage, the pineal body. In some of the "lower" vertebrates, including the reptile *Sphenodon,* the pineal body is light-sensitive and is often referred to as the third eye. Many reptiles, including *Sphenodon,* have a hole in the roof of the skull that transmits light to the pineal body.

The midbrain, the smallest of the three divisions of the brain, gives rise to a pair of swellings, the optic lobes, which are associated with vision. Birds usually have a very well developed sense of sight, and the optic lobes are correspondingly large. These lobes are also fairly well developed in reptiles, and even more so in the bony fishes. Most mammals have keen vision too, but the visual impulses are interpreted in the cerebrum rather than in the optic lobes, which therefore tend to be fairly small. In tetrapods there is another pair of swellings from the midbrain, the auditory colliculi, which are associated with hearing.

The third division of the brain, the hindbrain, is largely associated with balance and muscular coordination and is therefore something of a control center for locomotion. The specific center of balance is the cerebellum, a convoluted swelling from the roof of the hindbrain. As might be expected, the cerebellum is particularly well developed in animals that move in the three planes of space—the birds, the bats, and most aquatic animals, especially the cetaceans.

Because the brain holds so many clues to the sensory and behavioral potentials of living animals, it is of considerable interest to paleontologists to see whether its structure can be inferred from the skull. The brain is encased in a bony capsule, the cranium, and in mammals and birds the cranium conforms so closely to its contours that even impressions of blood vessels can be seen imprinted on the surface of the bone. The reptilian brain, however, usually does not fill the cranium. Furthermore, it is not always completely encased in bone. Consequently, there is a less precise correspondence between the shape of the brain and that of the cranial cavity in reptiles.

For mammals and birds and, to a lesser extent, for reptiles, the structure of the brain can be reconstructed simply by making an internal cast of the cranial cavity. The procedure is quite simple. The first step is to close all the holes that lead into the cranial cavity (except for the foramen magnum). In life these holes transmitted blood vessels and nerves to and from the brain, and they can easily be plugged with Plasticine. With the skull held vertically, liquid rubber latex is poured into the foramen magnum (the hole through which the spinal cord passes on its way to the brain). Enough latex is used to form a skin on the endocranial surfaces, and the latex is swirled around to ensure an even coating. When the latex has dried and cured, which takes a few days, it is carefully peeled away and pulled out through the foramen magnum. The outside surface of the latex cast faithfully reproduces the contours of the endocranial surfaces, and the whole structure, which has the shape of the brain, is called an endocast. The endocast can be filled with plaster to make it rigid, and a mould can be formed around it so that it can finally be cast in plaster.

Sometimes a skull becomes filled with sediments during preservation, and the consolidation of the material in the cranial cavity produces a natural endocast. One such rare event appears to have occurred during the preservation of the British Museum specimen of *Archaeopteryx,* the oldest known bird. Some time after

Natural brain casts are sometimes formed by the consolidation of minerals within the endocranial cavity. This rare event occurred during the preservation of the British Museum (Natural History) specimen of ARCHAEOPTERYX.

the consolidation of the endocast, the cranium became exposed to the environment and was partly weathered away, exposing a natural endocast.

Endocasts have been made for a large number of extinct animals, including many dinosaurs (Jerison, 1973; Hopson, 1977), but as the reptilian brain does not completely fill the cranium, allowances have to made when we deduce brain sizes from them. An estimate of the discrepancy has been made by comparing endocast volumes with actual brains for some present-day reptiles (Jerison, 1973). The endocasts were found to be about twice the size of the actual brains. Consequently, reptilian endocasts are taken to represent twice the volume of the actual brains. This is only a rough estimate, however, because in some cases the reptilian brain fills the cranium, while in others it may occupy less than half of the cranial volume.

Dinosaur endocasts are not only larger than the brains were in life but are lacking in much of the original detail as well. Consequently they do not always look very brain-like so we cannot obtain a precise picture of the actual brain's shape, which limits what can be said of the behavioral potentials of the dinosaurs.

The most obvious feature of a dinosaur brain is that it was typically reptilian, generally lacking any regional enlargements. The most prominent region was the cerebrum, but this was not particularly large, except in certain small theropods like *Troodon* (formerly called *Stenonychosaurus*). The pituitary gland was usually well developed. The cerebellum, which is largely concerned with the coordination of body movements, was moderately well developed. Some dinosaurs had reasonably well developed olfactory bulbs— they were especially large in the dome-headed dinosaurs (Giffin, 1989)— suggesting that they may have had a good sense of smell. The optic lobes appear to be small but their extent cannot be determined. This is because they were covered by large blood vessels in life and their structure is therefore obscured in endocasts.

The size of the brain was often quite large. An endocast obtained for the hadrosaur *Edmontosaurus*, for example, is nearly one foot long (27 cm), but we have already seen that absolute brain size is a poor guide to an animal's intellectual abilities, and we have yet to discuss how we can measure relative brain size.

Some of the early investigations into the relationship between brain size and body size merely divided the weight of the brain by the weight of the body, but the results were not very helpful. The brain of the domestic house cat, for example, represents 0.94 percent of its body weight, whereas that of the lion is only 0.18 percent. This would suggest that domestic cats have far greater intellectual potentials than lions, but this is obviously an erroneous conclusion. Even before the turn of the century it was realized that the relationship between brain weight and body weight was exponential—that is, one set of variables, in this case

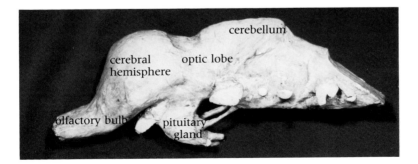

A plaster endocast of EDMONTOSAURUS. *Because the brain did not fill the cranial cavity, the endocast is larger than the actual brain and lacks detail.*

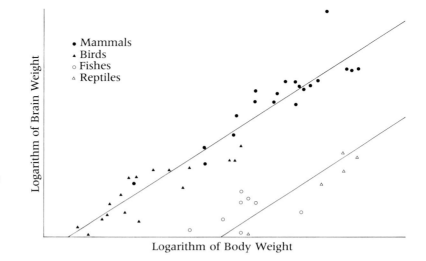

Logarithm of Brain Weight

• Mammals
▲ Birds
○ Fishes
△ Reptiles

Logarithm of Body Weight

Logarithmic graph of brain mass plotted against body mass for a wide range of vertebrates.

body weights, increases more rapidly than the other (brain weights)—but relatively few data were available for verification.

If exponential data are graphed, a curved line is obtained, but if logarithms of the original data are used instead, a straight line is obtained.[1] The gradient, or slope, of the line is easily measured, and this gives the rate of change of one variable relative to the other. When Harry Jerison, a pioneer in the field of assessing relative brain sizes, plotted data logarithmically for brain weights and body weights for a wide variety of modern vertebrates, he obtained two non-overlapping clusters of points, to which straight lines were then added (1969).[2] The upper line represented birds and mammals, the lower line the reptiles, amphibians, and bony fishes.[3] At any given body size, the brain of a bony fish or reptile is smaller than that of a bird or mammal. This dichotomy of the vertebrates is of much evolutionary significance.

Jerison obtained a slope of $\frac{2}{3}$ for the graphs and found that, for birds and mammals, the equation for this line was

$$\text{brain weight} = 0.07 \times (\text{body weight})^{2/3} \qquad (1)$$

while for bony fishes and reptiles the equation was

$$\text{brain weight} = 0.007 \times (\text{body weight})^{2/3} \qquad (2)$$

At any given body weight, the brain weight of a bony fish or reptile is therefore about one-tenth that of a bird or mammal.

With these equations it is possible to calculate the size of the brain of an animal of known body size, provided that it has an average-sized brain for its kind. For example, suppose that we wanted to calculate the brain weight of a lizard whose body weight is 540 grams. Substituting this value into equation 2, we get

$$\text{brain weight} = 0.007 \times (540)^{2/3} \text{ g}$$
$$= 0.007 \times (66.03) \text{ g}$$
$$= 0.46 \text{ g}$$

In Jerison's graphs some animals, like man, the great apes, and the whales, lie well above the line, which means that their brains are larger than would be predicted from the equation. Others lie below the line and have smaller brains than would be predicted. For example, from equation 1, the brain weight of a squirrel monkey weighing 1,000 grams is

$$\text{brain weight} = 0.07 \times (1,000)^{2/3} \text{ g}$$
$$= 0.07 \times (100) \text{ g}$$
$$= 7 \text{ g}$$

But the actual brain weight is 24 grams. The monkey's brain therefore exceeds the predicted weight[4] by the ratio 24/7 = 3.4. Jerison used the ratio of the actual brain weight to the predicted weight as an index of the relative development of the brain, calling the ratio the *encephalization quotient.* For our own species the encephalization quotient is about seven; that is, our brain is seven times larger than the brain of an "average" mammal of our body size. It is about three for the porpoise and dolphin, two for the chimpanzee, just over one for the African elephant, and just under one for the lion. The encephalization quotient appears to be a useful index of relative brain size, and this in turn appears to give an indication of intellectual potential. Equipped with the endocast technique, and the encephalization quotient, we are now able to assess the behavioral potentials of dinosaurs and their contemporaries.

Mesozoic brains

Is there any truth in the widely held belief that dinosaurs had such small brains that they were intellectually deficient? Jerison (1973) estimated endocast volumes for a number of different dinosaurs and then compared these with the values predicted using equation 2. His resulting encephalization quotients, and other pertinent figures, are listed in the table on the opposite page. Notice that since the density of living animals is close to unity, volume and mass can be used interchangeably.

Before we discuss the implications of these results, it must be pointed out that the estimates of body and endocast volumes are only approximate. Furthermore, halving the endocast volume to obtain an estimate for the brain volume is a gross approximation that is probably the source of much error. The values obtained for the encephalization quotient can therefore serve only as a rough guide to relative brain size.

Even by reptilian standards the giant sauropods *Brachiosaurus* and *Diplodocus*, the armored dinosaur *Stegosaurus*, and the horned dinosaur *Triceratops* had very small brains. On the other hand, *Troodon*, a small theropod that was only 6.5 feet long (2 m), had a remarkably large brain. The remainder had brains that were typically reptilian in size. Dinosaurs, it appears, had a wide range of relative brain sizes.

What sort of behavior might be expected of a dinosaur with a typically reptilian-sized brain? How limiting might a small brain have been to the complexity of its life-style? We will never have answers to these questions, but we can make some educated guesses on the basis of our knowledge of living reptiles. The closest reptilian relatives of the dinosaurs are crocodiles, a logical group upon which to model our ideas of dinosaurian behavior. A digression into the world of crocodiles will therefore provide some useful insights into the lives of large reptiles.

Crocodiles have traditionally been pictured as relatively inactive animals that spend most of their lives floating in the water like so many logs. But observations made on the Nile crocodile over a period of several years have revealed a surprisingly varied range of behavioral patterns, including social behaviors, some of them complex (Pooley and Gans, 1976). A crocodile will frequently capture a large prey, which it dismembers by seizing and jerking movements, sometimes twisting its whole body over and over in the water until a suitable piece is torn off. When a carcass is too

Brain and body sizes for dinosaurs

Genus	Estimated body volume (10^6 ml)	Estimated endocast volume (ml)	Estimated brain volume (ml)	Predicted brain volume (ml)	Encephalization quotient
Diplodocus[a]	11.7	100	50	363	0.1
Brachiosaurus[a]	87.0	309	154.5	1383	0.1
Triceratops	9.4	140	70	329	0.2
Stegosaurus	2.0	56	28	112	0.2
Protocertops	0.2	30	15	24	0.6
Iguanodon	5.0	250	125	206	0.6
Camptosaurus	0.4	46	23	38	0.6
Anatosaurus	3.4	300	150	160	0.9
Tyrannosaurus[b]	7.7	530	265	288	0.9
Allosaurus	2.3	335	167.5	122	1.4
Troodon (formerly Stenonychosaurus)[c]	0.045	49	49	9.2	5.3

Source: Modified from Jerison (1973).

a. If Hopson (1977) is correct in concluding that the brain probably filled the cranial cavity in sauropods, the brain volumes and encephalization quotients given here should be doubled.

b. Data from Osborn (1912).

c. Data from Russell (1969). The brain probably filled the cranial cavity in Troodon, as it does in birds and mammals.

awkward for one crocodile, the assistance of a second is then obtained. The second crocodile holds the carcass still while the first rotates, or they may both rotate in opposite directions. Each individual swallows the parts it rips off, without any indication of aggression toward the other. In early spring, when river levels are rising, sub-adults (individuals that are not yet fully mature) often cooperate in catching fish by forming a semicircle around the mouth of a channel that is being flooded. Each individual keeps its own station, even though it might mean losing a fish to a neighbor, and there is no fighting over the fish.

Male crocodiles reach sexual maturity after twelve to fifteen years, and the competition for females results in the establishment of a hierarchical system similar to that seen in many species of birds and mammals. Males participate in ritualized displays of aggression, and if a defeated male is too slow in taking flight, he is able to appease the victor by the submissive gesture of exposing the vulnerable throat region. A similar submissive gesture is seen in a number of mammals, including wolves. Once pairing has taken place the male and female appear to remain together for a long time. An elaborate courtship ritual precedes copulation, and when the time comes to lay eggs the female deposits them into a pit excavated in the river bank. When all the eggs have been laid, eighty or so, the burrow is covered up and she begins a vigil of guard duty lasting about three months. The female appears to fast during this period. The male remains in the vicinity and makes the occasional visit, but he does not approach the nest site.

Just prior to hatching time the young crocodiles—still inside the eggs buried in the bank—begin piping, just like unhatched chicks. Their chirping sounds are apparently so loud that they can be heard from sixty feet away (20 m). When the female hears them she starts uncovering the nest, and with the same huge jaws that can dismember an ox she gently picks up the hatchlings and carries them to the water, where they are bathed and then allowed to swim back to shore. The male returns to the nest at this time and he may also help to ferry the hatchlings to the water. Sometimes a youngster has difficulty in breaking through the egg shell, in which case the egg will be taken into an adult's mouth and rolled back and forth until the shell breaks. The hatchlings are looked after for about two months, after which time they leave the parents and join up with other youngsters, who migrate to isolated streams away from the parental rivers and lakes. Communal burrows are dug into the stream bank, into which they can retreat when threatened with danger. They live in this secluded environment for about five years and then, when grown, return to the main adult group.

If crocodiles, with their typically reptilian brains, are capable of such complex behavioral patterns, there is no reason to suppose that dinosaurs with similarly developed brains should not have led similarly complex lives. But what about those dinosaurs, like the sauropods, that had relatively small brains? As we saw in Chapter 5, the sauropods probably lived their lives at a leisurely pace, and this interpretation accords well with their relatively small brains. *Stegosaurus* and *Triceratops* also had small brains, suggesting that they too probably had unhurried and uncomplicated life-styles.

Stegosaurus, and several other dinosaurs, have long been credited with having had two brains:

> Behold the mighty dinosaur,
> Famous in prehistoric lore,
> Not only for his power and strength

But for his intellectual length.
You will observe by these remains
The creature had two sets of brains—
One in his head (the usual place),
The other at his spinal base.
No problem bothered him a bit
He made both head and tail of it.
If something slipped his forward mind
'Twas rescued by the one behind.
And if in error he was caught
He had a saving afterthought.
Thus he could think without congestion
Upon both sides of every question.
(Excerpt from a poem by B. L. Taylor)

Bizarre as these lines may sound, there is an element of truth in them. Dinosaurs, like all the vertebrates, had only one brain, but the spinal cord, the large nerve trunk running from the base of the brain to the tail, was considerably dilated in the sacral region (near the pelvis). We know this because the neural canal (the space in the vertebral column in which the spinal cord lies) is considerably expanded in the sacral vertebrae. In *Stegosaurus* the adjacent vertebrae are partially fused together, forming a bony cavity that has a volume about twenty times greater than that of the endocast. All tetrapods have a spinal enlargement in the sacral region[5]—another in the shoulder region too—but it is unusual for it to be as big as it is in some dinosaurs.

The significance of this enlargement is usually explained in terms of the role of the spinal cord. When an animal moves its limbs or its tail, the actions are not controlled solely by the brain—the spinal cord serves as a local center of coordination. We have all experienced the phenomenon of touching something hot or treading on a pin and seeing the affected limb withdraw automatically, even before we are aware of what has happened. Such automatic responses, termed reflex actions, involve three sets of nerves: sensory nerves, which transmit impulses from the limb to the spinal cord; motor nerves, which transmit impulses from the spinal cord to the muscles of the limb; and relay nerves within the cord linking the two sets of nerves. Dinosaurs are characterized by their large hindlimbs and long tails. *Stegosaurus* had the additional complication of having a spiked tail, which was probably used for defense. Spinal enlargement in dinosaurs therefore seems

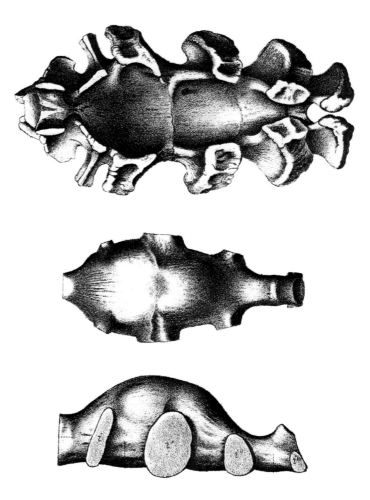

STEGOSAURUS *was unusual in having a spinal cord that was seemingly considerably enlarged in the sacral region.* Top: *The neural arches in the sacral vertebrae are expanded and form a large chamber.* Middle and bottom: *An internal cast, shown here in dorsal and lateral views, gives an approximate idea of the shape of the spinal enlargement. This was probably largely occupied by an enlarged glycogen body rather than by a greatly enlarged spinal cord.*

attributable to enhanced neuro-muscular control, but careful comparisons with living animals show that the spinal cord cannot account for such large neural spaces (Giffin, 1990). The most likely explanation is that the spaces housed a large glycogen body, a structure of uncertain function found today in birds

At the other end of the scale from these lumbering giants, both in stature and brain development, were small theropods, like *Ornithomimus* and *Troodon*, which were among the intellectually élite of dinosaurs. *Troodon* had an encephalization quotient of 5.3, which is probably higher than that of any other dinosaur and even exceeds that of *Archaeopteryx*. This interesting little dinosaur, from the Late Cretaceous of western North America, is unusual for its enormous orbits, which are directed forward, suggesting that it possessed very large eyes and had well-developed binocular vision. Remains of its skeleton are fragmentary, but it was obviously fairly small and lightly built. The hindlimbs appear to have been long and slender, and its well-developed grasping hands suggest a predatory mode of life. Perhaps it preyed on large insects, small reptiles, and mammals.

The intellectual potentials of dinosaurs probably covered a broad spectrum, but how did they compare with their reptilian contemporaries? Unfortunately our knowledge of the brain structure of other Mesozoic animals is rather scant. We have some information for the pterosaurs, for *Archaeopteryx* and for the ichthyosaurs, and also for the early mammals, but we will not be concerned with them. The pterosaurs, like all flying animals, were subjected to a high selection pressure for weight reduction and its bones were thin and light. As a consequence the bones of the cranium were thin and conformed closely to the contours of the brain. Endocasts taken from pterosaurs therefore look very brainlike, as they do for birds and mammals, and the volume of the cast probably corresponds closely to the volume of the actual brain. The avian appearance of the brain—well-developed cerebral hemispheres, a prominent cerebellum, small olfactory bulbs, and fairly large optic lobes—suggests high levels of body coordination and balance, which are essential prerequisites for flight. The large optic lobe correlates with the large orbit and sclerotic ring, both of which indicate the possession of a large eye. Estimates of the encephalization quotient are difficult to make, not because of problems of estimating the volume of the brain but because of the difficulty of obtaining a reliable estimate of the weight of the body.

Archaeopteryx, the earliest known bird, lived during the late Jurassic Period. It is the only Mesozoic bird for which we have reliable endocast data.[6] The natural endocast of the British Museum specimen, mentioned earlier, reveals a typically avian organization, with fairly large and unconvoluted cerebral hemispheres, a well-developed cerebellum, and moderately well-developed optic lobes. Estimates for the volume of the brain[7] are not as problematic as those for the weight of the body, and there is considerable disagreement among specialists on this point. By reptilian standards, *Archaeopteryx* had a large brain, and its relative brain size was probably intermediate between those of reptiles and those of modern birds. There is some evidence that brain

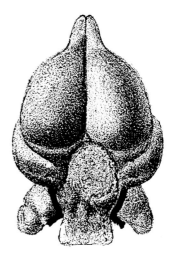

A natural endocast of the pterosaur SCAPHOGNA-THUS. *The appearance is remarkably avian, with large, smooth, cerebral hemispheres, large optic lobes, and a prominent cerebellum.*

enlargement was not completed during the evolutionary history of the birds until after the close of the Mesozoic Era (Jerison, 1973).

We have but a single endocast for the ichthyosaurs, and it is only partial, so our data are incomplete (McGowan, 1973b). The evidence suggests a well-developed cerebellum, very large optic lobes, moderate cerebral hemispheres, and fairly well-developed olfactory bulbs. These features suggest that sight was an important sense, and this is supported by the evidence for enormously large eyes. Olfaction was probably also important for ichthyosaurs, in contrast with whales and dolphins, which have lost their olfactory bulbs and, presumably, their sense of smell.

Not Wholly a Fish

T H E Cobb, a solid wall of stone that curves out to sea like a beckoning finger, has protected the harbor and town of Lyme Regis for over four hundred years. A winter storm can hurl the sea against the wall with enough force to send spray skyward with the roar of thunder. But on summer days the sea is all blue innocence, lapping gently against the gray stone on a fitful breeze. Lyme Regis stands on the south coast of England, about one hundred and forty miles southwest of London, in the county of Dorset. It is the archetype of the English seaside town—pebble beach, gull cliffs, seaweed, crabs, and ice cream. Visitors have been coming to Lyme since the seventeen hundreds, many of them for the safe bathing and sea air. But for others the main attraction has always been the fossils.

No visit to Lyme Regis would be complete without a stroll along the curving stone wall of the Cobb. Jane Austen walked this way, so did the frivolous Miss Musgrove in her novel *Persuasion*—we are in good company. The Cobb gives us an excellent view of the town and surrounding countryside. Behind us, over on the other side of the English Channel, lies France, and there, stretched out before us like a Norman tapestry are some of the finest exposures of marine Jurassic rocks to be seen anywhere in the world. These exposures, rich in fossils, are the result of the relentless action of wind and waves and sun on the very fabric of the land. The cliffs, mostly marls and shales, interspersed by limestone slabs, are soft and crumbling so they weather very fast. They are also prone to slumping, especially after heavy rains, and there have been some impressive land slips over the years. New exposures are contin-

The town of Lyme Regis, on the southwest coast of England.

ually being revealed, attracting fossilists to the Dorset coast like birds to a freshly ploughed field.

Most of the rocks along this stretch of the Dorset coast are from the Lower Jurassic, which is usually referred to as the Liassic, derived from the Gaelic word *leac,* meaning a flat stone. Since the rocks belong to the earliest part of the Liassic, they are said to be Lower Liassic in age. To the east of the town is Church Cliffs. The crescent-shaped band of rock that outcrops at its base belongs to the Blue Lias, so-called because of its blue-gray color. The Blue Lias, which is rich in fossils, is the earliest part of the Lower Liassic and is therefore geologically the oldest part of the Jurassic. There are similar exposures of Blue Lias to the west of the town; these are more extensive than the Church Cliffs exposures.

Among the most important fossils found in the Blue Lias are the ichthyosaurs ("fish-lizards"). This group of Mesozoic reptiles—not obviously related to any other—lived in the sea and were named fish-lizards because of their remarkable resemblance to fishes. They were first described in the scientific literature in 1814, by Sir Everard Home. This was not the first time that ichthyosaurian remains had ever been reported, but the earlier accounts had mistaken them for fish bones.[1] Other types of fossil reptiles have also been found in the Dorset Lias, and these were often referred to, collectively, as saurians.

The Blue Lias, exposed at Church Cliffs, just east of Lyme Regis. Notice the alternating bands of limestone and shale.

Beyond Church Cliffs is Black Ven, a somber cliff of crumbling dark marls. Hazardous to climb when dry, this clay material becomes a glutinous and impassable mass when it is wet. Fossils are found there too, but they are geologically slightly younger than the ones at Church Cliffs; that is, they are slightly higher up in the rock sequence than the Blue Lias. Indeed, the rocks become progressively younger to the east, older to the west, because of the slope of the strata. Let's stroll along the beach, toward the west, and take a closer look at the Blue Lias.

Crunching over the pebbles is hard work, so we move up the beach and walk on the large stones and boulders that have weathered away from the cliffs. Some boulders are of limestone, which is a pale blue-gray, while the darker-colored ones, which weather more rapidly, are of shale. Many of our stepping stones contain fossils, and some are crammed full of ammonites. These extinct molluscs, distantly related to present-day squids, appear as thin whorls of shell.

At places along the beach the tide has scoured away the pebbles, revealing the underlying limestone. This was once part of the seabed of a tropical lagoon, and we are placing our feet where ancient life once teemed. Here is a huge calcareous spiral, the remains of a giant ammonite. It is difficult to imagine a mollusc whose shell was more than a yard in diameter. It is equally difficult

An ammonite. These shelled fossils, related to present-day squid, are very common along the foreshore of Lyme Regis.

to imagine that it lived and died in those tranquil waters over two hundred million years ago.

The cliffs are not very high, fifty feet (15 m), possibly more, of alternating bands of shale and limestone. A steady dripping of groundwater can be heard somewhere above our heads, but the source remains hidden. The limestone ledges vary in thickness, as do the bands of shale that separate them; some are a little wider than my hand but others are no thicker than my thumb. Some of the shale layers are of similar thickness, but others are considerably thicker—half my height at least. We can easily break off a handful of shale; they are called paper shales because they are layered, just like pages of a book. The pages, which are about as thick as five or six sheets of paper, cleave apart easily and cleanly. How many years does each layer represent? One hundred years? One thousand years? We keep on turning the pages of the book—earth book, time book—our eyes peering where human eyes have never peered before. And then we see something: an ammonite no bigger than a fingernail and squashed as flat as a pressed flower. It is a trivial find, but it forges a tangible link with the remote past and for a few moments it is the most precious object on Earth. The ammonite is slipped into a pocket and we carry on along the beach.

Almost two centuries have passed since Mary Anning searched this beach for fossils. She is one of the most widely known of the

An ichthyosaur skeleton from the Lower Liassic of southwest England. The skeleton, which is almost six feet (2 m) long, is in the British Museum (Natural History), London.

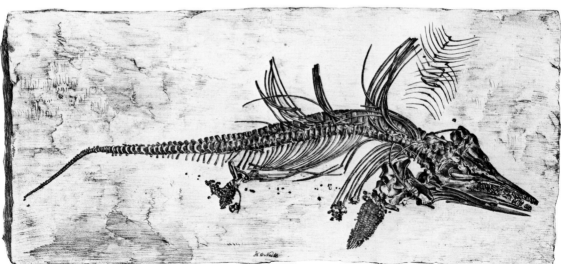

NOT WHOLLY A FISH

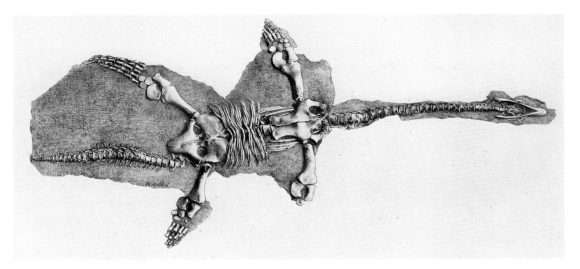

The skeleton of a plesiosaur (PLESIOSAURUS), as figured by Hawkins in his book of sea dragons.

earliest fossilists, and for good reason. She was only eleven or twelve when she helped her brother excavate the first ichthyosaur—a magnificent skull with associated vertebrae and pectoral girdle—the one described by Sir Everard Home in 1814.[2] The money she received from its sale went a long way to relieving the hardship that had befallen her family with her father's death the previous year. She discovered a second saurian that was new to science in 1821. It was the first specimen of a new group of reptiles for which the name *plesiosaur* was chosen ("near lizard"; De La Beche and Conybeare, 1821). These were marine reptiles, like ichthyosaurs, but they were anatomically quite distinct. Instead of having a body that was narrowed from side to side, like a fish, their bodies were narrow from top to bottom, like a turtle. And, like turtles, they had large, oar-like limbs for propelling themselves through the water, instead of the shark-like tails of ichthyosaurs. Seven years after discovering the first plesiosaur Mary Anning collected Britain's first pterosaur (Buckland, 1829). She became widely known as a collector of importance, and some of the most influential names in the new science of paleontology visited her fossil shop in Lyme, seeking information and buying her wares.

The great importance of her discoveries can only be appreciated when considered in the context of what was happening in the scientific world at that time. Paleontology was still a relatively new branch of science—as molecular genetics is today—and far fewer fossils, especially Mesozoic ones, were known then. The first dinosaur, *Iguanodon* ("iguana tooth"), was not discovered until about 1822 (described three years later by Mantell, 1825). Even then the significance of this rather unimpressive find (some teeth from a stone quarry in Sussex, England) was not fully appreciated, and it was not until 1841 that the name *dinosaur* ("terrible lizard") was coined—by Richard Owen. Only three well-established genera of dinosaurs were known at that time, and these were all fragmentary: *Iguanodon* (an ornithopod), *Megalosaurus* ("big lizard"; a theropod), and *Hylaeosaurus* ("woodland lizard"; an ankylosaur). With astonishing foresight Richard Owen (1841) recognized that these three fossils represented a new group of Mesozoic reptiles, a group characterized by gigantic size and terrestrial habit.

How were these various Mesozoic reptiles—primarily the dinosaurs, ichthyosaurs, plesiosaurs, and pterosaurs—viewed by the scientific community of the day? They were certainly recognized as being extinct, a concept that had been established by the work of the great French anatomist Georges Cuvier (1769–1832). For some people, though, extinction was unacceptable on religious grounds: God had created all species and to suggest that some had become extinct was to question the perfection of the Creation and of the Creator. Cuvier attributed extinctions to changes in the physical world—floods, climatic changes and the like—which he referred to as revolutions. The Reverend William Buckland (1784–1856), first professor of geology at Oxford University, identified Cuvier's last "revolution" with the Great Flood of Noah, and was thereby able to reconcile the fact of extinction with the conviction of his faith. This view no doubt gave comfort to others who saw conflict between their religious beliefs and the fossil record.

Cuvier believed that species were stable entities that did not undergo change with the passage of time. However, his fellow countryman, Jean-Baptiste Lamarck (1744–1829), discounted both the idea of the stability of species and the concept that some had become extinct. Lamarck believed that species were continuously changing and that the species represented by fossils had gradually become transformed into modern species. Somewhat similar evolutionary concepts had been discussed by others, including Darwin's own grandfather, Erasmus Darwin (1731–1802), but most authorities were inclined to dismiss such ideas. For them dinosaurs and other Mesozoic reptiles were not the

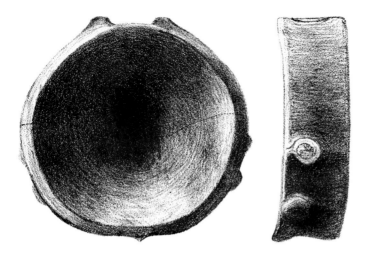

An isolated ichthyosaur centrum, seen from the front (left) and from the side. These bony discs are easily identified by the deep concavities of their anterior and posterior surfaces.

modified descendants of earlier animals, but were reptiles of the original Creation that had not survived beyond the Mesozoic. This viewpoint was not to change significantly until well after the appearance of Darwin's *Origin of Species* in 1859.

The appellation *ichthyosaur*, "fish-lizard," for Mary Anning's discovery is easy enough to understand: the bony remains of these creatures *do* look more like fishes than reptiles. The fore- and hindlimbs are highly modified as fins, and the individual bones, save for the humerus and femur, have lost the usual shape of limb and finger elements. Instead, they are polygonal in shape, though they become more rounded distally, and they all fit together like a mosaic. The forefin is often much larger than the hindfin, it often has more than the usual five digits, and there are many more individual bones—phalanges—in each one.

The skull looks very fishy, too, with its long, pointed jaws and sharply pointed teeth. But the most prominent feature of the skull is its enormous orbit. And in the middle of the orbit, almost filling it, is a large sclerotic ring. The external naris lies immediately in front of the orbit, low on the skull, rather than high on top of the head as it is in present-day cetaceans (dolphins and whales).

Although many complete skeletons have been collected since Anning's day, they are still fairly uncommon and it is more usual to find isolated parts, like teeth and vertebrae. Ichthyosaurian vertebrae are distinctive because the centra are deeply bi-concave

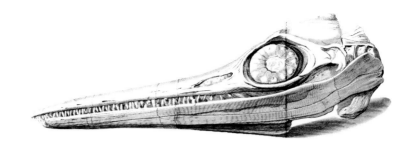

This ichthyosaur skull, which is well over a yard (1 m) long, was collected by Mary Anning in 1811 and subsequently described as a new animal by Sir Everard Home in 1814. Specimen in the British Museum (Natural History), London.

and are described as amphicoelous ("double-hollowed"). Fishes similarly have amphicoelous vertebrae, but they are easily distinguished from ichthyosaurian centra because the latter have smooth sides whereas those of fishes are ridged.

The vertebral column is gently arched between the fore- and hindfins, then straightens out into a fairly long tail. A singular feature of the tail is an abrupt downward bend, which occurs about two-thirds of the way along its length. Owen, the most respected anatomist of his day, believed that the bend was a postmortem effect, possibly brought about by the contraction of a tail ligament (Owen, 1840b). He concluded that the vertebral column was essentially straight in life, but that there was some sort of tail fin in the region of the bend. The same year that he published these findings he also published a report on ichthyosaurs and plesiosaurs that he read at the 1839 meeting of the British Association for the Advancement of Science (Owen, 1840a). These were Owen's first contributions on ichthyosaurs, most previous writings having come from the pens of Sir Everard Home, Henry De La Beche, the Reverend William Conybeare (the vicar of Lyme's neighboring town of Axminster), and the eccentric Thomas Hawkins of Somerset.[3] Henry De La Beche, who was also a resident of Lyme and prominent in geological circles, was only three years older than Mary Anning. They had always been very good friends, and it has been said that they might have married had it not been for her lowly working-class background.

Fish or lizard? A brief history of research

Mary Anning's first ichthyosaur skull caused Sir Everard Home much perplexity. He was initially impressed by its resemblance to the crocodile, which led him to the premature conclusion that it was indeed a new species of crocodile (Home, 1814). When he examined the vertebrae, however, he saw that they were amphicoelous, a condition hitherto found only in fishes. He concluded that the new discovery must have been a fish, but, he wrote, "I by no means consider it as wholly a fish." He believed that it formed a connecting link between fishes and crocodiles. Home wrestled with the problem for the next five years, finally coming to the conclusion that the new fossil was intermediate between lizards and salamanders. The salamander that he considered its closest relative was the genus *Proteus,* and he therefore coined the name *Proteosaurus* for the new fossil. He noted, however, that the name *Ichthyosaurus* was already in use, and it is the latter name that has since been adopted.[4]

Richard Owen (1804–1892), who was probably the greatest of all anatomists, was still in his teens when all of this was happening. By the 1840s, when he became actively interested in ichthyosaurs, it was already well established that they were reptiles, and several different species had been recognized and described. Most of the specimens had been collected from the Lower Jurassic outcrops of Dorset and Somerset, many of them by Mary Anning. Thomas Hawkins was also an avid collector, and while he did do some collecting at Lyme Regis, most of his specimens were taken from the adjoining county of Somerset, particularly from the inland quarries of the town of Street and its environs. Ichthyosaurs were also being unearthed in Europe, notably in southern Germany. Owen, the well-travelled ambassador of British science, had the opportunity of examining this material.

Although their marine habit had been recognized since the earliest days, it was believed that ichthyosaurs used to haul themselves up on land, as seals do. Nineteenth-century restorations frequently depict ichthyosaurs basking on rocks, in the pose that Owen selected for the life-sized models that were built on the grounds of the Crystal Palace, in the London suburb of Sydenham. The Crystal Palace, an impressive edifice in iron and glass, was originally erected in 1851 in Hyde Park to house the Great Exhibition. This Victorian extravaganza in celebration of science and technology was organized by Queen Victoria's husband, Albert, the Prince Consort, and Owen was one of his scientific advisers. When the Crystal Palace was relocated in Sydenham, Owen supervised the reconstruction of a number of extinct animals to adorn the grounds. Completion of the models was celebrated, with typical Victorian eccentricity, with a banquet held inside the

A Jurassic lagoon as seen by Hawkins in his book of "great sea dragons." Notice that ichthyosaurs have been given a straight tail and a predilection for sunbathing.

hollow body of the dinosaur *Iguanodon*. According to the report in the *Illustrated London News*, twenty-one gentlemen were accommodated within the beast, with another seven seated at a side table.

The straight-tailed, rock-basking image of ichthyosaurs persisted until about the time of Owen's death. Indeed, the straight-tailed concept was so deeply entrenched that there is at least one ichthyosaur whose natural tailbend appears to have been straightened out during preparation. This makes an interesting exception to the more usual and more recent trend of exaggerating the tailbend during preparation (McGowan, 1989d). The reversal in thinking regarding the straight tail was brought about by some unusual specimens that were discovered in southern Germany. In these skeletons the body outline was preserved as a thin film of carbon, which showed that the tailbend was a natural feature and that the down-turned segment of the vertebral column formed the lower lobe of a deep and lunate tail fin. This new information was difficult to reconcile with the idea that ichthyosaurs hauled themselves up on land. Not only would such a deep-bodied animal have had difficulties balancing itself on its belly, but the tail fin would have got in the way too.

The quarries of southern Germany—mostly in and around the town of Holzmaden (Hauff, 1954)—have been worked for slate for several hundred years, and ichthyosaur remains were found there over two centuries ago (Ziegler, 1986). As in England, their significance was not realized until much later. The quarrymen, alert to the fact that well-preserved specimens could be sold for relatively large sums of money, took a special interest in finding more material. And, because these quarries were larger and more productive than their English counterparts, the numbers of ichthyosaurs recovered from the Holzmaden area soon surpassed those of southwestern England. Large collections of Holzmaden ichthyosaurs have been built up, and the center for research on ichthyosaurs therefore shifted to Germany. Whereas the ichthyosaurs from southern England are Lower Liassic in age, those from Holzmaden are Upper Liassic, about 12 million years younger.

The end of the nineteenth century and the first half of the twentieth century witnessed a proliferation of German publications, with major contributions from Eberhard Fraas of the Royal Natural History Museum in Stuttgart, which no longer exists, and Friedrich von Huene of the University of Tübingen. Von Huene wrote extensively on many different fossil vertebrates, but ichthy-

These Lower Jurassic quarries, near the town of Holzmaden in southern Germany, are still being worked for slate today.

osaurs were his first love and he published more papers on them than any other paleontologist. He was largely concerned with the interrelationships among ichthyosaurs, especially the German species, and also with the relationships between them and other reptiles. Other German paleontologists interested themselves in interpreting the conditions under which the skeletons had become preserved and in extracting the maximum amount of information from the remarkably well preserved specimens at their disposal. These specimens, especially those in which the body outline has been preserved, have allowed us to fill in some of the gaps left by the Lyme Regis material.

Germany, like so many other countries, is dependent on the Middle East for oil, and when these supplies were cut off during the two world wars she desperately sought other sources of supply. The slate quarried at Holzmaden had long been known to contain oil and bitumen, up to 20 percent, and this resource was hurriedly developed as a vital part of the war effort. The layers that are rich in bitumen are also those where the best-preserved ichthyosaurs are found, the ones for which the body outline has been preserved. This was obviously no mere coincidence. Experiments with present-day animals have shown that bitumen does have preservative properties (Heller, 1966).

The usual interpretation of the film is that it is a carbonaceous residue of the original skin, but David Martill of the Open University in England has raised serious doubts on this (1987). His studies have shown that the film is partly formed of mats of bacteria that have replaced the original skin. Although these mats have retained the broad outlines of the body, Martill considered it unlikely that the mats would have retained outlines as sharply defined as are seen in most Holzmaden specimens. This led him to suspect that the original specimens may have been embellished by the preparators who extricated the fossils from the rock. I have encountered other examples of "improvements" being made on nature and therefore share Martill's concerns.

Only the lower surface of the body has the carbonaceous film—there is no similar preservation of the upper surface. This suggests that when a dead ichthyosaur settled on the soft seabed its underside became embedded in a bituminous slime and was preserved, while the soft tissues of the upper surface rotted and were swept away. The bacterial mats were formed by anaerobic bacteria, that is, those that thrive in oxygen-free conditions. Anaerobic bacteria are commonly found in muddy sediments and are re-

sponsible for the black color and bad smells associated with the mud of some seashores.

Samples of the film have been analyzed and found to consist mainly of hydrocarbons (compounds formed of hydrogen, carbon, and oxygen) and amino acids, together with small amounts of fatty acids (Heller, 1966). The latter component indicates the original presence of lipids (fats), which suggests that there might have been a layer of fat beneath the ichthyosaur's skin. Reference was made to this fatty layer in Chapter 6 with regard to thermal insulation.

Preservation of the body outline is exceedingly rare in fossil vertebrates. The ichthyosaurian specimens, which are uncommon but not rare, provide us with information that would otherwise be unavailable. That ichthyosaurs were very fish-like was self-evident from the earliest days, but the full extent of the resemblance became apparent only with the discovery of the skin specimens. The presence of some sort of caudal fin had already been suspected because of the tailbend, but the discovery of a dorsal fin was a complete surprise because there is no internal skeleton to hint at its existence. The superficial resemblance to the dorsal fin of the swordfish is quite striking. Martill (1987) has questioned the authenticity of these particular impressions, warning that the existence of a dorsal fin has yet to be proved. I do not doubt that some dorsal fins have been fabricated, but I am inclined to believe that others are genuine. The paired pectoral and pelvic fins both appear to have been wide at the base, indicating a limited mobility. This suggests that they were primarily used as hydroplanes, for adjusting the swimming level, rather than as paddles for rowing and maneuvering the animal through the water.

The cast of the forefin of a Jurassic ichthyosaur (left), *when sectioned* (center), *reveals a streamlined shape. Compare this with the sectioned porpoise fin* (right). *The leading edge is to the left in each case.*

The body outline is beautifully streamlined, with the deepest part of the body occurring a little less than halfway along its length, at about the same level as the dorsal fin. Most vertebrates that swim or fly well have streamlined bodies. The paired fins are also streamlined, in profile, as can be seen by cutting sections through a fin or, rather, through a plastic cast of a fin. Each section has a streamlined shape, and although they are not as pronouncedly teardrop-shaped as the sections in a similar series cut through a porpoise fin, it must be remembered that we are looking only at the bony parts. In life the bone would have been covered with soft tissue, as in the porpoise fin, and may then have been more markedly streamlined. An examination of specimens where the body outline is preserved reveals that the paired fins have extensive trailing edges that are unsupported by the internal skeleton. Very occasionally, however, traces of rays are found. Owen (1841), who first described them, suggested that these were rare probably because they were cartilaginous or horny. It is concluded that the trailing edges of the paired fins were continued as gently tapering blades, probably stiffened internally by non-bony rays. There is also a narrow unsupported area on the leading edge of the paired fins. Both features contributed to the streamlining of the fins.

The lower lobe of the tail similarly had a wide and unsupported trailing edge, immediately posterior to the down-turned vertebral column, and a narrow, unsupported leading edge immediately anterior to it. This suggests that the lower lobe of the tail had a streamlined cross section. It is possible that the trailing edge of the tail, like the trailing edges of the paired fins, was supported by rays, but there is no evidence for this.

As there is no internal skeleton for the upper lobe of the caudal fin, nor for the dorsal fin, their cross-sectional shapes cannot be determined. However, it is likely that both would have been streamlined, like the tail flukes and the dorsal fins of cetaceans, which similarly lack internal skeletons. Man has utilized the streamlined shape in a wide variety of applications, from ships' hulls to turbine blades. Understanding how the streamlined shape works requires some knowledge of the forces that act on a body moving through a fluid, that is, a knowledge of fluid mechanics.

Until now our primary focus has been upon the terrestrial reptiles of the Mesozoic—the dinosaurs. Many of their features, we have seen, are best understood as solutions to the problem of supporting their bodies against the force of gravity. Now that our attention is turned to reptiles that moved in fluids—specifically to ichthyosaurs that swam in the sea and, later (Chapter 12), to pterosaurs that flew in the air—we shall see that fluid forces, not gravitational forces, predominate. In order to understand these forces, and how they have shaped the bodies of the animals that cope with them, we need to spend some time on fluid mechanics. This is the subject of the next chapter.

NINE

The Mechanics of Swimming

A body moving through a fluid, whether air or water, experiences a resistance to its forward motion because of its interaction with the surrounding fluid. This resistance, usually referred to as *drag*, can be a remarkably large force, as was graphically illustrated for me one summer when I was towing some surveying equipment behind a boat. The equipment, a yard-long (one meter) sonar transducer shaped like a cigar, did not generate much drag on its own—it was like having a good-sized salmon on the line. But I was amazed at the additional drag generated by the 100-yard (92 m) cable, no thicker than my thumb. If we had been using several hundred yards of cable for a deep survey we would have needed a trawler and power winch to handle the additional drag forces.

Since drag forces rob a moving object of such large quantities of energy, anything that can be done to reduce these forces is advantageous. A widely used mechanism for reducing drag is the streamlined shape, familiar to everyone from a variety of man-made objects, like aircraft and submarines, as well as from living organisms, like fishes and birds. Understanding how the streamlined shape works requires some knowledge of the forces that act on a body moving in a fluid.

When a fluid flows gently and smoothly so that its layers slip past one another without any swirling movements, its flow is described as being *laminar.* The airflow over the windshield of a car is usually laminar. The path air takes is usually invisible, but it can readily be seen when driving during a snowfall. Orderly

Left: *When a thin plate is placed perpendicular to the direction of flow of a fluid, it generates considerable turbulence and drag forces are maximal.* Right: *When it is placed parallel to the direction of flow, the drag forces acting on the plate are minimal.*

streams of snowflakes rush toward the windshield, swooping up in smooth curves as they follow the contours of the glass, but once they have passed the windshield the flakes break up into a tumbling, swirling maelstrom. A disturbed flow of fluid, like that demonstrated by the swirling snowflakes, is described as *turbulent*.

Most bodies generate swirling movements when they move through a fluid. Since the energy that sets the fluid particles into motion comes from the body itself, this swirling is the cause of most of the resistance to forward motion. If a body could move through a fluid without causing such disturbances, the drag forces would be greatly reduced. A useful way of examining drag forces is to visualize what happens when a thin plate is moved through a fluid. (Actually, most experiments on fluid flow have the object stationary and the fluid in motion, as in experiments conducted in wind-tunnels. This is purely a matter of convenience; it makes no difference to the end result—all that matters is that there is a flow of fluid *relative* to the object being tested.)

The maximum amount of turbulence is achieved when the plate is placed perpendicular to the direction of flow. The fluid can be thought of as being made up of numerous small particles, some of which strike the leading surface of the plate and oppose forward motion head-on. Many more get diverted around the edges before they can reach the plate, and some of these move inward and create rotating areas of flow called vortices. Extensive vortices can be seen downstream from the plate. Energy is required to move each fluid particle around the plate, and also to change the motion of the particles from being laminar to being turbulent. The amount of energy required depends on the inertia of the particle—the tendency for it to stay where it is—which is proportional to its mass and therefore to the density of the fluid (mass per unit volume). This component of drag is referred to as *pressure drag* or *form drag*. The latter term, which emphasizes that this component

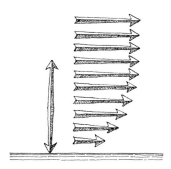

When a fluid flows over a body the layer in contact with the surface is stationary. Adjacent layers move at increasingly higher velocities until the full velocity of the free stream is reached. The zone of increasing velocities is called the boundary layer (vertical arrow).

of drag is determined by the shape or form of the body, will be used in preference to *pressure drag*. Almost all of the resistance experienced by the perpendicular plate is due to form drag.

A second drag component, called *friction drag* (or skin friction), acts on the surface of the body. Friction drag is attributable to a property of fluid called *viscosity*. Viscosity may be thought of as a measure of how readily one layer of a fluid slips past an adjacent layer; in general, the lower the viscosity, the easier the flow. Pancake syrup, for example, is far more viscous than water, which is why it takes longer to pour syrup out of a jug than water.

When a fluid flows over a body, the layer in contact with the outside "sticks" to the surface and is therefore stationary. Above this stationary layer is a zone of transition where the velocity of adjacent layers increases until it reaches that of the fluid that is undisturbed by the body, called the free stream. This whole region of transition, from the surface to the free-flowing fluid, is called the *boundary layer*.

Adjacent layers of fluid in the boundary layer are slipping past one another, and each layer is moving relatively faster than the one nearer to the body. The ease with which adjacent layers slip past one another—that is, the ease with which they shear—varies with the viscosity of the fluid. If we had an imaginary fluid which had no viscosity at all, there would be no boundary layer, because there would be no resistance to shearing between the stationary layer of fluid that is in contact with the skin and the layer adjacent to it. The latter would therefore be travelling at the full speed of the free stream. Fluids with high viscosities, in contrast, have thick boundary layers.[1] Needless to say, if there were no viscosity there would be no friction drag.

To return to the experiment with the thin plate: when the plate is placed perpendicular to the fluid flow, friction drag is negligible. This is because there is barely any surface area parallel to the free stream, and almost all of the resistance to forward motion is due to form drag. The reverse holds true when the plate is placed parallel to the flow. The reason is that the fluid flow is now almost entirely laminar, and almost all of the resistance to foward motion is attributable to friction drag.

The total drag acting on a body is a combination of form drag and friction drag, attributable to the density and to the viscosity of the fluid, respectively. The relative sizes of these two forces depend on the size of the body and the speed of the fluid flow. A large and a small body therefore experience different forces even though they are moving in the same fluid. The relationship between these variables is expressed by a ratio called *Reynolds number*, after Osborne Reynolds, abbreviated Re:

$$\text{Re} = \frac{\text{length} \times \text{speed} \times \text{density}}{\text{viscosity}}$$

where density and viscosity refer to the fluid and speed and length refer to the body moving through it. When calculating a Reynolds number, you must be sure that all of the variables are measured in the same units. (Because the units in the equation cancel out, the Reynolds number itself has no units.)[2] Although we have been using imperial, or English, units of measure in this book (and metric units parenthetically), metric units are almost universally used in science and will accordingly be used here to determine Reynolds numbers. In metric units, the density of water is 1,000 kg/m^3 and the viscosity of water (at 10° C) is 1.304×10^{-3} N · s/m^2 (newton-second per meter squared).

Suppose we wanted to calculate the Reynolds number of a two-meter-long (6.5 ft) dolphin moving along at five meters per second (11 mph). The Reynolds number of the dolphin at this speed is:

$$\frac{2 \times 5 \times 1,000}{1.304 \times 10^{-3}} = 7.67 \times 10^6$$

which is quite high. How about a smaller animal, say a twenty-millimeter (0.75 inch) guppy swimming in a fish tank? We know from a previous chapter that small animals move more slowly than large animals, and a reasonable speed for a guppy would be about forty millimeters per second (1.5 inches per second).

$$\text{Re} = \frac{0.02 \times 0.04 \times 1,000}{1.304 \times 10^{-3}} = 613.5$$

which is quite low. Very small bodies, such as planktonic organisms, move at Reynolds numbers that are very much less than ten. Movements at low Reynolds numbers are dominated by viscosity forces, and the greatest drag component is therefore friction drag. This is irrespective of whether the fluid has a high or low viscosity. Water, for example, has a fairly low viscosity, but the microscopic animals we see swimming in a droplet of water appear to us as if they are swimming in a thick syrup. At high

The seahorse (left), *a slow swimmer, moves at low Reynolds numbers and is not streamlined. The salmon is a fairly fast swimmer that moves at high Reynolds numbers and is streamlined.*

Reynolds numbers the situation is reversed; form drag is the main source of resistance, and viscosity, hence friction drag, is of secondary importance.

In addition to the sources of resistance changing with Reynolds number, the thickness of the boundary layer also varies with Reynolds number. We can see this relationship every time we drive a car in the rain. At low speeds the Reynolds number is low and the boundary layer is correspondingly thick. The beads of water that form on the windows lie within the boundary layer and are therefore not exposed to the full force of the free flow of air—they are not displaced by the air flowing over the car. As the speed of the car increases, the thickness of the boundary layer decreases until it is thinner than the beads of water. At this point those parts of the beads projecting beyond the boundary layer are exposed to the full force of the free airflow and are accordingly swept back by it. The largest droplets are the first to move, because they project farthest, and the car has to be travelling very fast before the smallest ones are swept away.

The streamlined body

If a round plate is placed in a fluid, perpendicular to the direction of flow, the vortices generated—hence the form drag—are maximal, as we saw earlier. Replacing the disc with a ball of the same diameter reduces the resistance to about 50 percent (Kermode, 1972). The flow is essentially laminar around the front or *leading surface* of the ball, and continues like this beyond the widest part of the ball, called the *shoulder,* toward the back or *trailing surface.* Behind the ball the fluid flow becomes turbulent and numerous vortices are generated.

If the leading surface is drawn out into a rounded point and

Left: *A sphere placed in a flow of fluid generates less turbulence than a flat plate of the same diameter and therefore experiences smaller drag forces.* Right: *A streamlined body form generates even less turbulence and therefore experiences even smaller drag forces.*

the trailing surface is gently tapered, the vortices can be essentially eliminated and the resistance is reduced to about 5 percent that of the perpendicular disc. This torpedo shape is called a *streamlined body.* The optimum streamlined shape for minimizing total drag has the shoulder placed between about one-third and one-half of the way back from the front of the body.

A useful way of expressing the relative slenderness of a streamlined body is the ratio of the length of the body divided by its diameter, a ratio known as the *fineness ratio* (sometimes the slenderness ratio). Increasing the fineness ratio decreases the form drag, but since it also increases the surface area the friction drag is correspondingly increased. There is an optimum fineness ratio at which the total drag is minimal, and this has been found to correspond to a value between three and seven (Weihs and Webb, 1983).

The major drag component of a streamlined body is attributed to friction, which, we have seen, involves the boundary layer. The fluid flow in the boundary layer, like that in the free-flowing fluid beyond it, can be either laminar or turbulent. If laminar flow could be maintained in the boundary layer, friction drag would be reduced to about 10 percent of what it would be if the flow were turbulent (Kermode, 1972). This ideal is probably never achieved, but it can be approached.

The flow in the boundary layer usually starts off being laminar over the first part of the leading surface (because the local Reynolds number is still low there), but a transition region is eventually reached beyond which the flow becomes turbulent. The position of the transition region moves forward as the object moves with increasing speed, so as the body moves faster more of the boundary layer becomes turbulent, therefore increasing the magnitude of the friction drag. As surface roughness is one of the factors that causes the boundary layer to become turbulent, having a smooth body is one effective way of reducing friction drag.

Not only does the boundary layer become turbulent with in-

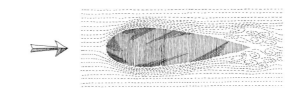

creasing speed, but it also begins to separate from the surface of the body. Separation begins at trailing surfaces and moves forward as the body increases in speed. Large vortices are generated in the fluid by the separation, and this reduces, then eliminates, the beneficial effect of the streamline shape. Separation, like turbulence, can be caused by rough surfaces, which is another reason for having a smooth body (though under some circumstances, described later, roughness can be beneficial).

Just how smooth the surface of a body needs to be to promote laminar flow and delay separation depends on the thickness of the boundary layer, which is determined by the Reynolds number. Modern aircraft, for example, move at such high Reynolds numbers that the boundary layer is only about as thick as a paperclip. Even small projections extend well into the boundary layer, causing turbulence and promoting separation, which explains why the rivets that secure the metal skin to the airframe have to be flush with the surface—take a look next time you board a plane. Separation on wing surfaces is especially critical because it reduces the lifting capacity (see Chapter 11). A build-up of ice on wings can therefore have disastrous consequences.

As noted, however, turbulent flow can be more beneficial than laminar flow. At very high Reynolds numbers, for example, a turbulent fluid may actually remain in contact with a surface for longer than it would if its flow were laminar. This is why golf balls have dimples. The turbulence the dimples generate in the boundary layer delays separation so that the air stays in contact with the surface longer, actually reaching the trailing surface before breaking away. Consequently the width of the wake, hence the drag, is much smaller. It has been estimated that a swing that would drive a regular ball for 230 yards (212 m) would drive a smooth ball for only about 50 yards (46 m—Shapiro, 1961).

So far we have considered resistance to surfaces that are either parallel or perpendicular to the flow of a fluid. What of surfaces that are between these two extremes—what forces are experienced by objects that meet the flow, or move through a fluid, at an angle? When a plane (a flat surface, that is, not an aircraft) is inclined at a small angle to a moving fluid, such that the leading edge is higher than the trailing edge, it experiences an upthrust or lift force. This is because the plane deflects fluid away from its lower surface and the force applied to the fluid has an equal and opposite reaction on the plane. This can readily be demonstrated by making one's hand into an inclined plane and holding it out the window of a moving car. If you tilt the hand from the horizontal position, you will feel it being pushed upward by the air.

If the angle of an inclined plane to the flow, called the *angle of attack,* is gradually increased, the magnitude of the lift force increases—but so too does the drag force. The lift force does not keep on increasing, however; it reaches a maximum—at an angle of attack of 10–20°—after which it begins to decrease. The drag force increases all the time, reaching a maximum when the angle of attack reaches 90°. If the angle of attack is increased beyond 90°, that is, if the plane is inclined downward, the plane experiences a downthrust instead of a lift.

The rudder of a ship is a vertical inclined plane that can be inclined to one side or the other. When the rudder is moved so that its trailing edge moves to the right, the thrust is to the left. This pushes the stern to the left, which in turn swings the vessel around and points the bow to the right.

The effectiveness of an inclined plane can be expressed by the ratio of the lift force to the drag force. The lift force increases with

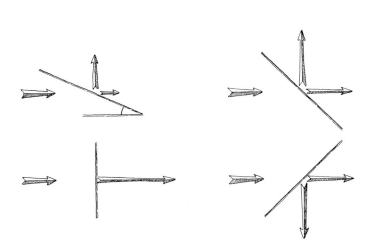

Top left: *An inclined plane placed in a flowing fluid at a small angle of attack experiences an upthrust and drag forces.* Top right: *With increasing angle of attack, the magnitudes of these forces increases.* Bottom left: *When the plane is perpendicular, only drag forces are experienced.* Bottom right: *At angles greater than 90°, however, the drag forces are accompanied by a downthrust. The magnitude of the forces are indicated by the lengths of the arrows.*

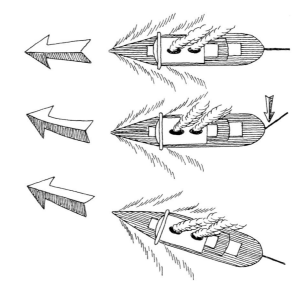

increased area, but so does the drag force, so we cannot improve the lift-to-drag ratio simply by increasing the size of the plane. A long, narrow plane, with its long axis across the flow, however, will have a better lift-to-drag ratio than a square plane. That is, of two planes of equal area, one square and the other long and narrow, the second plane would generate the greater lift for a given drag force. The relative narrowness of an inclined plane is expressed by the *aspect ratio,* which is the ratio of the length to the width. A plane 10 units long and 10 units wide has an aspect ratio of one, whereas one that is 20 units long and 5 wide has an aspect ratio of four. One of the reasons inclined planes (and wings) with high aspect ratios have a higher lift-to-drag ratio is that they generate less of a vortex at the tips. The lift-to-drag ratio of a plane is also increased by its having a streamlined profile, because, at a given, high, Reynolds number, a streamlined body experiences lower drag forces than one with a rectangular profile.

The motion we are concerned with—the movement of animals through water and air—occurs at high Reynolds numbers. These movements are dominated by inertial forces, forces that are attributed to the density of the fluid. Viscosity forces are smaller, but their effects cannot ignored. Most of our knowledge of the drag forces acting on a body moving at high Reynolds number has been obtained empirically, that is, by observation and experiment. It has been found that the drag force increases approximately with the square of the speed, with the density of the fluid, and with the area of the body. The drag force can therefore be expressed as: drag $\propto$ (velocity)^2 $\times$ density $\times$ area. This is not a precise relationship, however; it gives only an approximate fit with real data. Notice that there is no viscosity term, which contributes to the imprecision of the equation. Imperfections aside, what are the implications of this relationship?

Because the velocity term is squared, the drag force increases very rapidly with increasing speed. Doubling the speed quadruples the drag, and since energy has to be supplied to overcome drag it follows that moving at high speed is very expensive. That is why in these times of diminishing oil supplies we are legislated into driving our cars more slowly.

Drag increases with the density of the fluid. Since water is over eight hundred times denser than air, the implication is that swimming is much more expensive than flying. Actually this is not the case; as we saw in Chapter 6, swimming is relatively inexpensive (Schmidt-Nielsen, 1972). Part of the reason for this is that swimmers are buoyed up by the water, whereas flying and terrestrial animals have always to counteract the force of gravity when they are moving. Swimmers can do little to decrease the density of the fluid that surrounds them, but fliers have some control because air density decreases with altitude. Migratory birds are known to fly at relatively high altitudes. Some shorebirds have been recorded at altitudes of five or six thousand feet (1,500–1,800 m), geese at nine thousand (2,800 m), and cranes probably up to fifteen thousand (4,600 m—Van Tyne and Berger, 1976). How birds deal with the problems of reduced oxygen levels and lowered temperatures at these altitudes is not immediately apparent.

Drag increases with the frontal area of the body (the maximum cross-sectional area) and with external projections from the body, and anything that can be done to minimize these projections reduces the overall drag. This principle is well exemplified in the tunas and their relatives, which are among the most highly modified fishes for high-performance cruising. The pectoral fins can be flattened against the body, and although they are thin they have a shallow depression in each side of the body to allow the fins to lie absolutely flush with the surface. The anterior dorsal fin can be completely collapsed into a longitudinal groove, and the paired pelvic fins can similarly be tucked into recesses. The gill cover (the operculum) also fits flush, and even the eyeball is modified so that it faithfully conforms to the shape of the surface. Because tunas keep the fins flush with the body during swimming, extending them only when changing directions, the drag forces are much reduced. The surface of the body is remarkably smooth, which reduces friction drag. Smoothness, together with the very gentle curvatures of the streamline shape, delays the separation of the boundary layer. It is likely that much of the fluid flow over the surface of the body is laminar.

In Chapter 10 we will return to the Mesozoic Era, but first we will take a look at how these principles of fluid mechanics have shaped the bodies of present-day fishes. They will be our guides to the body forms and swimming habits of ichthyosaurs.

Fishes: form and function

The fishes are the largest vertebrate group, with more species than any other, and present us with a dazzling array of forms. There are long-bodied fishes, like eels, that swim by flexing their bodies so that waves of undulation pass from head to tail, while others, like the seahorse, have inflexible bodies and rely upon the movements of their fins to propel them through the water. Paul Webb, one of the most prolific authorities on fish locomotion, helped make sense of this otherwise confusing array of body forms by pointing out three major specializations (there are others): cruising, maneuvering, and accelerating (1984). Most fishes are generalists, performing moderately well in each of the three basic areas, but some specialize in one area, almost to the exclusion of the other two. The pike, for example, with its long, slender body, broad tail, and large, posteriorly placed dorsal and anal fins, is specialized for acceleration. It can lunge from its hiding place among the reeds and snatch a passing fish in a flash of silver but, like the lion, it can only sprint for short distances. The butterfly fish cannot sprint like a pike, nor can it cruise—that is, keep up continuous swimming for an hour or more. But when it comes to maneuvering in and out of the coral it has few equals. The butterfly fish achieves such maneuverability by having a laterally compressed body that is disk-shaped when viewed from the side. The tuna, in contrast, is not very adept at maneuvering, nor does it accelerate as well as a pike does, but it can cruise at high speeds for hours on end, covering considerable distances in the process. Among the features that contribute to its outstanding cruising performance are its streamlined body and its narrow, crescent-shaped tail. The maintenance of a relatively high body temperature is probably also a factor. Many sharks are also specialized for cruising and they similarly have streamlined bodies and crescentic tails. But they are not in the same league as the tuna, because the shark's flexible body produces more drag than the tuna's stiff form does. (We will return to the differences between tunas and sharks later in this chapter.)

Fishes must create the thrust that moves them through water. Thrust can be generated by wave-like undulations and by back-and-forth oscillatory movements. Undulatory movements, one of the most common methods used by fishes, usually involve the body and tail. The salmon, for example, throws its whole body into undulatory movements. The waveform that moves down the length of the body, from the head to the tail, reacts against the resistance of the water to produce a forward thrust. Some fishes have fins that undulate and produce thrust. This is best seen by visiting a pet shop and looking at the median fins (the dorsal and anal fins) of certain tropical fishes.

Oscillatory movements sometimes involve side-to-side movements of a stiff tail, as in the tuna, but some fishes generate thrust by the oscillatory movements of other fins—the paired fins (the pectoral and pelvic fins). Fins that flap in this way need to be moved in different directions and consequently are connected to the body by a narrow base. In contrast, fins that undulate and fins that are used as hydroplanes—that is, for adjusting the swimming level—have a wide base. The undulatory fins can be distinguished from the others because they are supple rather than stiff. Of all these methods of producing thrust, we will focus on tail movements because they are of particular significance in interpreting ichthyosaurs.

Tails, tunas, and sharks

The tuna and it relatives, which include the mackerel, are collectively called scombroid fishes. Scombroids are characterized by having a stiff, lunate tail. The aspect ratio of the tail is remarkably high—about seven compared with only about two for the salmon. Other fast swimmers, including many sharks and cetaceans, have similarly shaped tails, but they do not have such a high aspect ratio. It has been suggested that this is because they lack the stiffness and strength of the scombroid tail (Van Dam, 1987).

The tuna and its relatives have streamlined bodies and their tails have high aspect ratios. This basic body plan is described as thunniform.

Because of its stiffness, achieved by a series of ossified fin rays, the scombroid tail moves as a rigid unit, without any appreciable bending.

Animals whose tails have a high aspect ratio and whose bodies are streamlined are described as *thunniform,* meaning tuna-shaped. Scombroid fishes are thunniform, and so are many sharks and cetaceans (Webb, 1988). One of the features of thunniform animals is that they have a particularly narrow caudal peduncle—this is the constriction at the base of the tail where it is attached to the body. A flexible joint is formed at the caudal peduncle that allows the tail to be moved freely from side to side (or up and down in the case of cetaceans). When the tail is moved from side to side, the posterior edge always trails behind the leading edge. The tail therefore acts as an inclined plane moving with a small angle of attack on every stroke. Each stroke is therefore a power stroke and generates a lift that is directed obliquely toward the head.

The side-to-side movements of the tail, because of the oblique direction of the thrust, tends to cause the body to move from side to side—the tail wagging the fish. This yawing movement is undesirable because it increases the total drag on the body, mainly because it interferes with the boundary layer and therefore increases friction drag. A strategy for dampening out these movements is to have a deep body, ideally one that is deepest at the center of mass (Webb, 1988). The possession of a tall dorsal fin at this same position also helps to increase the depth of the body.

Although tunas and certain sharks (lamnid sharks) are both described as being thunniform, there are important differences between them, both morphological differences (pertaining to body shape) and physiological ones (pertaining to body function). The most significant morphological difference, besides the tail, is that the body of the shark is flexible whereas that of the tuna is stiff. Since some of the shark's forward thrust of swimming is derived from the undulatory movements of the whole body, being flexible would seem to be an advantage. But the shark pays dearly for its flexibility because flexible bodies generate considerably more drag than stiff bodies—perhaps as much as between two and five times more (Webb, 1982).

The additional drag that accompanies lateral movements of the body is greatest at the tail end, where the displacement is the highest, and this explains why fast endurance swimmers have a narrow caudal peduncle. It also explains why having a stiff tail is more efficient than having a flexible one. The most efficient swimming strategy is therefore to have a stiff body and tail and to allow only the tail to move, as in the tuna. Why, then, have not the fastest sharks adopted this strategy? The reason probably has much to do with Young's modulus; cartilage, which comprises the shark's skeleton, is not as stiff as bone, therefore the shark is unable to have a stiff body.

While the stiffness of the tuna's body is attributable to the stiffness of bone and to the way that the skeleton is constructed, muscle action may also play a role. The tuna has both red and white muscle fibers and some of the red fibers, which we know are resistant to fatigue (Chapter 6), are attached to the vertebral column. It has been suggested that these fibers may be used to reduce skeletal mobility, thereby stiffening the body (Graham, Koehrn, and Dickson, 1983).

The significant physiological difference between sharks and tunas is that tunas are endothermic whereas sharks rarely elevate their temperatures above ambient levels. Indeed, the metabolic rates of sharks were generally believed to be lower than those of most other fishes (Webb, 1984), but the low rates may only reflect that most shark studies have been conducted upon the less energetic species (Webb, personal communication).[3] The higher activity levels of tunas give them higher food requirements, which means they spend most of their time cruising the seas in search of food. Sharks, in contrast, are more opportunistic predators, adopting a sit-and-wait strategy.

The Jurassic ichthyosaurs whose body outlines have been preserved show all the features of the thunniform body form: a lunate tail with a fairly high aspect ratio (above five in some species), a narrow peduncle, and a streamlined body form. Most other Jurassic ichthyosaurs have similar skeletal proportions, suggesting that they too were probably thunniform. We may infer from this that most Jurassic ichthyosaurs were specialized for fast cruising, and indeed this conclusion has been drawn by fish specialists (Webb, 1988) and by paleontologists (Massare, 1988). The same may have been true for Cretaceous ichthyosaurs, but most Triassic ones, as we will see later, were not thunniform.

A striking feature of the ichthyosaurian tail is that although it is externally symmetrical, it is markedly asymmetrical internally. This is because the lower lobe is supported by the down-turned portion of the vertebral column, while the upper lobe has no skeletal support. Such a condition is the exact reverse of that found in the shark, whose vertebral column extends into the upper lobe of the tail. The shark tail is said to be *heterocercal* ("other tail"), and the ichthyosaurian tail has accordingly been interpreted as functioning like a reversed heterocercal tail.

The traditional understanding of the heterocercal tail, one that has been with us for almost a century, is that it generates an upward as well as a forward thrust. The upthrust on the tail tends to rotate the shark about its center of mass, causing the head to pitch downward. This tendency is balanced by a lift force generated in front of the center of mass by the paired pectoral fins acting as inclined planes. These two forces balance one another, and the shark is therefore able to maintain a horizontal swimming level.

One of the reasons this mechanism made such good sense was that sharks were believed to be denser than seawater. Inactive sharks, deprived of the lift generated by the tail and pectoral fins, would therefore sink to the bottom, and sharks are often seen resting on the seabed. However, Keith Thomson, a paleontologist at the Academy of Natural Science in Philadelphia, has questioned this conventional wisdom (Thomson, 1976; Thomson and Simanek, 1977). One of his arguments was that most sharks are probably not denser than water but are, instead, neutrally buoyant. He also pointed out an anomaly in the mechanism. According to the traditional interpretation (which has yet to be proven), as a shark swims faster the upthrust on the tail increases, requiring a greater correcting force to be generated by the pectoral fins. This correcting force is achieved by increasing the angle of attack on the pectoral fins, but putting the fins in this position would also generate a larger drag force. It seemed illogical to Thomson that the drag on the pectoral fins should be increasing at the very time the shark is trying to increase its swimming performance, and he sought an alternate mechanism. The alternative that he arrived at has influenced subsequent studies on sharks and also on ichthyosaurs, and it therefore warrants our attention. But it must be pointed out that there are problems with the mechanism (Blake, 1983) and it has been rejected by many specialists in the field of swimming (Webb, personal communication).

By reviewing the results of other investigators, and by making some of his own observations on living sharks, Thomson concluded that the heterocercal tail does not generate a lift force that causes rotation about the center of mass. Instead, the heterocercal tail produces a forward thrust that is inclined downward and through (or close to) the center of mass. He reasoned that there would accordingly be no tendency for the shark to rotate about its center of mass. Instead there would be just an overall sinking effect, which would be counterbalanced by the lift generated by the pectoral fins. Be that as it may, the sinking effect generated by the tail would still increase with increased swimming speed. Counteracting the sinking effect would require a greater uplift to be generated by the pectoral fins, which was one of Thomson's objections to the traditional interpretation of swimming in sharks.

To understand how the shark's tail may work (this explanation is based largely on Thomson's work) let us first imagine that the vertebral column is not upturned but is, instead, perfectly straight. When the tail moves from side to side the stiff main portion of the tail (the dorsal lobe) would be bent back—partly by muscular action and partly by the pressure of the water—forming an inclined plane. The dorsal lobe would therefore generate a forward thrust directed toward the head. The ventral lobe of the tail, which is smaller, is at right angles to the dorsal lobe, and hence at right angles to the vertebral column. According to the usual textbook account, the ventral lobe, being more flexible than the dorsal lobe, is deflected by the pressure of the water when the tail is moved from side to side. The ventral lobe would therefore form an inclined plane whose thrust is directed upward, at right angles to the vertebral column. This hypothetical tail would therefore generate a forward thrust, due to the dorsal lobe, accompanied by a small upthrust, generated by the ventral lobe.

Suppose the vertebral column is now tilted upward, as it is in sharks. The dorsal lobe of the tail still gives a thrust toward the head but, because this segment is now tilted at an angle to the horizontal, its thrust is directed forward and downward. Similarly,

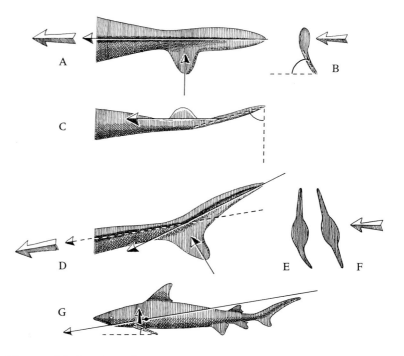

The possible function of the tail of the shark, explained in terms of inclined planes. These diagrams have been exaggerated for clarity. A, a hypothetical shark's tail, seen from the side, in which the vertebral column is straight. The ventral lobe of the tail is assumed to be more flexible that the (supported) main lobe. The arrow on the left shows the direction of the shark. B, in the posterior view, the ventral lobe is shown acting as an inclined plane. The arrow on the right shows the direction of movement of the tail. C, the view from above. D, the tail of a shark with the usual upward tilt of the vertebral column. The arrow on the left shows the direction of the shark. E, this posterior view shows the upthrust of the tail as a result of the deflection of the ventral lobe, which is the conventional explanation. The arrow on the right gives the direction of movement of the tail. F, Thomson proposes that upthrust is due to the twisting of the whole tail. G, the body of a shark depicting the net thrust of the tail as a downward and forward force acting through the center of mass. The pectoral fin acts as an inclined plane and generates an upthrust.

the ventral lobe still gives an upthrust, but this is reduced in size for it is no longer directed vertically upward but is inclined toward the head, because of the tilt on the tail. Thomson did not follow the usual interpretation and attribute the upthrust to the deflection of the ventral lobe. Instead, he attributed the upthrust to the twisting of the entire tail about it longitudinal axis. The twisting of the tail during each lateral sweep causes it to act as an inclined plane, generating an upthrust. According to Thomson, the net result of these two forces—generated by the dorsal lobe and by the entire tail—is a forward thrust that is directed diagonally downward, through, or close to, the center of mass.

Whether the ventral lobe generates an upthrust, according to the orthodox account of the heterocercal tail, probably depends on its size and flexibility. In the dogfish, which is often used in experiments because of its ready availability, the ventral lobe is fairly large and is more flexible than the dorsal lobe. As a consequence it generates an upthrust. I often demonstrate this to my students using a simple apparatus in which an amputated tail is wagged from side to side in a tank of water. The tail often rises so rapidly that it pops right out of the water. But the problem with experiments like this, which is similar to one described by Alexander (1965), is that it gives the impression that the heterocercal tail of all sharks functions in this simple manner. That this is not the case was shown by some experiments on amputated tails of the Port Jackson shark and the piked dogfish (Simons, 1970). These experiments clearly showed that the ventral lobe actually *reduced* the force of the upthrust. The explanation for this unexpected result is that the ventral lobe was not deflected by the water to form an inclined plane moving at an acute angle of attack. Instead, the ventral lobe was inclined in the opposite direction, and therefore it gave a downthrust. Although experiments such as these can be useful, one of their shortcomings is that amputated tails probably do not react in the same way as living tails do. This is because the shape of the tail appears to be modified by the action of muscles in the ventral lobe.

Much work needs to be done before the function of the heterocercal tail is properly understood, and it is obviously more complex than was formerly believed. There is, for example, a wide range of variation among sharks not only in the general shape of the tail, but also in the relative stiffness of the dorsal and ventral lobes. Dogfishes, for example, have fairly stiff dorsal lobes and flexible ventral lobes, but in other sharks, such as the hammerhead shark *(Sphyrna),* the ventral lobe is stiff while the skeletally supported dorsal lobe is compliant. There is also variation in the extent and direction of the vertical component that accompanies the horizontal swimming thrust. This may be an upthrust or a downthrust, or there may be no vertical component at all, and all this appears to be under the control of the individual shark (Thomson, 1976). The heterocercal tail therefore appears to be more versatile than was formerly considered.

The Sea Dragons

WE will begin this third and last chapter on things aquatic with a look at ichthyosaurs as living animals. When we have done with painting a picture of their life in the sea, we will turn our attention to a survey of their geographical and geological distribution, which spanned much of the Mesozoic Era.

Life in the water

Our understanding of swimming in living animals is far from complete, and we are able to observe them directly, so the possibility of understanding how ichthyosaurs moved through water seems remote. We can draw some tentative conclusions, but we must always remember that these are both speculative and open to revision.

In the past, for example, I took a traditional view of the reversed heterocercal tail, interpreting it as a device for generating a downthrust to overcome the natural buoyancy of the air-breathing ichthyosaur (McGowan, 1973c, 1983). However, Mike Taylor, who works on extinct marine reptiles at the Leicestershire Museum in England, has suggested that ichthyosaurs may not necessarily have been less dense than water (1987). We tend to think of air breathers as being less dense than water, and while this is certainly true of many vertebrates, ourselves included, it is probably not true for others. Many aquatic animals, including hippos, otters, and some seals, have densities that are similar to or slightly greater than that of water. Their heaviness is achieved primarily by their having very dense bones (Wall, 1983). The functional

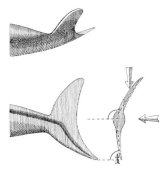

The possible function of the ichthyosaurian tail. Top, view from above. Bottom left, *side view.* Bottom right, *posterior view. The unsupported upper lobe is depicted here as being more flexible than the lower lobe. The horizontal arrow shows the direction of motion. The vertical arrows depict the vertical forces generated by the upper and lower lobes. Because the upper lobe is deflected more than the lower lobe, its vertical force is greatest and the tail generates a net downward thrust.*

significance of their high density is that they do not have to expend energy to remain submerged—they are neutrally buoyant. Cetaceans, surprisingly, do not similarly have dense bones. When they are swimming at the surface their density is less than that of the water, so they tend to float. They then expel most of the air from their lungs prior to diving and, because they tend to dive to great depths, the water pressure collapses the lungs so that the density of their bodies exceeds that of water. Although some ichthyosaurs may have dense bones, the Lower Jurassic ones I have examined do not, and in this regard they are similar to cetaceans. Taylor stressed the importance of the increasing water pressure as the ichthyosaur dived, and suggested that this would have soon destroyed any positive buoyancy and caused them to sink. Influenced by Thomson's re-analysis of the shark's tail, he concluded that the propulsive thrust of the ichthyosaurian tail was probably directed forward and slightly upward, through the center of mass. The thrust would have required minimal correctional adjustments of the pectoral fins—essentially only to correct for changes in buoyancy—and this would have reduced the drag generated by the fins. He argued that if an ichthyosaur flexed its body so that the tail was depressed, the line of thrust of the tail could have been raised above the level of the center of mass. This would have produced a powerful downward pitching of the head to initiate diving, of particular importance when the animal was at the surface.

Although Taylor's scenario for swimming and diving makes perfectly good sense, we must keep in mind that a reversed heterocercal tail is not an exact analogue for an ichthyosaur tail. This is partly because the skeletally unsupported lobe—ventral in sharks but dorsal in ichthyosaurs—is relatively much larger in ichthyosaurs than in sharks. Indeed, the two lobes have essentially the same area, whereas in sharks the unsupported lobe is only about half the size of the dorsal lobe. The other difference is that the skeletal support is cartilaginous in sharks but bony in ichthyosaurs and therefore likely to have been stiffer. The implication of these differences depends upon how the two lobes behaved in water. If the unsupported lobe of the ichthyosaur's tail was more compliant than the skeletally supported lower lobe, it would have been deflected so as to act as an inclined plane at an obtuse angle of attack. Under these circumstances the tail would have generated a net downthrust. The forward propulsive thrust of the ichthyosaurian tail would therefore not have been directed slightly up-

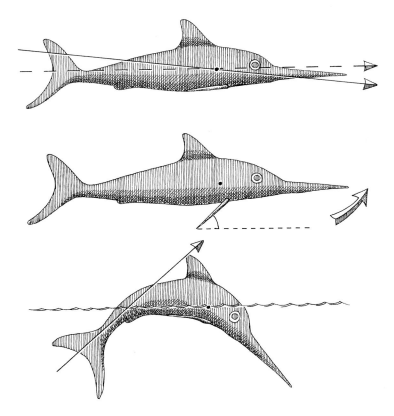

Top: *If the upper lobe of an ichthyosaur's tail was more flexible than the lower lobe, the tail would probably have generated a downward as well as a forward thrust (solid arrow). If the lobes were equally stiff, the tail would probably have generated only a forward thrust, with no vertical component (broken arrow).* Middle: *Changes in swimming level may have been effected by the pectoral fins acting as inclined planes.* Bottom: *Ichthyosaurs may have initiated a dive by flexing the body.*

ward, as Taylor suggested, but slightly downward instead. It is equally plausible that the unsupported lobe was stiff (as in the hammerhead shark), as stiff as the lower lobe. If it were, the tail would have generated no downward component at all and the forward propulsive thrust would have been directed horizontally. This is the situation in tunas, and most other bony fishes, and it seems the most likely alternative because once a diving ichthyosaur had lost its buoyancy, a tail that generated a downward thrust would be disadvantageous.

Taylor suggested that an ichthyosaur swimming at the surface may have initiated a dive by a downward flexure of the tail, which would have produced a strong downward pitching of the head. Once an ichthyosaur was below the surface, changes in its horizontal swimming level were probably effected by using the pectoral fins as inclined planes. The pelvic fins may also have been used for the same purpose, though a more likely function was that they acted as stabilizers. The pelvic fins, like the pectorals, were inclined obliquely downward from the horizontal, something like the flight feathers of a dart or arrow, and like flight feathers they were placed behind the center of mass. They would therefore have functioned as inclined planes when the body deviated from a straight path, generating correcting forces that would tend to bring the body back onto a straight course (see page 287). Since their orientation was oblique, they would have corrected for up-and-down movements as well as for side-to-side ones.

The thunniform shape—the body deep and streamlined and the tail with a high aspect ratio—was not typical of all ichthyosaurs. Some of the early ones, like the Middle Triassic genus *Cymbospondylus* ("hollow vertebra"), had a fairly long and narrow body and relatively long limbs. *Cymbospondylus*, obviously not adapted for cruising, may be visualized as an ambush predator. The Lower Jurassic species typically conformed to the thunniform pattern, but there was a fairly wide range of variation, as there is in modern sharks. There is now evidence, for example, that the long-snouted species *Leptopterygius tenuirostris*, which straddles the boundary between the latest Triassic and the earliest Jurassic, had only a modestly downturned tailbend, suggesting that its tail had a low aspect ratio (McGowan, 1989a). Perhaps *L. tenuirostris*, like *Cymbospondylus*, was more of a sit-and-wait predator.

Several Lower Jurassic species had particularly large tails, with fairly high aspect ratios. The aspect ratio of the tail of *Stenopterygius megacephalus*, for example, is much higher than that of the salmon but probably not as high as that of the tuna. In *S. hauffianus*, however, the tail is slender and has an aspect ratio comparable with that of a tuna. The forefin also has a relatively high aspect ratio, and it seems likely that *S. hauffianus* was a fast endurance swimmer. The possession of such a slender caudal fin makes me suspect that it may have had some internal supporting structure.

Eurhinosaurus ("broad-nosed lizard"), a large ichthyosaur that was a reptilian analogue of the modern swordfish, has exceedingly long and slender paired fins. Although no specimen has been found in which the body outline has been preserved, we can be fairly confident that these fins had high aspect ratios. The tail is also shown as being very steeply downturned, suggesting a high-aspect-ratio tail, but the veracity of this conclusion has been questioned by Jürgen Riess (1986). By reference to field photographs of specimens taken before and after they were assembled as skeletons, he has shown that the angle of the tailbend was a figment

Jurassic ichthyosaurs had large tails with high aspect ratios. The aspect ratio in STENOPTERYGIUS MEGACEPHALUS (top) *is higher than in the salmon* (bottom row, left) *but lower than in the swordfish* (bottom row, right). S. HAUFFIANUS (middle) *probably had an aspect ration as high as that of the swordfish.*

of the preparator's imagination! Unfortunately, this kind of problem may not be rare (McGowan, 1990). But there is a solution as far as the tail is concerned. The tailbend is formed by half a dozen or so vertebrae with wedge-shaped centra. If these wedge-shaped centra can be examined and measured in side view, it is a matter of simple geometry to calculate their angle of camber. An estimate of the angle of the tailbend can then be obtained by

EURHINOSAURUS, *one of the largest Holzmaden ichthyosaurs, is unusual for its abbreviated lower jaw. Research has shown that the tail is not as steeply down-turned as depicted in the skeleton* (top) *and in the restoration.*

adding these individual angles. These centra are seldom seen in lateral view, but I have been able to overcome the problem by X-raying the relevant section of the vertebral column using a CT-scanner of the kind used in hospitals to give three-dimensional images of the body (1989c). By using this technique I have been able to confirm that *Eurhinosaurus* did have a tailbend, but it was not so steeply downturned as it is usually shown in restorations.

The Middle and Upper Jurassic genus *Ophthalmosaurus*, like most of the Lower Jurassic ichthyosaurs, had a thunniform body, and the same appears to have been true for the Cretaceous genus *Platypterygius*. Most ichthyosaurs, then, appear to have been adapted for fast cruising, and it has previously been suggested that some may have been endothermic, like tunas (Chapter 6).

Knowing that most ichthyosaurs were fast swimmers gives us some clues about their feeding habits, but of course their teeth tell us more. Almost all ichthyosaurs have a dentition consisting of numerous conical teeth that are fairly sharply pointed. There was, however, a range of forms, from relatively slender and sharply pointed teeth to fairly stout, somewhat blunted ones. Massare (1987) has discussed how these different types of teeth may have been used for different prey species.

The teeth are so closely set that when the jaws are closed they mesh together, leaving no gaps. This type of dentition—found today in dolphins and seals—suggests a diet of relatively small, fast-moving prey. Dolphins and seals feed primarily on fishes, though some dolphins prefer squid. Such prey require a fast snapping action of the jaws—there is no time for chewing so the prey are swallowed whole—and most ichthyosaurs are modified for such rapid movements. The upper and lower jaws are long and slender, and the jaw muscles did not have to overcome a large inertia when they snapped the lower jaw closed. While there are serious limitations to what can be inferred from presumed muscle scars on bones (Chapter 4), such areas that have been identified on the lower jaw of ichthyosaurs lie close to the jaw joint (McGowan, 1973b). This suggests that relatively small contractions of the muscles would have brought about large movements at the tips of the jaws. In the same way only small movements with the finger and thumb can bring about large movements at the tips of a pair of tweezers. When muscles do not have to work against a large load they are able to contract through short distances almost instantaneously, so the jaws were probably able to close very rapidly.

Our suspicion that ichthyosaurs were active predators is confirmed by the remains of the gut contents that have been found in the abdominal regions of many specimens (Pollard, 1968). Coprolites, or fossilized feces, found in the same localities as ichthyosaurs have also been used to interpret their diets, but there is no way of confirming that the feces were left behind by ichthyosaurs—and not, say, by plesiosaurs or sharks. The Reverend William Buckland reported finding fish scales in the abdominal cavity of an ichthyosaur as long ago as 1836, and he identified them as belonging to the genus *Pholidophorus*. *Pholidophorus*, similar in size and appearance to the herring, was probably a fairly fast swimmer. More common than fish scales are the small hooklets that were once part of belemnites. These animals, classified as cephalopods, are relatives of the modern squid and had tentacles armed with suckers; the hooklets were part of these suckers. Cephalopods, like fishes, are wary animals that avoid capture, which brings us to the question of how ichthyosaurs may have located their prey.

Most ichthyosaurs have a remarkably large orbit, usually with a well-preserved sclerotic ring, which gives a good idea of the actual size of the eye. Since the sclerotic ring is large, filling most of the orbit, it is concluded that the eye was large too.

The function of the sclerotic ring is obscure, though the usual explanation given is that it protected the ichthyosaur's eye from compression during diving. But cetaceans lack sclerotic rings and many of them dive to considerable depths, and birds have sclerotic rings but very few of them dive underwater at all. Furthermore, the vertebrate eye is fluid-filled, and since fluids are incompressible, there is no need for any additional device to prevent the eye from being compressed. The primary function of the sclerotic ring may simply have been to give some rigidity to the eye; perhaps fluid pressure alone was not sufficient to maintain the spherical shape of such a large eye.

The large size of the eyes strongly suggests that sight was the predominant sense, and there is some evidence to show that the optic lobes of the brain were well developed. Cetaceans have good sight, as experiments with captive dolphins have shown, but their sense of hearing is far more important for finding food and for avoiding obstacles in the water. Their acute hearing is part of an elaborate echolocation system that involves the transmission, and reception, of short bursts of high-frequency sounds. Echolocation is also used by bats, and both systems have been subjected to intensive investigation. The question naturally arises whether ichthyosaurs echolocated too, but before we can hope to answer it we must understand how the system works.

When the sound waves emitted by a dolphin strike an object in the water they rebound, and the returning waves are received by the dolphin's ears. The impulses from the ears are then transmitted to the brain, where they are analyzed. Humans cannot echolocate, but we can determine the direction from which airborne sounds are coming. We probably do this by detecting the slight differences in signal strengths reaching the left and right ears, and also perhaps by the slight time difference between reception of the sound at the two ears. A sound to the right of our head, for example, not only sounds slightly louder to the right ear, but also reaches that ear slightly before it reaches the left ear. Our ability to detect these slight differences between right and left sides hinges on the fact that the two sound receptors, the inner ears, are isolated from one another, each being surrounded by the bony otic capsule.

When sound waves strike an object they may be completely reflected, completely absorbed, or a combination of the two, depending on the difference in densities between the object and the medium in which the sound waves travel. The greater the density difference, the greater the extent of reflection over absorption. This is why hotel rooms in buildings with concrete floors and walls are so much quieter than those in wooden buildings. Most of the sound waves falling on our heads are reflected away, because of the large difference in density between the flesh and bone of our heads and air. The only waves that reach the inner ear structures are therefore those that pass down the bony canal running from the earflap to the eardrum. Because water has about the same density as the body, most of the sound waves that reach an underwater swimmer now pass straight through the head. There is now barely any difference in signal strengths, or arrival times, between the two inner ears, and the direction of sound is difficult, if not impossible, to detect. This can be verified by conducting some simple experiments in a swimming pool.

Cetaceans have overcome this problem by surrounding each otic capsule with a layer of spongy tissue, insulating it from the rest of the skull so that there is no direct bone-to-bone contact. The bone of the otic capsule is also very dense and thick, which probably enhances its auditory isolation. The fact that it is so robust, and that it is not firmly attached to the rest of the skull, explains why the otic capsules of cetaceans are commonly found as fossils.

Examination of an ichthyosaur skull reveals that the two otic capsules are firmly braced against the cranium, and it appears that there was no room for an intervening layer of insulation (McGowan, 1973b). It is therefore doubtful that the otic capsules were insulated from one another, which makes directional hearing, hence echolocation, very unlikely.

The ability to echolocate requires not only directional hearing but also ears that are sensitive to sound. Sound travels in the form of waves—like the movement of ripples across a pond—and the speed of the waves is fixed for a particular fluid at a particular temperature. The speed of sound in air is about 730 mph or 1,070 feet per second (330 m/s), and the speed of sound in water is about 3,200 mph or 4,700 feet per second (1,450 m/s). The distance between wave crests, the wavelength, decreases with increased frequency, so high-pitched sounds have shorter wavelengths. The wavelength also increases with increased velocity, which means that the wavelength of a given sound would be longer in water than in air. The wavelength of a sound can be calculated by dividing the velocity of the sound by the frequency of the sound. Middle C, for example, has a frequency of 256 cycles per second (hertz, or Hz), so the wavelength of this sound, in air, is 1,070/256 = 4.2 feet (1.3 m). In water, middle C would have a wavelength of 4,700/256 = 18.4 feet (5.7 m).

When sound waves strike an object whose density is different from that of the surrounding fluid, they rebound, forming an echo, but only if the object is sufficiently large. It is easy enough to obtain an echo from a large brick wall, but a parked car or a lamp post would probably not produce an echo. To obtain an echo the object has to be about as wide as the wavelength of the sound falling upon it. Stated another way, high-frequency sounds are reflected by objects that are too small to reflect low-frequency sounds. This is why animals that are able to echolocate use high-frequency sounds. For example, to detect a herring-sized fish (0.5 ft; 15 cm) requires a sound frequency that is at least as high as 9,400 Hz (frequency = velocity/wavelength, therefore frequency = 4,700/0.5 = 9,400). Obviously an animal that echolocates has to be able to receive the high-frequency signals as well as transmitting them, and we have already seen that reptiles, in lacking a cochlea, are generally insensitive to high-frequency sounds (Chapter 4). This, taken with the fact that ichthyosaurs had their two otic capsules braced against the cranium, makes it very unlikely that they were able to echolocate.

Parent or cannibal?

The most remarkable examples of fine preservation among vertebrate fossils must surely be those ichthyosaurs that have been preserved with embryos within the abdominal cavity. In one rare instance a female seemingly died at the very instant of giving birth. Most ichthyosaurs with preserved embryos have been collected from the Holzmaden area of southern Germany, but there has been at least one similar find in England. The English discovery was made at about the same time as those in Germany and was reported in 1846 by Chaning Pearce. He described how he was removing some layers of shale from the pelvic region of an ichthyosaur when he came upon series of small vertebrae. He was astonished at what he saw, and when he removed more of the overlying shale he came upon a tiny lower jaw, together with other parts of a small skull. Such a singular discovery had to be treated with extreme caution, and Pearce wrote to the highest authorities in the land: Richard Owen and William Buckland. Both men gave their assurance that there was no reason ichthyosaurs should not have been viviparous (live-bearing), and Pearce proceeded, with caution, to share his findings with the scientific community.

Twelve years later the great German paleontologist, Friedrich Quenstedt, made a similar discovery but, instead of concluding that the smaller individual was an embryo, he decided that it had been devoured by the adult. As the quarries of Holzmaden continued to yield more and more ichthyosaurs, additional skeletons were discovered that contained small inclusions. Parent or predator? A controversy sprang up that continued well into the present century. In 1907 the German geologist Wilhelm Branca made the valid point that the two alternatives were not mutually exclusive. He suggested that those small inclusions that lay with their heads facing forward had been grasped by the tail and swallowed tail first, whereas those that faced toward the adult's pelvis were embryos, their long snouts being particularly suited for engaging the narrow birth canal. Branca's fellow countryman, Fritz Drevermann (1926), agreed in principle, but he argued that, although a fleeing youngster would initially be grasped by the tail, it would then be flipped around and swallowed head first, which is just how dolphins eat fishes. Inclusions that faced toward the pelvis of the adult were consequently interpreted as food items, whereas those that faced the other way were believed to be embryos, birth normally occurring tail first, as in the cetaceans. The debate, largely between German paleontologists, continued for many years. Wilhelm Liepmann (1926) questioned the significance of the orientation of small individuals within the bodies of adults, pointing out that, in domestic animals, tail-first births were almost as common as head-first ones. Drawing from human obstetrics, he also noted that embryos can be displaced into the abdominal region after the death of the mother and that birth can even occur

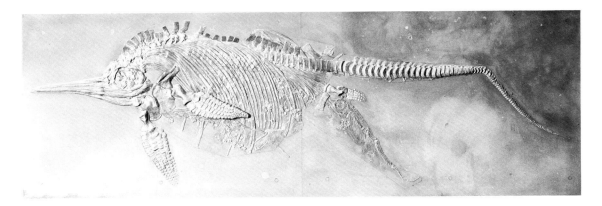

A female ichthyosaur with her offspring. She may have died at the very moment of giving birth, but it is more likely that the embryo was expelled after her death.

after death. He believed this last possibility to be the most plausible explanation for the occurrence of ichthyosaur skeletons that had been preserved during the actual process of birth.

Birth after death has been observed in whales that have been stranded on shore. In 1974, for example, a school of false killer whales was stranded on a sandy beach on the north coast of Tasmania (Scott and Green, 1975). After initial examination, the bodies were buried in the sand, but when the site was revisited some months later the bodies were found to be lying on the surface again, because they had not been buried deeply enough. Four of the females were pregnant, and in two cases a fetus was seen, partly protruding from the genital slit, presumably because of the mounting gas pressure in the carcass. Whales are normally born tail first, but in three of the dead females the offspring were in the head-first position. The relative position and orientation of a small individual within the body cavity of another is therefore not an infallible guide to whether it is an embryo.

How, then, can we reach a decision in the case of ichthyosaurs? Each case must be judged on its own merits after all factors have been assessed. The size of an inclusion can be a useful guide; young ichthyosaurs would normally be available as food items only after they had been born, therefore very small individuals are almost certain to be embryos. If the inclusion is very scattered it is likely to have been devoured, especially if it is found in association with stomach contents such as cephalopod hooklets. The orientation is probably of little significance, but the engagement of the head or tail in the birth canal, or the close proximity of either to the pelvic region, suggests an embryo. Conversely, an inclusion that lies toward the front of the body cavity of the adult, especially if its head is pointing backward, is more likely to be an item of food than an offspring. Although several skeletons containing inclusions have been found since Chaning Pearce's day, there are probably not more than a few dozen of them.[1]

The identification of embryos within the body cavity of an extinct animal is not only extremely unusual, it is also very useful. First, it establishes the sex of the adult beyond any reasonable doubt. On the basis of skeletal differences male dinosaurs have sometimes been distinguished from females; the identification of male hadrosaurs on the basis of their more elaborate crests is a notable example. But while these differences may well be due to sexual dimorphism (differences between the sexes), they cannot be confirmed as sexual differences. Female ichthyosaurs can be positively identified when embryos are present, and they can give some clues to sexual dimorphism. With the possible exception of *Eurhinosaurus,* it appears that ichthyosaurs, at least those from the Lower Jurassic of Germany, were not sexually dimorphic (McGowan, 1979c).

Pregnant ichthyosaurs also provide us with a unique opportunity to identify individuals that unquestionably belong to the same biological species. By comparing the skeletal anatomy of parent and offspring, we can obtain valuable information about the changes that occur during growth. For example, young ichthyosaurs, like the young of most other animals, including ourselves, have heads that look far too big for their bodies.

Because of their small size, young ichthyosaurs were born into a different ecological niche than that occupied by their parents. Immature ichthyosaurs therefore have more in common with the young of other species than they do with adult members of their own species. For example, mature individuals of the Lower Jurassic species *Stenopterygius quadriscissus* are characterized by being toothless, but their offspring have teeth, just like the offspring of other species (McGowan, 1979c).

Most of our attention has been focused on the Lower Jurassic ichthyosaurs, primarily because they are the most numerous and best preserved, and therefore the best known. But ichthyosaurs are found throughout most of the Mesozoic. A survey of their distribution not only give us an overview of their diversity, but also provides us with some insights into life in the Mesozoic seas.

Ichthyosaurs of the Triassic

Ichthyosaurs first appeared during the Lower Triassic, and although they extended into the Upper Cretaceous, they did not survive to the end of that period.[2] Their geological history spanned a considerable period—about 150 million years—but our knowledge of the group is rather patchy, partly because of the vagaries of preservation and partly because good exposures of sedimentary rock are uncommon. We therefore catch only glimpses of the evolutionary history of the ichthyosaurs, as if viewed through a series of rents in a curtain. Each glimpse, which sometimes represents a single fossil locality, spans but an instant of geological time and covers the smallest speck on the globe. Some glimpses are more revealing than others, and we get our first good view in the Middle Triassic exposures of Switzerland and the adjoining areas of northern Italy. Numerous well-preserved skeletons have been found in both localities, but, as Martin Sander (1989) of the University of Bonn has pointed out, they have not yet been adequately described. Almost all of the material belongs to a single species, *Mixosaurus cornalianus,* a small ichthyosaur that barely exceeded a yard (1 m) in length. The skeletons are beautifully preserved, some have traces of skin impressions, and there is one that appears to have embryos (Bürgin et al., 1989). The vertebral column was not downturned to form a tailbend as it is in Jurassic and Cretaceous ichthyosaurs. Instead, it is gently curved in the caudal region, and the curve corresponds in position with changes in the height and slope of the neural arches. The inference to be drawn is that there probably was a caudal fin, but it was not crescentic as in post-Triassic forms. The fore- and hindlimbs are already modified as fins, but they still retain the primitive five-fingered (pentadactyl) condition that is found in so many vertebrates, including ourselves. The forefin is about twice the length of the hindfin, and both are relatively broad. The slender rostrum is armed with numerous teeth, each one set in its own socket. The anterior teeth are sharply pointed, but they become blunt and quite peg-like toward the back of the jaws. In post-Triassic ichthyosaurs the teeth lie in a continuous groove instead of being set in individual sockets, and they have the same shape throughout the entire jaw, though they are often sharper toward the tip.

Mixosaurus ("mingling lizard"), which has also been found in the Lower Triassic (Callaway and Brinkman, 1989), is one of the earliest of ichthyosaurs. It is somewhat transitional between the

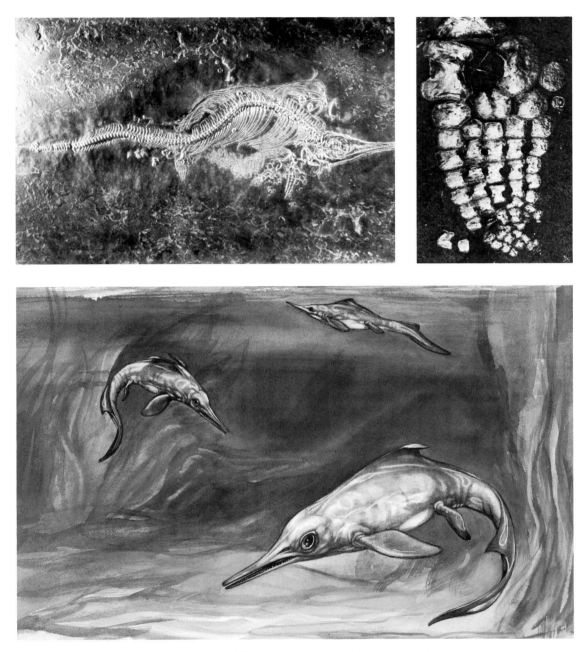

MIXOSAURUS, *from the Lower and Middle Triassic, is one of the earliest ichthyosaurs. It is small, lacks a tailbend, and retained the primitive five-fingered limb. As the restoration shows, however, the body was fish-like and the eye large, as in later forms.*

fully terrestrial ancestor that gave rise to the group and the highly specialized swimmers that appeared later in the fossil record. Ancestral (or primitive) features include the pentadactyl limb, the straight tail, and the socketed teeth. The advanced features that are associated with the aquatic way of life include limbs modified as fins, the enormous orbit, and the streamlined body shape.

So far we have been unable to identify the group from which the ichthyosaurs evolved, but this has not discouraged speculation, and most reptilian groups have been proposed at one time or another as the most likely ancestral stock. Until fairly recently the ichthyosaurs, which have a single (upper) temporal opening in their skull, were placed in the subclass Euryapsida, together with certain other reptiles. There is a growing consensus, however, that ichthyosaurs should be placed into the subclass Diapsida, together with lizards, snakes, and other related forms. This conclusion is partly based on some evidence that they possess the remnants of a second (lower) opening in the side of the skull. A discussion of this proposal, which was first made over a century ago (Baur, 1887), lies beyond the scope of this book. Suffice to say that the proposal does have some merit and, although it has not been universally accepted, it is becoming more widely used (Tarsitano, 1983).

Most of the specimens of *Mixosaurus* were collected from Switzerland and northern Italy, but the genus was very widespread. Specimens have been found in Middle Triassic deposits of Spitsbergen and Exmouth Island in the Arctic, Alaska, British Columbia, Nevada, China, Germany, Turkey, and Timor in the Indonesian Archipelago, and also in the Lower Triassic of British Columbia (Sander and Bucher, 1990; Callaway and Brinkman, 1989).

The Lower Triassic ichthyosaurs are neither as plentiful nor as well preserved as the skeletons from the Middle Triassic of Switzerland and Italy, and they are therefore not as well known. Indeed, it is largely because of their incompleteness that there have been such disagreements regarding their relationships. Some of the specimens have been identified as *Mixosaurus,* but others have been placed into genera of their own. One of these genera, *Grippia,* is from the Lower Triassic of Spitsbergen. I have followed von Huene and considered *Grippia* to represent another species of *Mixosaurus,* but the French paleontologist Jean-Michel Mazin, who has made a study of the ichthyosaurs from Spitsbergen, has argued that it does represent a distinct genus (1981). A somewhat similar situation pertains for *Utatsusaurus hataii,* from the Lower Triassic of Japan. Like *Mixosaurus* this is a small ichthyosaur, with a five-fingered forefin and a vertebral column that shows no evidence of having been downturned to form a tailbend (Shikama, Tadao, and Masafumi, 1978). Thirteen specimens have been found, but none are complete, so, as in the case of *Grippia,* it is not possible to make detailed comparisons with *Mixosaurus.* Whether these Lower Triassic ichthyosaurs really do belong to separate genera, or merely represent different species of *Mixosaurus,* will not be satisfactorily resolved until more material is found. But one thing they do have in common is their small size. By Middle Triassic times, though, some very large ichthyosaurs had evolved. Bearing in mind the short duration of the Lower Triassic—only five million years—this represents a rapid rate of evolutionary change. These giants lived and died in the seas that once covered western North America, and finding out more about them will take us on a visit to Nevada, one of their last resting places on Earth.

Nevada, a vast and desolate state in the American southwest, can give a traveller a feeling of utter loneliness. Much of the land is dry and dusty and will support little vegetation besides cacti and sagebrush. You can drive for hours without seeing another car, and one of the few signs that man has ever ventured beyond the road are the barbed-wire fences. In places they stretch for miles and are marked by ominous warnings that the area is used for weapons testing. Ichthyosaurs were first found in Nevada during the latter half of the nineteenth century (Leidy, 1868), but it was not until the beginning of this century that an intensive collecting effort was made. The central figure in these expeditions, from the University of California, was Annie Alexander, an enthusiastic and capable amateur. Her father, a pioneer of the Hawaiian sugarcane industry, moved his family to California when Annie was fifteen. She completed her schooling at a public high school in Oakland. The Alexander family travelled extensively during this period, spending relatively little time in their Oakland home.

The winter of 1900 saw her parents enjoying the sunshine of India, but Annie preferred to stay in Oakland. To amuse herself she attended some lectures given by John C. Merriam, assistant professor of paleontology at the University of California at Berkeley. She soon became deeply interested in the subject, and her love of the outdoors led to her participation in a paleontological field trip the following year. This was the first of many trips, and it was not long before she was organizing expeditions herself. Her sound financial position enabled her to sponsor expeditions to various parts of California and Nevada, and her continued generosity to the university led to the founding of the Museum of Paleontology at Berkeley in 1921.

Since her early twenties Annie had suffered from eye problems, but this did not seem to affect her in the field. She collected hundreds of specimens, including ichthyosaurs that were new to science. Her contributions to paleontology were considerable, and one is reminded of Mary Anning, who lived a century before.

The most complete ichthyosaur collected from Nevada during this period was *Cymbospondylus petrinus*, represented by several

CYMBOSPONDYLUS, *from the Middle Triassic of Nevada, reached an estimated length of thirty-two feet (10 m). The skull, which exceeds one yard (1 m) in length, is armed with sharp teeth. There is some evidence of a slight tailbend, and there may have been a caudal fin as in* MIXOSAURUS. *The tail may therefore not have been as straight and featureless as depicted in the restoration.*

specimens, including an almost complete skeleton and a well-preserved skull (Merriam, 1908). Indeed, *Cymbospondylus petrinus* is one of the best known of Triassic ichthyosaurs. It was a large and robustly built animal, with an estimated length of thirty-two feet (10 m)—about the length of a killer whale (*Orcinus*). The skull has a very long rostrum armed with conical teeth, which, unlike those of *Mixosaurus,* are all about the same size and shape. The teeth do not extend throughout the entire length of the rostrum; there appear to be none posterior of the level of the external nares. As in other Triassic ichthyosaurs, the teeth are set in individual sockets rather than in a groove. The orbit is not set as far back as it is in *Mixosaurus,* and there is consequently a relatively wide post-orbital region; the orbit is relatively small. An unusual feature of *C. petrinus* is the crest on top of the skull, at the level of the orbits. Another unusual feature, which is in fact unique, is that the occipital condyle, the knob at the back of the skull that articulates with the first neck vertebra, is concave rather than convex, as it is in other ichthyosaurs and in most reptiles.

Unfortunately, the post-cranial skeleton is not as complete as the skull. The fore- and hindfins are incomplete, but, judging from the few (proximal) elements that have been preserved, they appear to have been of approximately equal size. The body was very long, with more than sixty presacral vertebrae (the vertebrae between the skull and the pelvic girdle), which is probably more than in any other ichthyosaur (Merriam, 1908). Although the skull is large, about four feet (1.2 m), it is relatively small compared with the body. It is not known whether *C. petrinus* had a well-developed caudal fin, but there is evidence of a slight tailbend: Merriam (1908) reported that some of the centra of the caudal vertebrae are wedge-shaped. These vertebrae occur at a point where there is a change in slope of the neural spines. It seems likely that there was some sort of caudal fin, perhaps like that in *Mixosaurus.* Until quite recently *C. petrinus* was the only well-known species belonging to the genus, but Martin Sander (1989) has now described a large, moderately complete skeleton from the Middle Triassic of Switzerland as a new species of *Cymbospondylus, C. buchseri.* His report firmly establishes the occurrence of *Cymbospondylus* in Europe as well as in North America.

The early expeditions to Nevada uncovered only ichthyosaurs that were Middle Triassic in age, but Upper Triassic ones have since been found there, and these turned out to be the largest of all ichthyosaurs. The site where they were found can be visited by driving west across Nevada on Highway 50. The highway brings a traveller to the old silver-mining town of Austin, and a little beyond this is a small side road that runs south. This road follows the course of the Reese River, which has carved a wide, flat valley through the Shoshone Mountains. Fifty dusty miles down this road stands the tired little town of Ione—just a few wooden buildings and looking like a ghost town.

A few miles beyond Ione, the road begins to climb into the foothills. Cactus and sagebrush now give way to juniper and piñon pine, and the tumbledown buildings of a real ghost town come into view. This was once the silver-mining town of Berlin. There are dozens of other ghost towns in this part of the West, but Berlin is unique because it stands within the boundaries of a singular park—Berlin Ichthyosaur State Park.

Ichthyosaur State Park came into being largely through the efforts of Charles L. Camp, who was the director of the Museum of Paleontology at Berkeley. Fossils were first recognized in the area in 1928, and when Camp paid his first visit, in 1953, he was impressed by the richness of the site and staggered by the immense size of the ichthyosaur skeletons that had been found there. This was unquestionably an important paleontological area, but it also had the potential to be of considerable interest to the public. Some years previously a rich dinosaur quarry in Utah had been developed as a visitor center and named Dinosaur National Monument. Camp no doubt had this successful venture in mind when he discussed his ideas for an ichthyosaur park with the authorities of the State of Nevada. He volunteered his services to develop the site and spent several summers there with his family and students from the University of California. The remains of more than thirty ichthyosaurs were located, and a number of these were excavated for exhibition. A large building was erected over one of the quarries, to protect the skeletons from weathering and to provide visitors with a convenient facility for viewing the material.

The skeletons are Upper Triassic in age and therefore geologically younger than *Cymbospondylus*. While sharing some features with *Cymbospondylus*, these ichthyosaurs were obviously quite distinct, and Camp (1976) erected a new genus, *Shonisaurus*, for them, recognizing three distinct species. The commonest species, *Shonisaurus popularis*, is represented by thirty-seven individuals, but descriptions of the other two species are based upon very little material. *Shonisaurus*, named for the Shoshone Mountains, is the

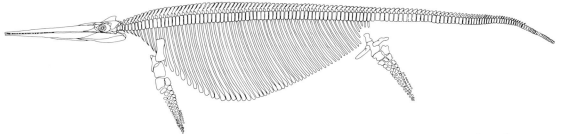

largest of all ichthyosaurs; it reached an estimated length of fifty feet (15 m), which is about half-again as large as *Cymbospondylus*. The skeleton is robust, like that of *Cymbospondylus*, the individual bones being thick and heavily built. The fore- and hindfins, which are of similar size, compare in length with the height of a man. Both are slender and have undergone reduction from the primitive five-fingered condition seen in *Mixosaurus* to having three main fingers and remnants of a fourth. There is uncertainty regarding the number of presacral vertebrae. Camp (1980) originally estimated that there were over fifty but revised the estimate to approximately ninety.[3] Massare believes that the number may be closer to that of *Cymbospondylus*, that is, about 60,[4] which agrees with the estimate of 64 given by Kosch (in press). There appears to have been a slight tailbend, perhaps like that of *Cymbospondylus*, suggesting that *Shonisaurus* probably had some kind of caudal fin. Although there are many skeletons, not one has a complete skull, and our knowledge of this region is based upon three incomplete specimens. As in *Cymbospondylus*, the rostrum is long and slender, and the orbit is small and is not set very far back along the length of the skull. There is a crest on the top of the skull, similar to that seen in *Cymbospondylus*. The teeth are set in sockets and are relatively small, as in *Cymbospondylus;* teeth appear to be restricted to the anterior portion of the skull. The occipital condyle is convex, as it is in all other ichthyosaurs (except *Cymbospondylus*). An unusual feature of *Shonisaurus* is that the ends of the ribs are expanded, a feature that has not been seen in any other ichthyosaur.

During the excavations, Camp became aware of something unusual: the majority of the skeletons were lying in the same direction. The skeletons that were close together were obviously parallel to one another, and compass bearings taken from the other skeletons showed that they, too, had the same north-south orientation. Camp concluded that the animals had probably been stranded on a sandbar, just as sometimes happens today with whales. He modified this view in later years, however, concluding that the evidence for a stranding was very weak—not only were relatively few of the skeletons buried at the same time but the sediments appear to have been deposited in deep water (Camp, 1980).

Three of the ichthyosaur genera so far considered— *Mixosaurus, Cymbospondylus,* and *Shonisaurus*—are represented by relatively large amounts of material and are consequently fairly well known, but the same is not true for most of the other Triassic genera, which includes *Shastasaurus* (after Shasta County, California); *Californosaurus* ("California lizard"), formerly referred to as *Delphinosaurus,* the "dolphin lizard" (Callaway, 1989); *Merriamia,* named for J. C. Merriam; and *Toretocnemus.* These four genera are from the Upper Triassic of California, though *Shastasaurus* also occurs elsewhere, including the USSR and Mexico (Callaway and Massare, 1989a). *Shastasaurus,* the largest of the four, is robustly built, like *Cymbospondylus,* and probably reached lengths of between thirteen and sixteen feet (4 and 5 m), though lengths of up to thirty-nine feet (12 m) have been reported (Callaway and Massare, 1989a), which is slightly larger than *Cymbospondylus.* The fore- and hindfins are both incomplete, as is the skull, so we do not have a clear picture of its anatomy. However, Callaway and Massare have given a composite restoration of the skeleton. They depict an ichthyosaur with a fairly deep and sharply tapering skull, fairly narrow fins, each with only three digits, and a slight tailbend.

The next largest genus, *Californosaurus,* appears to have been a much smaller species, reaching lengths of between six and ten feet (2 and 3 m). The body proportions appear to be similar to those of Jurassic ichthyosaurs, with about forty-five presacral vertebrae, but the individual vertebrae had relatively longer centra and were therefore more cylindrical than discoidal, so the body was probably relatively longer than in most Jurassic species. The vertebral column is thrown into a gentle curvature in the caudal region, similar to that seen in *Mixosaurus,* and there is good evidence of a tailbend (Callaway and Massare, 1989a). The forefin is not well preserved but is obviously narrow and appears to have only three digits; the hindfin is fragmentary.

Merriamia and *Toretocnemus* are both small, about a yard (1 m) long, which is about the size of *Mixosaurus. Merriamia,* the better-known genus, has narrow fore- and hindfins, each with only three digits and remnants of a fourth. The hindfin is only about half the length of the forefin, as in most Jurassic ichthyosaurs. From what remains of the skull it seems that the teeth may have been set in an open groove rather than in individual sockets, and this again is characteristic of Jurassic forms. The teeth are sharply pointed and, in contrast to *Cymbospondylus* and *Shonisaurus* but like *Mixosaurus,* they extend all the way back to the level of the orbit. There is little that can be said about *Toretocnemus* because the material is very incomplete. The fore- and hindfins are both narrow and appear similar to those of *Merriamia,* but there is little difference in length between the two fins.

A problematic and poorly known Triassic ichthyosaur that deserves mention is *Omphalosaurus,* first found from the Middle Triassic of Nevada and described by Merriam in 1906. The material is incomplete, comprising little more than a badly crushed skull, and it is so unusual, with its massive jaws, that Merriam placed it in a group of its own. In more recent times it became widely accepted as an ichthyosaur, but I have always had problems with this identification because the massive jaws are so atypical of the group. There are two vertebrae attached to the skull that are typically ichthyosaurian, however, and it is therefore difficult to deny that the specimen is an ichthyosaur. The heavy jaws, which would be suitable for crushing hard materials, suggest a mollusc-eating diet.

Leptopterygius tenuirostris, which has been mentioned earlier and which is usually regarded as being Jurassic, actually extends from the uppermost Triassic to the Lower Jurassic and should therefore be included here. In most regards this species, which is from southwest England, is typical of other Lower Jurassic ichthyosaurs, but unlike other post-Triassic ichthyosaurs, it appears to have had only a slightly downturned tailbend (McGowan, 1989c).

What conclusions can we draw from this wide and perhaps confusing array of Triassic ichthyosaurs? The first point is that early ichthyosaurs were not only highly specialized but were widely diversified, with more genera than in the Jurassic and Cretaceous Periods combined (Callaway and Massare, 1989b). Consider, for example, the differences between the small and relatively short-bodied *Mixosaurus,* with its wide fins and large orbit, and the long-bodied giant *Cymbospondylus,* with its narrow fins and relatively small orbit. A second point is that within each genus we can see some features that are primitive, that is, that

LEPTOPTERYGIUS TENUI-
ROSTRIS, *originally referred
to as* ICHTHYOSAURUS
TENUIROSTRIS, *extends from
the uppermost Triassic to the
Lower Jurassic. Named for
its long and slender snout,
this species appears to have
had only a slight tailbend.
One of the most complete
skeletons* (top) *is from
Street, Somerset. This isolated skull* (bottom), *which
is very well preserved, once
belonged to Thomas Hawk-
ins.*

are in a condition typical of many early reptiles, and some features that are advanced, that is, in a condition typical of some later ichthyosaurs. As we saw earlier, the five-fingered fin of *Mixosaurus* is a primitive feature, because early reptiles have five fingers, whereas the enormous orbit is an advanced feature, few reptiles having such a large orbit relative to the length of the skull. This occurrence of primitive and advanced characters within the same animal is referred to as mosaic evolution. It shows us that some features change fairly rapidly during the evolutionary history of a group, while others remain essentially unaltered. Mosaic evolution appears to be the rule rather than the exception.

Although Triassic ichthyosaurs are diverse, two basic types can be recognized: those with narrow fins and those with broad fins. A similar division can be seen during later geological periods, and the distinction has led most investigators to recognize two separate groups: the longipinnates (narrow fins) and the latipinnates (wide fins). While this classification has been useful in the past, it is probably of limited value and will not be used here (McGowan, 1978b, 1979c). For our next glimpse of the evolutionary history of the ichthyosaurs we have to return to the place of their discovery, the Lower Jurassic sediments of southwestern England.

Ichthyosaurs of the Jurassic

The Lower Jurassic provides us with our best view of the history of the group. The majority of ichthyosaurs that have been found are of this age, and we are presented with a wide array of species from both England and Germany.

Most of the ichthyosaurs found in England are from the Lower Liassic, though some Upper Liassic ones have also been found. The Upper Liassic ones are referred to *Stenopterygius* ("narrow-fin") and to *Eurhinosaurus* (McGowan, 1974a, 1974b, 1989a). The reverse is true for southern Germany, and the rich strata of the Holzmaden area are all Upper Liassic in age. The commonest English genus is *Ichthyosaurus,* a modest-sized ichthyosaur from the Lower Liassic. *Leptopterygius tenuirostris,* which ranges from the uppermost Triassic to the Lower Liassic, is of similar size, and it is also common, but few complete specimens are known. *Temnodontosaurus* ("cutting-tooth lizard"), a veritable giant, is quite rare, but complete skeletons have been found.

In contrast to their Triassic predecessors, the Jurassic ichthyosaurs have a distinct tailbend, and their teeth, which are pointed and numerous, are set in a continuous groove instead of in individual sockets. *Temnodontosaurus* has a relatively long and slender body compared with *Ichthyosaurus,* and the fore- and hindfins, which are approximately of equal length, are relatively narrow, having undergone reduction to three or four digits. *Ichthyosaurus,* in contrast, has a forefin that is usually twice the length of the hindfin, and the fins are relatively wide, having at least five digits and usually six or seven. There are three species of *Ichthyosaurus,* and the commonest, appropriately named *I. communis,* is known from dozens of skeletons that have been collected from Lyme Regis and Street. Examples of this dolphin-sized ichthyosaur, which seldom exceeds six feet (2 m) in length, may be seen in many museums. The smallest species, named *I. conybeari* in honor of the Reverend William Conybeare, was similar in size to the Triassic *Mixosaurus,* barely exceeding three feet (1 m) in length. This uncommon little ichthyosaur has a thin, pointed rostrum,

ICHTHYOSAURUS COMMU-
NIS, *the most common Eng-*
lish species, seldom exceeds
six feet (2 m) in length. The
skeleton (above) *was col-*
lected from Somerset by
Thomas Hawkins. Notice
the large tail fin and the
shark-like dorsal fin in the
restoration.

armed with numerous needle-sharp teeth, and has a relatively large orbit, like most ichthyosaurs. In contrast is the third species, *I. breviceps,* whose short rostrum gives it a bird-like appearance and makes the orbit appear enormous. *L. tenuirostris* used to be included in the genus *Ichthyosaurus,* with which it has much in common.

The fact that the rocks along the southwest coast of England have been searched for fossils for almost two centuries could lead

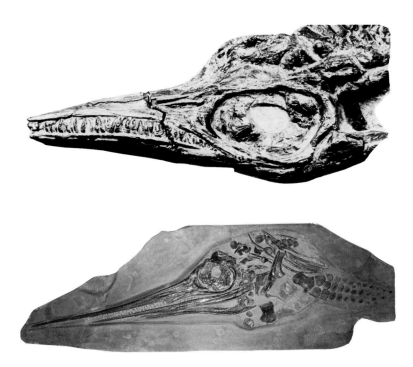

This complete skeleton of the rare species ICHTHYOSAURUS CONYBEARI, *from the Lower Jurassic of Lyme Regis, is about three feet (1 m) long (top). The skull of the short-snouted species,* ICHTHYOSAURUS BREVICEPS, *was also recovered from Lyme Regis, Dorset* (middle). EXCALIBOSAURUS, *from the Lower Jurassic of Somerset, England, was discovered in 1984 (bottom). The unusual feature is that the long, slender snout extends well beyond the lower jaw.*

one to believe that everything that could be found already has been found. However, new fossils do show up from time to time, and an especially interesting ichthyosaur was discovered in 1984. What was unusual about this particular specimen was that its lower jaw was only about three-quarters the length of the skull, giving it an extensive overbite reminiscent of the Upper Liassic ichthyosaur *Eurhinosaurus,* whose lower jaw is only about half the length of the skull. This specimen was sufficiently different from all other ichthyosaurs that a new genus, *Excalibosaurus,* was erected for its reception (McGowan, 1986a; 1989b). The name was derived from Arthurian legend, Excalibur being the name given to King Arthur's sword, which, like the sword-bearing ichthyosaur, was taken from the waters of the West Country. The possibility exists that *Excalibosaurus* may have been ancestral to *Eurhinosaurus.*

Temnodontosaurus is represented by several species, the commonest of which is *T. platyodon,* which is also the largest. This is the species to which Mary Anning's impressive first ichthyosaur belongs. *T. platyodon* reaches lengths of up to thirty feet (9 m) and is therefore only slightly smaller than the Triassic genus *Cymbospondylus.* Like *Cymbospondylus* it has a relatively long and robust body, and its long rostrum is armed with a battery of conical teeth. One of the largest ichthyosaurs of the Jurassic, it was probably a formidable predator. We have no direct evidence of what *T. platyodon* fed upon, but a wide range of prey animals were available, including the various squid-like molluscs, fishes, and other marine reptiles. Perhaps, too, they preyed upon plesiosaurs, or even upon other ichthyosaurs.

The most likely candidate as a hunter of other ichthyosaurs is *T. eurycephalus,* a species that is unfortunately represented by a single skull. The rostrum is very short, very much like that of *Ichthyosaurus breviceps,* and the skull is correspondingly deep. The teeth are large and well spaced and have bulbous roots, which suggests that they were firmly anchored in the jaws. These features, indicative of a powerful crushing apparatus, remind us of similar adaptations that we have seen in the giant flesh-eating dinosaurs. This particular individual has clenched between its teeth a basisphenoid bone, which is one of the bones from the base of the skull. It is likely that this element was displaced from its natural position during preservation, but it is possible that the bone does not belong to this animal but to some unfortunate victim.

While the majority of Lower Jurassic ichthyosaurs found in England are from the southwest, a few other localities have been productive. All of these lie upon a band of Liassic rocks that runs diagonally across the country from the Dorset coast in the southwest to the Yorkshire coast in the northeast. One of the more important localities is near the town of Whitby, in Yorkshire, where the cliffs look over the somber North Sea. These exposures belong to the upper division of the Liassic and are therefore geo-

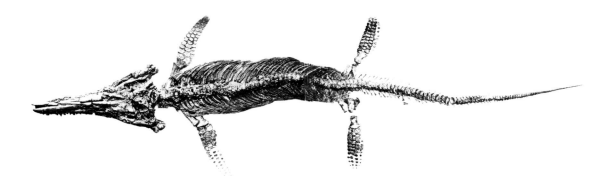

Top: *A complete skeleton of the giant English species* TEMNODONTOSAURUS PLATYODON *from Lyme Regis.* Middle: *The fins are relatively larger than in* ICHTHYOSAURUS COMMUNIS, *and the tail was probably smaller, as shown in the restoration.* Bottom: *The deep jaws of* TEMNODONTO-SAURUS EURYCEPHALUS, *also from Lyme Regis, suggest a powerful bite. It is possible that this species may have hunted other ichthyosaurs.*

logically younger than the Lower Liassic strata of Lyme Regis and Street. Ichthyosaurs are not the only fossils that have been found along this stretch of the Yorkshire coast. Ever since Roman times, people have searched the cliffs for jet, a semi-precious form of coal once prized for making jewelry. There was also an industry for the quarrying of alum (an aluminium salt), which may have also dated back to Roman times (Benton and Taylor, 1984). Limestone was also quarried for making cement, and it was during these operations that most of the ichthyosaurs were discovered. Quarrying activities ended around the 1860s and ichthyosaurs are seldom found there now.

Whitby has yielded relatively few ichthyosaurs compared with Lyme Regis and, although there are several complete skeletons, the preservation is rather poor and the material is consequently not well known. The bone is hard, black, and glossy, much like Whitby jet, and the specimens all appear to belong to the same species, *Stenopterygius acutirostris* (sometimes referred to as *Leptopterygius acutirostris*). These are large ichthyosaurs, at least as large as *Temnodontosaurus platyodon*. More research is needed on the relationships among these large ichthyosaurs; perhaps we will find out that the differences between them are relatively minor.

Quarrying has long since ceased in the English localities, but many of the quarries in southern Germany are still active. My colleague Rupert Wild, of the Staatliches Museum für Naturkunde in Stuttgart, estimates that about thirty-five ichthyosaurs are found each year, and that the total number of specimens that have ever been found from the area is in the order of three thousand. These remarkable ichthyosaurs have been widely distributed throughout the world, and most major natural history museums have at least one Holzmaden ichthyosaur in their collection.

In many regards the German species are much like their English counterparts, and some species actually also occur in the Upper Liassic deposits of England. Three genera are represented (possibly more): *Stenopterygius*, *Leptopterygius* ("slender fin"), and *Eurhinosaurus*. *Stenopterygius* is the most diverse genus, and several species are recognized (McGowan, 1979c). The most common species, the dolphin-sized *S. quadriscissus*, is unusual because although the juvenile specimens have numerous well-developed teeth, the adults have undergone considerable tooth reduction. Not only are the teeth few in number, but they are so small that they barely project beyond the jawline. The adults were therefore essentially toothless. Another species, *S. hauffianus* (which occurs in Eng-

STENOPTERYGIUS QUADRISCISSUS *is the most common German species; the specimen at the top is from the Holzmaden area. Adults are unusual in having very small teeth, which are just visible in the photograph of the skull. The caudal fin is especially large, as depicted in the restoration.*

land), has an abbreviated rostrum, and in this respect it is similar to the English species *Ichthyosaurus breviceps*.

The largest ichthyosaur from the German Liassic belongs to the species *Leptopterygius burgundiae*. Reaching lengths of over thirty feet (9 m), this species is comparable in size with some of the giants of the Triassic. There are some obvious similarities with *T. platyodon*, from the English Lower Liassic. Both species are relatively long-bodied with large fore- and hindfins, which are of about equal size and which may have been used for generating some of the propulsive thrust of swimming. The genus *Leptopterygius* has become something of a catchall for a number of species and further research is needed.

Eurhinosaurus, which is as large as *Leptopterygius*, is unusual for its shortened lower jaw, which is only about half the length of the skull. This gives it a striking resemblance to the modern swordfish, *Xiphias*. The fore- and hindfins are both long and narrow, and the tail, we have seen, is usually depicted as being steeply downturned, which is not the case. Although the tail was not so markedly downturned, the caudal fin that it supported was nevertheless large because the vertebral column posterior to the bend is quite long. The rostrum is armed with teeth throughout its length, so too is the lower jaw. The modern swordfish, in contrast, lacks teeth, but the rostrum is flattened dorso-ventrally (from above to below) and, like a sword, it has sharp cutting edges. The swordfish apparently uses its long rostrum for making sideways slashes as it swims through a school of fishes. Then, having injured some of them, it swims back and swallows them. Perhaps *Eurhinosaurus* used its long rostrum in a somewhat similar fashion, but with an up-and-down instead of a lateral movement (McGowan, 1979c, 1988).

This latest glimpse of the evolutionary history of the ichthyosaurs, which spans from the Lower Liassic to the Upper Liassic, an interval of about 25 million years, presents us with a wide variety of species. We do not know whether these species were restricted to England and Germany or were widely distributed, but the available evidence suggests that ichthyosaurs, like many present-day whales and dolphins, were widespread (McGowan, 1978b). It is fortunate that this view of ichthyosaurian life was so extensive because we do not get another glimpse until the Middle and Upper Jurassic, some 20 million years later, and for this we have to journey back to the southeast of England.

The ancient kingdom of East Anglia has been swept by many waves of invasion and immigration during its long history. The Flemings brought their weaving skills. The marauding Vikings robbed and raped. But the pragmatic Romans built roads and dug drainage ditches. At the time of the Roman conquest much of East Anglia lay under water and settlements were confined to a few low hills that rose above the swamps. Several factors contributed to the poor drainage. The land, which is often referred to as the Fens, was flat, much of it lay below sea level, and the heavy clay soil prevented the water from passing through it. The drainage ditches and causeways that the Romans built to link their garrisons lowered water levels, but when the Romans returned home the works fell into disrepair. The water levels rose again, and no serious attempts were made to resolve the problem until the seventeenth century.

Today, the clay that was once the undoing of the land is used in the manufacture of bricks. Referred to as the Oxford Clay, it was formed in the sea that covered Britain during the later part of the Jurassic and is rich in fossils. Peterborough, a sprawling town sixty miles (100 km) to the north of London and one of the major centers of the brick industry, has yielded enormous quantities of vertebrate fossils, mainly ichthyosaurs and plesiosaurs, together with some crocodiles. A major portion of the material that has been collected has been acquired by the British Museum (Natural History). The brick-works in the adjacent county of Bedfordshire have also been productive, as have a number of quarries in the county of Cambridgeshire. Several other localities, including some in Dorset, have also yielded material, and there is consequently no shortage of late Jurassic ichthyosaurs. But this wealth of material is rather disappointing because it comprises largely isolated vertebrae, fin elements, and skull bones, with relatively few associations (bones belonging to one individual) and probably no complete skeletons.

Although the Oxford Clay is often referred to as being Upper Jurassic in age, most of the sediments were formed during the last part of the Middle Jurassic, during the Callovian. The Callovian, once included with the Upper Jurassic, lasted for only about six million years, which is about half the duration of the Lower Liassic (Harland et al., 1982).

Middle and Upper Jurassic ichthyosaurs are not confined to England, their remains having been found in many other parts of the world, including Europe and the Americas, but again, the material is largely incomplete. The incomplete nature of the specimens, however, was no deterrent to the paleontologists who first studied them, and more than twenty species have been described. The practice of naming new species on the basis of isolated scraps of bone has cluttered the literature with hundreds of meaningless names. This practice, which was very common in the nineteenth century, unfortunately survives to the present.

A study of Middle and Upper Jurassic ichthyosaurs has shown that almost all of the specimens belong to the same genus, *Ophthalmosaurus* ("eye lizard"). The North American ophthalmosaurs have traditionally been referred to a separate genus, *Baptanodon*, but the differences are not major and I am inclined to refer the North American material to *Ophthalmosaurus*. A number of species of this genus have been described, but the differences between them are minor. The best-known species are the English *Ophthalmosaurus icenicus* and the American *O. discus*.

Like many of its Lower Jurassic predecessors, *Ophthalmosaurus* is a dolphin-sized ichthyosaur, reaching a length of about thirteen feet (4 m). The fore- and hindfins are both broad and much rounder than in earlier forms. The individual bones of the fins are round rather than angular, and they barely make contact with one another. This is in marked contrast to the closely packed arrangements of fin elements in most Lower Jurassic fins. As a consequence it is almost impossible to put a disarticulated fin back together again, and very few have been found in an articulated state.

One of the striking features of *Ophthalmosaurus* is the enormous size of its orbit, which inspired the generic name. The rostrum is fairly long and slender and there appear to be few teeth. Judged from the many hundreds of bones that have been identified, *Ophthalmosaurus* was a common ichthyosaur. It was also widespread, both geographically and geologically, having been found in France, Argentina, the United States, and the Canadian Arctic and spanning from the Middle Jurassic to the Lower Cretaceous.

Ophthalmosaurus was not the only ichthyosaur to swim in the late Jurassic seas but we have managed to catch only the occasional glimpse of its contemporaries. One such sighting occurred, quite by chance, in 1958. An excavator that was digging a drainage ditch in East Anglia happened to strike a large solid object, and this turned out to be a large ichthyosaur skull, about four feet (1.2 m) in length. What was left of the specimen was collected

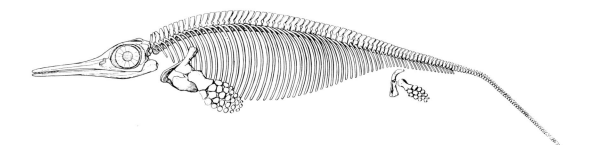

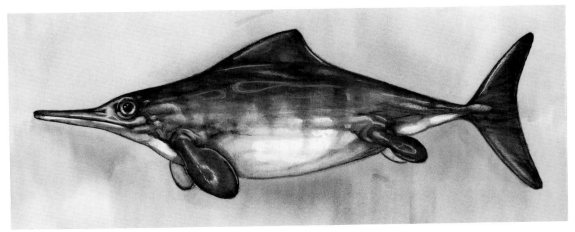

by the University of Cambridge, and I subsequently had an opportunity to take the specimen to Canada for preparation and study. In marked contrast to *Ophthalmosaurus*, the new ichthyosaur had large teeth and a relatively small orbit. Aside from a few vertebral centra and rib fragments, little remained of the postcranial skeleton. Especially unfortunate was the fact that nothing remained of the forefins, because these are especially useful in comparing one species with another. The loss of the hand inspired the name *Grendelius*, Grendel being the monster from the Fens whom Beowulf slayed by chopping off its hand in the Old English poem (McGowan, 1976).

Our impression of ichthyosaurian life during the latest Middle Jurassic and Upper Jurassic, which also lasted for about 25 million years, is one of abundance but little diversity. This contrasts with the Lower Jurassic, which also lasted for about 25 million years, when the seas thronged with a wide variety of species. Were

Although isolated bones of the Middle and Upper Jurassic ichthyosaur OPHTHALMOSAURUS *are very common, complete and even partial skeletons are rare. In the restoration,* OPHTHALMOSAURUS *is depicted as having a short body, a large tail, and rounded fins.*

ichthyosaurs really less diverse at this stage in their evolutionary history, or is this lack of diversity merely a reflection of the vagaries of the fossil record? We can never be sure, of course, but it looks as though the reduction in diversity is real because our samples have been drawn from a number of different localities. Furthermore, when we move up the geological series to the Cretaceous Period for our last glimpse of the ichthyosaurs, we find even less diversity. So it seems that the ichthyosaurs were at their peak during the early part of the Jurassic.

Ichthyosaurs of the Cretaceous

Ophthalmosaurus survived into the Lower Cretaceous and is represented by a single species, *Ophthalmosaurus cantabrigiensis.* This species appears to have been smaller than its Jurassic predecessors, but the material is incomplete and our knowledge correspondingly limited.

Immortalized in song by Vera Lynn during World War II, the white cliffs of Dover are part of an extensive wall of chalk that forms part of the coastline of southeastern England. These cliffs are actually the exposed edge of a large block of chalk that reaches a thickness of 1,600 feet (500 m) and gives rise to the well-drained and gently undulating countryside of southern England. The word *Cretaceous* comes from the Latin word for chalk. Thomas Henry Huxley, defender of the Darwinian faith, once delivered a famous lecture, "On a piece of chalk." He used the lecture to describe how this familiar type of rock was formed in the seas by the accumulation of the skeletons of microscopic organisms. Chalk cliffs have always been popular with fossil hunters and a number of ichthyosaurs have been found in them over the years, but these are mainly fragmentary. The time period covered by the chalk, which is Upper Cretaceous in age, is about 30 million years. Ichthyosaurs also occur in the Lower Cretaceous, of course, and many specimens, mainly isolated bones of the cranium and fins, have been found in fossiliferous sediments referred to as the Greensand.

Most Cretaceous ichthyosaurs can be identified with a single genus, *Platypterygius* ("broad fin"), which is widely distributed in both space and time. *Platypterygius* extends from the Lower Cretaceous to the early part of the Upper Cretaceous and has been found in North and South America, England, continental Europe, Russia, India, and Australia. A complete fossil record would probably confirm a world-wide distribution. At least fifteen species have been described, but most of these were founded upon inadequate material and there are no more than five recognizable species (McGowan, 1972). Even these are based largely on incomplete material, and the most adequately known species, *P. australis,* is represented by only a few skeletons. If we had several complete skeletons for each of the five species, we might find so much overlap among them that we would not want to recognize all of them as separate species. For the present, though, we must content ourselves with noting that *Platypterygius* was probably not a very variable genus, certainly not as variable as many of the Lower Jurassic genera.

What sort of an ichthyosaur is *Platypterygius?* First it is fairly large—about twenty-three feet (7 m) long, which is larger than *Ophthalmosaurus.* The skull is long and slender and, unlike that of *Ophthalmosaurus,* is armed with fairly large teeth. The orbit is relatively much smaller than in *Ophthalmosaurus* and the cheek is wider, both of these being primitive characters for ichthyosaurs. Until fairly recently it was thought that *Platypterygius* had broad fins, like *Ophthalmosaurus.* More material is now available, however, and a study made by Mary Wade (1984) of the Queensland Museum shows that the free part of the fin (that which lies distal to the humerus) is about three times longer than I had described, and probably a little wider too. The individual bones are tightly packed together, as they are in the Lower Jurassic forms, not widely spaced as in *Ophthalmosaurus.* Proximally they have a distinctive rectangular shape, like house bricks, and are arranged in regular columns, or finger rows, numbering up to ten. Distally, however, the individual elements vary in shape, some being rectangular, others polygonal, and others still are rounded, just like fins from the Lower Liassic. Indeed, the distal end of the forefin that Wade illustrates (from Boree Park) looks so much like that of a Lower Liassic ichthyosaur that if I were shown only that section of the photograph I would have identified it as having belonged to *Ichthyosaurus.* This is all the more remarkable when it is remembered that the Lower Liassic ichthyosaurs predate *Platypterygius* by about 100 million years. The fins are robust and the large size of the pectoral girdles with which they articulated suggests that they may have been used for paddling as well as for steering. The downturned segment of the tail appears to be relatively short compared with the length of the body, which suggests that the caudal fin was only of modest size, compared with the

PLATYPTERYGIUS, *the last of the ichthyosaurs, is found in Cretaceous rocks throughout most of the world.* Top right, *the skull of* P. AMERICANUS, *from the Upper Cretaceous of Wyoming.* Left, *an incomplete fin of* P. AUSTRALIS, *from the Upper Cretaceous of Northern Queensland.* Bottom right, *restoration based upon several specimens. The relatively small tail, which contrasts with Jurassic forms, may have been modified for cruising rather than for sprinting. Notice the remarkably long, narrow fins.*

relatively much larger caudal fins of the Lower Jurassic ichthyosaurs. But some caution is required here, because the vertebral column is rather poorly known and our knowledge of the tail is primarily based upon a single specimen.

Platypterygius seems to have been the sole representative of the ichthyosaurs during the latter part of the Cretaceous. It therefore had the dubious honor of being the last of its kind. The fact that it was not a very variable genus appears to confirm our suspicion that there had been a marked decrease in the diversity of the ichthyosaurs. This decline in diversity seems to have occurred sometime during the Middle Jurassic. The group continued on into the Upper Cretaceous but became extinct about 25 million years before the end of that period.

The Winged Phantom

DINOSAURS confound us with the enormity of their size. Ichthyosaurs beguile us with their beauty of form. But pterosaurs capture our imagination and carry it aloft on flights of fancy. They are the stuff of dreams . . . and of nightmares. They darkened the sky in Arthur Conan Doyle's *Lost World* and inspired the title *Dragons of the Air* for paleontologist H. G. Seeley's book on the subject (1901). The Reverend William Buckland wrote:

In external form, these animals somewhat resemble our modern Bats and Vampires. . . . Their eyes were of enormous size, apparently enabling them to fly by night. . . . It is probable also that the Pterodactyles had the power of swimming. . . . Thus, like Milton's fiend, all qualified for all services and all elements, the creature was fit companion for the kindred reptiles that swarmed in the seas, or crawled on the shores of a turbulent planet. (Buckland, 1836, pp. 223–224)[1]

Buckland was not the first to liken pterosaurs to bats—a reasonable comparison, given that both groups of animals fly on membranous wings formed of skin. But Kevin Padian of the University of California thinks the bat analogy has been taken too far, and that its pervading influence has distorted our view of the functional anatomy of pterosaurs (1983, 1987). The main point he makes is that whereas in bats the wing membrane extends to the hindlegs, which become incorporated into the wing, in pterosaurs the legs remained completely free from the wing. Not everyone agrees with this interpretation and, since the question has far-reaching consequences, especially for how pterosaurs

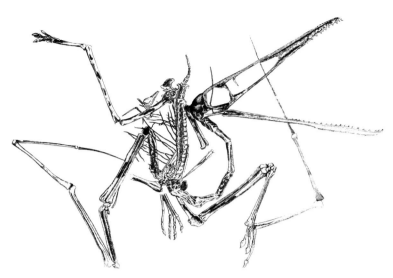

Left: *A reconstruction of the long-tailed pterosaur,* RHAMPHORHYNCHUS. *Notice that the wing membrane is supported by a single finger.* Right: *A short-tailed pterosaur,* PTERODACTYLUS, *from the Upper Jurassic of Germany.*

walked on land, we will have to weigh the evidence carefully. But first we need to become acquainted with our subjects.

The pterosaurs first appeared in the fossil record during the Upper Triassic Period (Wild, 1978), about 220 million years ago, and flourished until the close of the Cretaceous, some 155 million years later. Two separate groups are recognized, the Rhamphorhynchoidea, which have long, stiff tails, and the Pterodactyloidea, in which the tail is short. The long-tailed group evolved first, but it appears to have died out sometime during the latter half of the Jurassic, at about the time when the short-tailed group appeared. There was therefore an evolutionary succession from the long-tailed pterosaurs to the short-tailed ones.

Many of the early members of both groups were fairly small animals, about the size of pigeons, but some of the Cretaceous ones, all of which belonged to the short-tailed group, were gigantic. One of these, named *Quetzalcoatlus,* had an estimated wingspan of about thirty-five feet (11 m)—comparable to that of an executive jet—and is the largest flying animal known (Lawson, 1975). Appropriately enough, *Quetzalcoatlus* was discovered in Texas, where everything is said to be bigger.

Pterosaur remains have been found all over the world, but most skeletons have been collected from relatively few localities, and these are almost exclusively marine deposits (Wellnhofer, 1978; Langston, 1981). As was the case for the ichthyosaurs, it seems likely that the present-day distribution of pterosaurs merely reflects the distribution of good exposures of marine sediments. The fact that most pterosaurs are found in marine deposits suggests that most of them lived close to shore and earned their living from the sea, like present-day seabirds. Others, however, may have been completely terrestrial, and their burial in marine sediments may just have been coincidental. Some of the best-preserved specimens, including those with impressions of the wing membrane, have been collected from the fine-grained Upper Jurassic limestones of Bavaria, southern Germany. These same localities, near the towns of Solnhofen and Eichstätt, have yielded specimens of *Archaeopteryx,* the earliest bird; it was also from here that the first pterosaur was discovered. Pterosaurs have also been collected from the Lower Jurassic locality of Holzmaden and from the geologically slightly older coastal cliffs of Lyme Regis in England, but they are much rarer than ichthyosaurs at both of these localities.

An especially rich collecting ground for pterosaurs is the Niobrara Chalk of Kansas, which was laid down in the vast inland sea that covered most of North America for much of the Mesozoic. The Niobrara Chalk is Upper Cretaceous in age, and although it is richly fossiliferous most of the pterosaur material found there is fragmentary so there are no complete skeletons. Most of the material belongs to *Pteranodon,* a toothless giant whose wingspan reached about twenty-three feet (7 m) and which was the subject of a detailed study by the American paleontologist G. F. Eaton (1910). Kansas has been the most important collecting site for Cretaceous pterosaurs for over a century, but a new site that has recently been discovered in Brazil promises to be even more important. These deposits, which belong to the Santana Formation, are Lower Cretaceous in age and therefore geologically older than the Niobrara Chalk. This Brazilian locality has been known for many years for its remarkably well preserved fossil fishes, but its pterosaurs are no less remarkable. Many of the specimens are essentially uncrushed, which is most unusual for such fragile skeletons, and they represent a remarkable variety of specializations. So far a dozen new species have been described from this formation and we can look forward to more exciting discoveries in the future (Wellnhofer, 1985, 1987b).

Relatively few species of pterosaurs have been found, probably

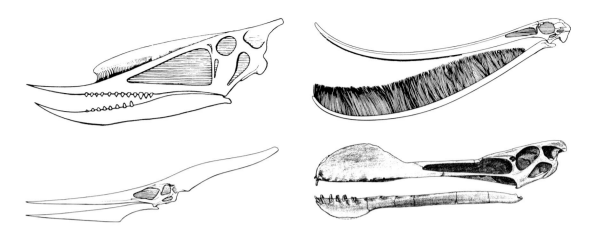

There is a wide variety of pterosaurs, some of which had remarkably specialized skulls. Clockwise, from top left: DSUNGARIPTERUS *from the Lower Cretaceous of China;* PTERODAUSTRO *from the Lower Cretaceous of Argentina;* TROPEOGNA- THUS, *from the Lower Cretaceous of Brazil; and* PTERANODON, *from the Upper Cretaceous of North America.*

fewer than a hundred, but they present us with a dazzling array of forms, from the unusual to the bizarre. *Dsungaripterus,* for example, from the Lower Cretaceous of Sinkiang, China, has pointed, upturned jaws that might have been used, like the bill of a shorebird, for prying open shells and other seashore animals. Even odder is *Pterodaustro,* from the Lower Cretaceous of Argentina, whose jaws support a fine-bristled straining device reminiscent of the baleen of whales. This sieve, like that of the whales, was probably used for feeding on plankton. Some pterosaurs were crested. In *Pteranodon,* from the Upper Cretaceous of Kansas, the crest is a huge structure that projects from the back of the skull, almost doubling its length, while in *Tropeognathus,* one of the remarkably well preserved pterosaurs from the Lower Cretaceous of Brazil, there is a crest on the snout and another on the mandible (Wellnhofer, 1987b).

There is occasionally some confusion as to which is the appropriate group name to be used for the flying reptiles. "Pterodactyl" is often used, especially in the older literature, because the first pterosaur to be found was named *Pterodactylus* ("wing finger"). This genus was actually described by the great French anatomist Georges Cuvier in 1809 (Wellnhofer, 1982, gives an excellent account of the early researches). The first name to be assigned to the entire group was the Order Pterosauria, so all the "dragons of the air" are properly referred to as pterosaurs rather than pterodactyls.

Bipeds or quadrupeds?

Pterosaurs, like dinosaurs and crocodiles, are archosaurian reptiles, all of which share several features, including the possession of two temporal openings in the skull. Pterosaurs also share many features with birds (also technically classified as archosaurs; see page 316), and the most significant of these are: an L-shaped shoulder girdle formed from a slender scapula (shoulder blade) and strut-like coracoid; a keeled sternum for the attachment of flight muscles (bats also have a keeled sternum, but it is very much smaller than it is in birds or in pterosaurs); a humerus with a prominent process at the proximal end, presumably for muscle attachments; and the possession of hollow bones. In some of the larger pterosaurs, like *Pteranodon,* there are openings in the bones called pneumatic foramina, as there are in birds.

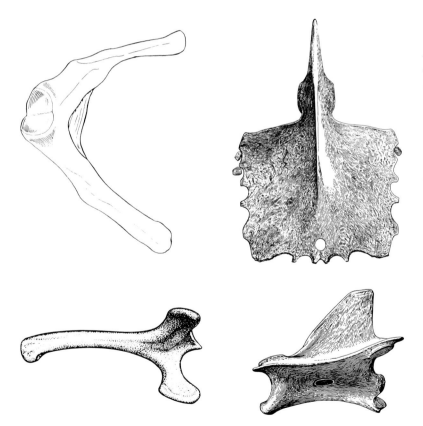

Many of the skeletal features of pterosaurs are similar to those seen in birds. Clockwise from top left, *the pectoral girdle of* PTERANODON; *the sternum of* PTERANODON; *a neck vertebra of* PTERANODON, *showing a prominent opening—a pneumatic foramen—in the side of the centrum; and the humerus of* RAMPHORHYNCHUS *(notice the prominent process at bottom right).*

The wing of a bat is supported by four of the fingers of the hand.

Bats and pterosaurs have a membranous wing, but in bats the wing is supported by most of the hand, whereas in pterosaurs it is only supported by one finger. This finger, the fourth, formed the leading edge of the wing and is remarkable for its great length. The fifth finger has been lost, and the other three—the thumb, index, and middle fingers—are of more normal proportions and are free of the wing membrane. All three fingers terminate in a sharp claw; they were probably used for grasping and perhaps also for climbing. A small bone arising from the wrist, the pteroid, was probably used to support a membrane at the leading edge of the wing. Birds and bats both have a leading-edge membrane (called the propatagium in birds), but in each case it is supported by an elastic ligament, not by a bone. You can see the propatagium for yourself the next time you prepare a bird for the oven—it is a particularly resilient web of skin that runs from the shoulder to the wrist. The propatagium serves to maintain the shape of the leading edge of the wing during flight, and since the leading edge of a wing is aerodynamically so critical, it is probably an important part of the flying apparatus.

Bats are excellent fliers, being remarkably maneuverable in the air, but once they are grounded they are usually fairly helpless.

The reason for this is that their legs are involved in the wing membrane and are consequently splayed out at the sides of the body. The hindlegs of other mammals, ourselves included, are placed vertically beneath the body, giving what is called an erect posture (see page 45): the head of the femur is offset from the shaft, and the hip socket faces downward as well as outward. In bats, the femoral head is essentially in line with the shaft, and the hip socket faces outward and slightly upward. Although some bats, including the vampire bats, are quite agile on the ground and can even jump, most bats do little better than a clumsy shuffle on all fours and therefore avoid being grounded at all costs. The toes end in sharp claws, which are used for clinging to footholds when roosting upside down.

What about pterosaurs? How well could they walk? This, as it happens, is one of the most contentious issues among pterosaur specialists.

The traditional view holds that in pterosaurs, as in bats, the hindlegs were incorporated into the wing membrane and were therefore directed laterally rather than vertically. This suggests that pterosaurs were obliged to walk on all fours, though some early restorations depicted a bipedal posture. At the other end of the spectrum is the viewpoint championed by Kevin Padian that the hindleg was completely free of the wing and that the posture was fully erect. Which is the correct view? As we tackle this question, bear in mind a few important points. First, if we had adequate data there would be no issue. This shortage of data is primarily attributable to the fragile nature of pterosaurs. Their bones are so thin-walled that they have usually undergone extensive crushing during preservation, which makes their interpretation difficult. Furthermore, although wing membranes have been preserved, their margins are not always obvious. They look nothing like the nice, clear-cut skin impressions seen in ichthyosaur remains, and they are always tantalizingly vague near the sides of the body. The last point is that pterosaurs probably did not all have the same hindlimb posture. All we can do here is make some educated guesses at what *might* have been; we cannot make definitive statements of what really *was*.

A central issue in the discussion of posture has been the anatomy of the pelvis, specifically the question of whether there was fusion between the left and right halves. In many animals, including theropod dinosaurs and mammals, the two halves of the pelvis are joined together ventrally and the hip sockets face outward and

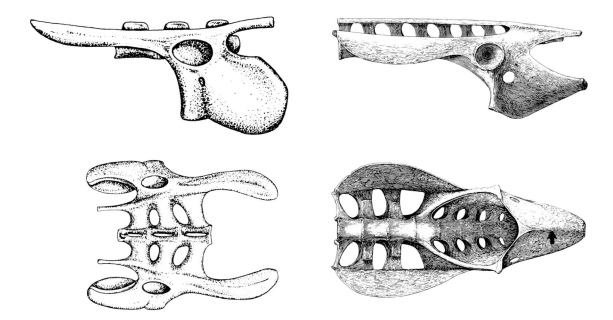

The pelvic girdle of the long-tailed pterosaur RHAMPHORHYNCHUS (left) *and of the short-tailed* PTERANODON (right), *shown in lateral view* (top) *and in ventral view* (bottom). *There is no ventral fusion in* RHAMPHORHYNCHUS, *but there is in* PTERANODON (bottom right; arrow).

somewhat downward. If the two halves became unjointed ventrally and swing slowly apart, the hip sockets would no longer face downward and outward but would come to face outward and upward. The leg posture would therefore change from erect to sprawled. The absence of ventral fusion in the pelvis of pterosaurs has therefore been interpreted as evidence for a sprawling posture.

Peter Wellnhofer of Munich, an authority on pterosaurs, believes that there was no ventral fusion—certainly in none of the long-tailed pterosaurs—and that the hip sockets therefore faced outward and upward (Wellnhofer, 1975, 1978). This was reinforced by his study of a particularly well preserved pelvis from the Lower Cretaceous of Brazil, belonging to the short-tailed genus *Anhanguera* (1988b). Here again there was no ventral fusion, the hip sockets faced outward and upward, and an erect posture seemed impossible. He noted that a similar orientation pertained in an incomplete pterosaur pelvis from the Lower Cretaceous of Queensland, Australia (Molnar, 1987). After making comparisons with other material, Wellnhofer concluded that the hindlimbs of pterosaurs were splayed out laterally, giving them a semi-erect gait.

On the other hand, Padian (1983) illustrated a pelvis of a long-tailed pterosaur in which the two halves were not separated ventrally, and from this he argued that the posture was erect. He also pointed out that in the short-tailed pterosaur *Pteranodon* the two halves are joined, a fact that nobody has questioned. Supporting evidence has recently been provided by yet another well-preserved pelvis of *Anhanguera* from the Lower Cretaceous of Brazil, described by Chris Bennett (1990). In this specimen there is evidence of ventral fusion and the hip sockets are described as being directed outward and backward. Bennett concluded that the animal could have walked bipedally, with the vertebral column inclined at about 60° to the horizontal. The shape of the sockets also suggested that the legs could have been held out laterally from the sides of the body, as during flight.

What about the position of the head of the femur? Does that help decide the matter? Happily there is no disagreement among specialists on this point—the femoral head is neither in line with the shaft, as in bats, nor set off at a right angle, as in birds. Instead, it is somewhat in between, though it is angled more nearly in line with the shaft than at right angles to it.

What should we make of all this? I think that too much has been made of the question of pelvic fusion. The equation that has been used is: no fusion = hip socket faces outward and upward = laterally splayed hindlegs = no bipedal locomotion. However, birds lack a fused pelvis and many have hip sockets that face outward and upward, but they are nevertheless bipedal (though, admittedly, the head of the femur is somewhat different from that of pterosaurs). There is actually a sound functional reason why birds lack a ventrally fused pelvis, and it has to do with the evolution of balance.

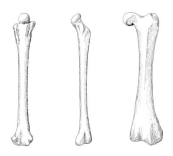

The left femur of a bat (left), a pterosaur (middle), and a bird (right), all shown in frontal view. The head of the femur is in line with the shaft in the bat and set off almost at right angles in the bird, but is intermediate in its position in the pterosaur.

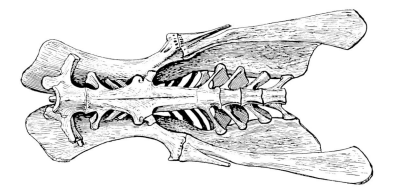

The pelvic girdle of a modern bird, seen from below. The left and right halves of the pelvis are splayed and there is no ventral fusion.

The earliest bird, *Archaeopteryx,* had a long, bony tail, a tail that would have been used to counterbalance the weight of the front part of the body about the hip joint, just as in bipedal dinosaurs (Chapter 4). Modern birds lack a long, bony tail and therefore do not have such a counterbalancing mechanism. They are able to balance because their heavy abdominal organs—primarily the gut—are centered over their legs (Young, 1962; Galton, 1970). The organs fit directly under the pelvis, the two halves of which are turned outward on the ventral aspect to make room for them.

One of the most remarkable features of a bird skeleton, which is also related to posture, is the shortness and stiffness of the trunk region. The backbone is so short that the pectoral and pelvic girdles almost touch, and most of the vertebrae are fused together and stiffened by ossified tendons. The functional significance of all this is that a short, stiff trunk makes it easier for the bird to balance its weight about the hip joint. The balancing movements are brought about by muscles that run between the pelvis and the femur. For example, the tendency for the bird to topple forward is resisted by muscles that run from the back of the pelvis to the back of the femur, the hip joint serving as a fulcrum. These postural movements are obviously easier to control if the body moves as a single (stiff) unit and if the balancing muscles are required to act over a small distance (the short trunk).

Pterosaur evolution seems to have paralleled that of birds, long-tailed forms having given rise to the short-tailed ones. There may also have been parallels in their respective postures. The rhamphorhynchoids had rigid tails, stiffened by long, bony processes of the vertebrae, as seen in certain dinosaurs like *Deinonychus.* Regardless of the role that the tail played in the air, it is difficult to conceive that it was not an important balancing organ on the ground. It therefore seems likely that at least the long-tailed pterosaurs were bipedal.

C. J. Pennycuick, a pioneer in the field of avian flight, rejected this line of reasoning on the grounds of the orientation of the hip sockets (1988), but we have already dealt with that objection. He then went on to discuss *Pteranodon,* pointing out that although it had a ventrally fused pelvis, which may have permitted bipedal posture, standing upright would have been impossible because it lacked the stiff backbone seen in birds. However, *Pteranodon did* have a stiff vertebral column, with extensive vertebral fusion, and it was also remarkably short.

There do not appear to be any good grounds for ruling out the

The skeleton of a modern bird has a short, stiff trunk and is almost tailless. Balance is maintained because the heavy abdominal organs are centered over the legs.

possibility that pterosaurs may have been bipedal. Padian (1983) went further and argued that the joints in the wing would not have permitted pterosaurs to walk on all fours. In any event, there was probably a wide range of locomotor abilities among pterosaurs, just as there is in living birds. This ranges from the ungainly belly shuffle of loons, through the waddling gait of geese, to the fleet-footed sprint of the roadrunner. So much for the legs of pterosaurs, but what of their wings?

Body plans and wingspans

Although a certain degree of flexibility is permissible in a wing, an animal cannot fly if its wings completely lack stiffness. Birds achieve stiffness through the rigidity of their feathers. Bats have a very resilient wing membrane, but it is stretched between the outspread fingers and the legs, so that it remains taut during flight. Pterosaurs also had a membranous wing but the only skeletal support was along the leading edge. If the hindlegs had been attached to the membrane they could have helped to keep the trailing edge taut, but what if the legs were completely free of the wing, as Padian advocates? This could present a problem, but Padian, referring to the work of Wellnhofer (1975c), pointed out that the membrane was stiffened by numerous fine fibers that spanned across the wing from the leading edge to the trailing edge, that is, across the chord of the wing. Pennycuick (1988) objected that what Padian thought were fibers were not structures at all but merely wrinkles, caused by the relaxation of the elastic wing membrane; he concluded that this explained why fibers could not be seen along the broken edges of specimens that showed the wing membrane.

When I started writing this chapter I thought that Pennycuick was right and that there were no fibers, just wrinkles. But then Jeremy Rayner, a specialist in animal flight at the University of Bristol, sent me some close-up photographs that changed my mind. The photographs showed the trailing edge of the wing membrane of a particular specimen of the long-tailed pterosaur *Rhamphorhynchus,* which has been illustrated elsewhere (Wellnhofer, 1975b, plate 17; Rayner, 1989, fig. 5a). Two features were of particular interest:

1. Some of the fibers were markedly curved as they approached the edge of the membrane. The curves are

difficult to reconcile with their being wrinkles, but they make sense if they are indeed fibers.

2. In some parts of the membrane the trailing edge has a slightly frayed appearance. Close inspection reveals that this seems to be due to some fine fibers that extend just beyond the edge.

While persuaded that the fibers are real and are not just due to wrinkling, I have difficulties visualizing how they could have stiffened the membrane because they are so very fine—not much thicker than human hair. And of what material might they have been made? Padian (1983, p. 220) suggested that they may have been cartilaginous, but cartilage lacks stiffness (Chapter 1). The only other likely materials are bone and keratin. Feathers are made of keratin, but their stiffness is largely due to their tubular construction and to their interlocking system of hooklets. Pterosaur fibers lack such complexity and appear to be simple rods. If these were of keratin, like hair, they would lack stiffness. Even if they were made of bone, which is stiffer than keratin, they would still lack stiffness because they are so thin.

Now it is true that composite materials, like fiberglass, which are made of fibers embedded in a second material, are tough, but they are not necessarily stiff too. Perhaps toughness (resistance to tearing) was the main function of the fibers. They were very closely packed together, though, and it seems that they would have increased the stiffness of the membrane in the direction of the chord, even though they were too fine to have made the whole membrane stiff, like the wing of a bird.

In 1989 a remarkable report was given of what appears to be a fragment of pterosaur skin in which it was possible to see microscopic details of its structure (Martill and Unwin, 1989). The skin was associated with the ulna, a bone in the forearm, of a short-tailed pterosaur from the Lower Cretaceous Santana Formation. There was no evidence of any stiffening fibers but there did appear to be muscle fibers and a rich blood supply. The latter feature suggested that the wing membrane may have been an important site for shedding excess heat, as it is in the bat, and this is consistent with the view that pterosaurs were warm-blooded. The fact that no fibers were found tends to open up the question of their existence again. Perhaps the fibers did not attach directly to the bones of the wing but to the muscles and connective tissue overlying them, which would explain their absence from the vi-

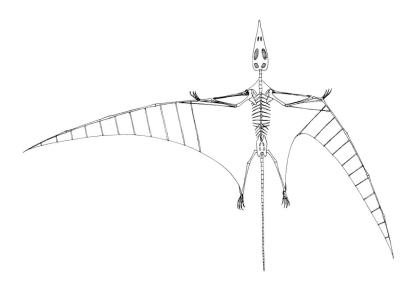

A reconstruction of the wing membrane of a long-tailed pterosaur, as proposed by Pennycuick. An elastic ligament is shown running along the trailing edge of the membrane, which is anchored to the ankle. Additional ligaments extend from this ligament to the leading edge. All of these details are conjectural.

cinity of the ulna. Perhaps fibers were not present in all pterosaurs—the skin came from a short-tailed pterosaur whereas the fibers discussed earlier were associated with the long-tailed *Rhamphorhynchus*.

We can conclude from all this that at least some pterosaurs had fibers embedded in the wing membrane and that the fibers may have afforded some stiffness in a chord-wise direction. Pennycuick (1988) suggested that an elastic ligament extending from the wing tip to the ankle served to stabilize the trailing edge and also to anchor a series of additional elastic ligaments that stretched across the wing to the wing finger. The tension in the trailing-edge ligament could be increased simply by moving the legs closer toward the body, and greater tension in the trailing edge, in turn, would have increased the tension in the additional ligaments. This all makes perfectly good sense, but is there any evidence for an ankle connection?

Pennycuick showed a photograph of the short-tailed pterosaur *Pterodactylus* in which the trailing edge of the wing membrane is clearly seen converging toward the lower leg region. Whether it actually attached to the ankle is not known, but it certainly appears to have attached somewhere on the lower leg. He also presented two pieces of circumstantial evidence to support an ankle connection, but this evidence pertains only to the long-tailed pterosaurs.

The outer toe of long-tailed pterosaurs differs from the others in being longer, lacking a claw, and being splayed out at an angle from the ankle region. These modifications suggest that it served a different function from that of the other toes. There is a somewhat similar structure in bats, though it is not a true toe but a single rod of bone, and it is on the inside rather than on the outside of the ankle. This bone, called the calcar, attaches to a segment of the membrane that runs between the legs. By analogy, it seems reasonable that the specialized pterosaur toe also attached to the wing membrane, but to its outer edge.

The second piece of circumstantial evidence is provided by a specimen of *Rhamphorhynchus*. In this specimen both legs have been drawn up toward the wings, ankles first, suggesting to Pennycuick (1988) that the ankles were attached to the wing membrane.

Wellnhofer (1987a) gave a detailed account of the wing membrane of the short-tailed genus *Pterodactylus*, based upon a particularly well preserved specimen in Vienna to which Pennycuick also referred. He concluded that the membrane attached to the lower leg rather than to the body, but probably not to the ankle. Rayner (1989) also discussed this specimen, but he concluded that the membrane attached to the femur, just above the level of the knee.

The evidence is not substantial but we can conclude that the wing membrane probably was attached to the hindleg and that it may have attached to the ankle in the long-tailed pterosaurs. It also seems reasonable that there was a system of elastic ligaments, as proposed by Pennycuick, though there is no direct evidence for such a system. Tension in these ligaments would have caused the wing to collapse toward the body, but this tendency could have been offset by the action of a wing-extensor muscle spanning between the shoulder and the wrist. As we will see later, there is evidence for a wing-extensor muscle in *Pteranodon*. This muscle probably attached to the prominent bony process that characterizes the pterosaurian humerus. When this muscle was fully contracted the wing would have been fully extended, whereas its relaxation would have permitted the wing to be folded toward the body.

Before proceeding to the basic principles of flight, let us review the conclusions we have reached about the pterosaurs. It seems that pterosaurs were probably capable of bipedal locomotion. This does not necessarily mean that none of them walked on all fours, and a wide range of locomotory abilities seems likely. The wing probably had fibers embedded in its membrane (at least in some pterosaurs), and these may have made it stiffer and less prone to tearing. There may have been an elastic ligament running along its trailing edge—though there is no evidence for this—which attached somewhere on the leg. In long-tailed pterosaurs, which have an elongated outer toe, the attachment was probably to the ankle. In short-tailed pterosaurs, which have a smaller outer toe, the attachment may have been higher than the ankle. The shape of the wing was probably maintained through the interaction among the tension in the elastic ligaments within the membrane, the tension generated by a wing-extensor muscle, and the movement of the hindlegs.

A flying lesson

Most flying machines are heavier than air and require a lift force to keep them airborne. This lift force is almost invariably provided by an airfoil, whether the flier be a locust or a Lear jet. The major differences among flying machines, both living and man-made, is in the power source used to make the airfoil work. Most aircraft have fixed wings that are driven forward through the air by engine power. Most flying animals, in contrast, have movable wings that are driven through the air by muscle power. However, there are some aircraft, called gliders or sailplanes, and some birds, called soarers, that fly on fixed wings, using moving air currents as their souce of power.

The terms *soaring* and *gliding* are often used interchangeably, but they do not have the same meaning and we need to distinguish between them. During soaring, which takes place in nonstationary air, energy is extracted from the atmosphere and this energy is used to keep the flier airborne. Gliding, in contrast, is passive in that no energy is extracted from the air and height is continuously lost during flight.[2] A vulture or a sailplane, for example, having gained height by soaring in a rising column of air, may leave the thermal and glide across country, losing height as it travels.

During the last decade a new approach has been offered to the study of the flight mechanics of animals, particularly flapping flight, that analyzes the vortices shed by the wings during flight. Pioneered by Jeremy Rayner, this methodology uses such ingenious devices as minute, helium-filled soap bubbles to track the path of the vortices left in the wake of a bird flying in a wind tunnel situation (Rayner, 1979; Spedding, Rayner, and Penny-cuick, 1984). This approach offers a promising research tool for quantifying the forces involved in animal flight, but it does not lend itself to ready comprehension. I will therefore use traditional methods for describing the principle of the airfoil.

Our knowledge of the way an airfoil works dates back more than two centuries, to the work of Daniel Bernoulli. Bernoulli found that when the velocity of a fluid increases there is a reciprocal decrease in pressure. This can easily be demonstrated using a pair of those colored glass balls used to decorate Christmas trees. If the balls are held about half an inch (1 cm) apart and air is blown through the gap between them, they will move closer together and touch. The explanation is that the air flowing over the two curved surfaces of the balls is speeded up, and so it experiences a reduction in pressure; since the rest of the air surrounding the two balls is at normal pressure, they are forced together. The airfoil, which usually takes the form of a wing, exploits Bernoulli's principle by having a convexly curved upper surface and a flat or concave lower surface. When the airfoil is moved through the air, the difference in speeds of flow over the two surfaces produces an area of reduced pressure next to the upper surface and an area of increased pressure next to the lower surface. As a consequence the airfoil experiences an upthrust, or lift. This lift increases with the speed and eventually the aircraft becomes airborne.

The lift force, which acts at right angles to the direction of motion of the airfoil, is always accompanied by a drag force, which acts in opposition to the direction of motion, that is, at right angles to the lift. The combination of the two forces is called the total reaction, and this acts through a point called the center of pressure. The lift force can be increased by increasing the camber of the airfoil—that is, the thickness relative to its width—and by increasing its surface area. Lift can also be increased by inclining the airfoil at a small angle of attack to the direction of the airflow. We saw in Chapter 9 that a flat plate can generate a lift force if it is inclined at a small angle of attack, but an airfoil generates a relatively greater lift than an inclined plane because of the Bernoulli effect. Increases in lift, however, are accompanied by increases in drag, and thus they require more energy to push the airfoil through the air. Modern airliners have a variety of high-lift devices to increase the camber and surface area of the wing prior

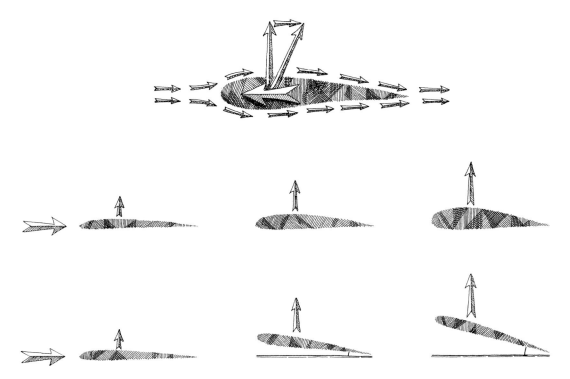

to takeoff and landing. These have the form of various leading-
and trailing-edge flaps that can be retracted when the aircraft is
in level flight.

As the angle of attack increases the lift increases but it soon
reaches a maximum value and then falls off very rapidly. At this
point the airfoil is said to have *stalled*. Stalling can also be brought
about by decreasing the speed of the air passing over the airfoil—
the *airspeed*. The airspeed at which stalling occurs is referred to as
the *stalling speed*. Stalling is caused by the separation of the airflow
from the surface of the airfoil. Separation begins at the trailing
edge and moves forward, and as it does so the lift decreases. Most
of the lift is lost at the stalling point, and when lift is no longer
large enough to support the weight of the aircraft, it loses height.
Pilots and flying animals alike therefore avoid allowing themselves
to get into a stalling situation during the normal course of flying.
The only time stalling is desirable is during landing, which is
essentially a controlled stall.[3]

Sometimes, if you are very lucky, you can catch a glimpse of

Top: *When air moves over
an airfoil the airfoil experiences lift and drag forces,
which act at right angles.
The resultant, the total reaction, acts through a point
called the center of pressure.
The lift generated by an airfoil can be increased both by
increasing the camber of the
airfoil* (middle) *and by inclining the airfoil at a small
angle of attack* (bottom). *Increases in lift, however, are
accompanied by increases in
drag. The magnitude of the
forces are indicated by the
lengths of the arrows.*

the wing of a bird actually stalling as it comes in to land. Pigeons are the most likely candidates, not only because they are common but because they have a low landing speed. The place to watch is the top surface of the wing. At the point of stalling any loose feathers that were being held in place by the airflow suddenly pop up as the airflow separates from the surface. This same method of capturing the instant of the stall is used during the testing of aircraft (tufts of wool are glued over the wing surfaces). In several years of pigeon-watching I have seen the feathers pop only once or twice, but I have seen the alula deployed on many occasions.

The alula, or bastard wing, is the thumb of the bird. It has the form of a short winglet with two or three short feathers, on the leading edge of the wing, about halfway along its length. It is generally thought that the alula functions as a leading-edge slot to delay the onset of separation, and hence to delay stalling. Similar devices are used on the leading edge of aircraft wings. In its simplest form, as used in aircraft built before World War II, the leading-edge slot was just a fixed rod that was attached along the leading edge of the wing, separated by a distance of about an inch or so. By redirecting the flow of air over the top surface of the wing, the leading-edge slot delayed the onset of separation. In later aircraft, including the Messerschmitt 109, the slot was flush-fitting but popped out automatically just before the wing stalled. The alula of the bird is similarly deployed automatically just before the wing stalls, which is why it can be seen when a pigeon lands. The alula is relatively small compared with the length of the wing, but it does, nevertheless, appear to serve a useful function. There is much variation in its relative development among birds, though, and in some species it is merely a vestige of a formerly important structure.

The wings of the bird at the top (a vulture) have a lower aspect ratio than the wings of the lower bird (an albatross).

Wing shape and wing loading

A useful measure of the efficiency of a wing is the ratio of lift generated to the accompanying drag. This, as noted in Chapter 9, is referred to as the lift-to-drag ratio. Long, narrow wings, that is, wings with high aspect ratios, have a higher lift-to-drag ratio than short, wide wings.[4] One of the consequences of narrow wing shape is that the gliding angle, that is, the minimum angle at which the bird can fly, is shallower. Gull wings have higher aspect ratios than pigeon wings, which explains why gulls can glide at much shallower angles. Having a low gliding angle means that longer distances can be travelled for a given loss of height. Therefore, if a gull and a pigeon both glided down to street level from the same height, the gull would travel furthest.

Wings with high aspect ratios have not only high lift-to-drag ratios but also high stalling speeds. This means that higher speeds have to be maintained to prevent stalling. The wing of the albatross has a high aspect ratio that enables it to remain aloft with the minimum of effort, but the relatively high stalling speed presents problems during takeoff and landing. Several years ago I visited a small albatross colony in New Zealand—a grassy hill beside the sea—and could see some of these problems firsthand. Just days before my visit one of the breeding males had struck a pole while coming in to land and had been killed. It if had been able to make a slower landing approach the disaster might have been averted. The warden of the colony had cut a grass runway for the birds that was headed toward the sea in the direction of the prevailing on-shore winds. During my visit a female made several attempts to take off. Each time she would line herself up on the runway, then start a waddling, flapping run toward the sea. But she never attained sufficient airspeed for takeoff and had to abort each attempt at the last moment.

Weight also has a considerable influence on flying performance. We take weight into account in terms of wing loading, that is, the total weight of the aircraft divided by the surface area of the wings. Aircraft with low wing loadings have lower stalling speeds and are therefore more maneuverable. They also consume less power and have shallower gliding angles.

The relationships among aspect ratio, wing loading, and flying performance may be demonstrated by comparing two famous fighter aircraft of World War II: the Messerschmitt 109 and the Supermarine Spitfire. There were many different versions of each aircraft during the war: the data used here were based upon the first operational models: the Messerschmitt 109 E-1 and the Spitfire I (Green, 1975). Apart from the Messerschmitt's slightly shorter wingspan (32 ft or 9.8 m compared with 37 ft or 11.3 m) and marginally greater weight (5,400 lbs or 2,455 kg compared with 5,280 lbs or 2,400 kg), the two aircraft were very similar in size. The Messerschmitt, however, did have a somewhat larger engine, its Daimler-Benz delivering 1,100 horsepower (hp) compared with the more modest 880 hp of the Spitfire's Rolls-Royce power plant. The most striking difference between them was the

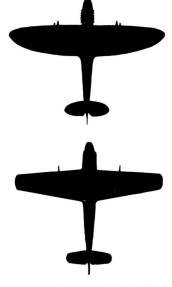

The Spitfire (top) *had relatively broader wings than the Messerschmitt, and they were slightly larger. Both aircraft weighed about the same. Consequently, the Spitfire had a lower wing loading and greater maneuverability.*

shape of the wings. The Spitfire had much broader wings, giving a lower aspect ratio and a wing area almost 40 percent greater than that of the Messerschmitt. As a consequence, the wing loading of the Messerschmitt was considerably higher (32 pounds per square foot; 154 kg/m^2) than that of the Spitfire (24 pounds per square foot; 115 kg/m^2) by 33 percent. The Messerschmitt could climb and dive faster than the Spitfire and had a higher operational ceiling, but it was much less maneuverable. Because the Spitfire could fly slower (its stalling speed was lower), it could make tighter turns. One of the defense strategies its pilots used was to put their machines into tight spirals, making it almost impossible for the faster and wider-turning adversary to get in a good shot.

Flapping and soaring

Most birds are able to glide, but since gliding is always accompanied by a loss in height they must also do some flapping or soaring. Some birds include both flying mechanisms in their repertoire, but others tend to be specialists. Most small birds are flappers, while larger birds tend toward soaring. The largest birds—such as the condor, albatross, vulture, and pelican—are all soarers, though they have to flap their wings to become airborne. They also have to flap their wings occasionally during flight, to make changes in course and to compensate for weakening air currents. Large birds are soarers primarily because of the relationship between body mass and power output. An animal's body mass increases approximately with the cube of its length, whereas its muscle power increases only with the square of its length— that is, with the area of cross section of its muscles. Large birds therefore tend to be relatively underpowered—they also have much higher wing loadings than their smaller relatives. Consequently they have to exert far more muscular energy to stay airborne by flapping, and the effort of becoming airborne probably involves anaerobic exercise. It is therefore unlikely that an animal as large as an albatross could maintain continuous flapping flight for any length of time.

Soaring aloft on unmoving wings is the least energetic form of flying and, provided a bird remains in an air current that is rising at least as rapidly as the bird's sinking rate, no height is lost. Each bird has its own sinking rate, which is determined by several factors including the aspect ratio and wing loading. Soaring birds are characterized by their low sinking speeds. An albatross sinks at about two feet (0.6 m) per second, which is about the same rate as that of a sailplane, whereas a pigeon sinks about four times faster than this (Brower, 1983).

The commonest sources of air currents are thermals, and these are usually caused by differential warming rates of the Earth's surface. Sailplane pilots learn to recognize where thermals are likely to occur, such as over rocky outcrops and dark patches of ground. I recall flying over Wyoming in a small aircraft on a bright spring afternoon. Much of the snow had melted, leaving dark patches of exposed ground. As the patches had absorbed more heat from the sun than the surrounding snowy areas had, they were the centers of thermals. As we flew over them the aircraft was lifted up, only to drop down again over the snow.

Thermals may be quite narrow, sometimes only a few yards or meters across, and if birds are to benefit from them they must be able to make tight enough turns to stay within them. The objective is to gain as much height as possible, then to leave the system and glide to the next one. Pennycuick (1972), who used a sail-plane to study thermal soaring in some African birds, was in no doubt that the birds were able to detect these systems because they selected flight paths that took them through more thermals than would have been expected if they had been flying randomly. Indeed, they were more experienced flyers than he was. He noted one particular flight in which he made much better progress on the way out, when he was following a vulture, than he did on the return trip, when he was on his own.

Thermals can be deflected by the prevailing winds. If they originate from a good source, such as an outcrop of rock warmed beneath a tropical sun, they may travel for many miles. These systems, which have been called thermal streets (Pennycuick, 1972), provide soaring birds with a means of soaring across country without having to circle to gain height. Thermals can also take the form of rising bubbles of air, and these may be a mile (2 km) in diameter. If a bird (or sailplane) stays within such a system it can fly in any direction within the bubble and still gain altitude.

Rising air currents can also be caused by updrafts, as when winds blow over hills or when onshore winds blow against sea cliffs—the familiar association between wheeling sea gulls and rugged cliffs brings this to mind. This type of flying is described as slope-soaring, for obvious reasons. As the zone of rising air follows the line of the coast, seabirds are able to patrol back and forth along the coast at will. A similar effect is also brought about at sea by the action of waves against the air. As the wave advances it pushes against the air, creating an updraft, and seabirds flying close to the surface are able to exploit the draft and slope-soar along the line of the wave front.

Yet another kind of soaring, described as dynamic soaring, involves making use of differences in horizontal wind velocities. As winds blow over the surface of the Earth they are slowed by friction, and the bottom layer of eight inches (20 cm) or so is essentially stationary. This is well known to sunbathers lying on the beach who only become aware of the wind when they sit up. There is, therefore, a velocity gradient, with the slowest moving air being at the bottom—we encountered this phenomenon of the boundary layer in Chapter 9. The albatross is said to specialize in dynamic soaring. A shallow dive is made into the wind, from a height of about sixty feet (20 m). As altitude is lost ground speed (speed relative to the ground) is gained. At the bottom of the descent, which is just above sea level, the albatross turns into the wind and the ground speed is so high that the bird ascends rapidly. As the altitude increases the ground speed decreases, but because the air encountered is moving progressively faster the air speed does not diminish very rapidly. As a consequence the bird continues to climb, even though its ground speed is very low, and it reaches a sufficient altitude to enable it to repeat the process.

Accounts such as this are available in most books, though Pennycuick's (1982) observations led him to conclude that the albatross's main soaring method was slope-soaring along the waves. His observations, however, were made from the land, and while albatrosses may slope-soar during short inshore flights, they probably do not do so during their long feeding journeys. Such long journeys are taken during the breeding season, one member of the pair flying out to sea while the other one remains ashore to incubate the eggs. Just how extensive these forages can be has been established for the Wandering Albatross using satellite telemetry (Jouventin and Weimerskirch, 1990). Radio transmitters were attached to five breeding males and their movements were tracked by satellite. The distances covered during a single trip, which lasted from two to almost five weeks, ranged from 2,257 miles (3,664 km) to a staggering 9,363 miles (15,200 km). Flying at speeds as high as 49 mph (80 kph), individuals travelled up to 554 miles (900 km) in a day. Most of the flying was done during daylight, but the birds were still active at night, especially by moonlight. They rested for only short times, alighting on the surface for periods that never exceeded about an hour and a half. Such flying performances are unsurpassed in any other species of bird.

Land birds are grounded at night because the earth rapidly loses its heat, and when there is no longer a marked difference in temperature with the overlying air, the thermals cease. The situation is different over the ocean in tropical latitudes because the trade winds, which blow toward the equator almost continuously, are relatively cooler than the sea. Consequently there are always convection currents and therefore always thermals, even at night. One of the birds that exploit these currents is the frigate bird, a familiar sight in the tropics with its sharply pointed V-wings, jet black against the sky. They are large birds—the wingspan is a little over seven feet (2.3 m)—and they are unusual in that their wings combine a low wing loading (0.77 1b/ft^2; 3.72 kg/m^2) with a high aspect ratio (12.8; data from Pennycuick, 1983). The low wing loading confers remarkable maneuverability because of the low stalling speed, which means that they can make tight turns and remain within narrow thermals. The high aspect ratio confers a high lift-to-drag ratio, giving improved gliding performance because of the slower sinking rate. Weak thermals can therefore be exploited, and when the thermals fail altogether the energy requirements for flapping flight are minimal. The disadvantage of having narrow wings is that takeoff performance is impaired. Frigate birds always take off from—and therefore must always land on—elevated perches such as trees, never from level ground or from the water (Pennycuick, 1983).

The frigate bird ranges far out to sea and is able to remain airborne for long periods of time, day and night. They do not alight on the water like pelicans and many other seabirds, but dive down and snatch fishes that are swimming at the surface. Very often they swoop on other birds and plunder their catch, and it is probably this marauding habit that has earned them their nickname, "man-o'-war birds."

Frigate birds, with their large size, low wing loadings, and high aspect ratios, are a close living analogue for pterosaurs. But we have much to learn about pterosaurs from birds of all sizes. For example, flapping flight requires high levels of continuous muscular activity, and therefore high metabolic rates. Even birds that spend most of their time soaring on stationary wings also need to be able to flap, not only to become airborne but also to stay aloft when air currents are unfavorable. It is therefore inconceivable that the pterosaurs were not endothermic, as all modern birds are.

Most of the smaller pterosaurs probably relied almost entirely on flapping flight to remain airborne, but the larger ones *(Pteranodon* is the best-known example) probably spent the greater portion of their time in soaring and gliding. Because flapping flight is difficult to analyze, especially when we have only skeletons to work with, and because so much has been written about the soaring abilities of *Pteranodon,* this crested giant is an obvious choice for a discussion of pterosaurian flight.

Pteranodon, the incredible flying machine

Few relics of the past have excited our imaginations or confounded our powers of reasoning more than *Pteranodon.* Ever since its discovery in the 1870s, paleontologists have speculated upon the possible life-style of such a large and seemingly fragile animal. And it was large too! With a wingspan of twenty-three feet (7 m), it was the largest known flying animal, a distinction it held for the next hundred years. Could they flap their wings? Did they launch themselves from cliffs? Would a grounded individual ever be able to become airborne again? These and many other questions went largely unanswered until the 1970s, when *Pteranodon* became the subject of a joint investigation between Cherrie Bramwell, a paleontologist, and G. R. Whitfield, an aeronautical engineer (1974). They chose *Pteranodon* as the subject of their aeronautical analysis for two reasons. First, because of its large size, they could assume that *Pteranodon* was a soarer, flying on fixed wings like a man-made aircraft. They could therefore assess its flying capabilities using methods that are applicable to aircraft. The second reason was that *Pteranodon* was the most completely known of the large pterosaurs. But before they could begin their analysis they had to familiarize themselves with the anatomy of the beast.

The most striking feature of *Pteranodon* is its enormous head— almost twice the length of its body—dominated by a huge crest. Such a massive head seems out of place in a flying animal but, like the rest of the skeleton, it is very lightly built. The bones are almost paper-thin and the crest, admittedly flattened during preservation, is only about an eight of an inch (3 mm) thick. Teeth

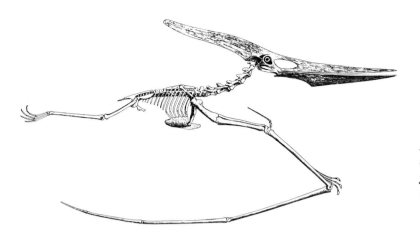

A reconstruction of the skeleton of PTERANODON, *as given by the American paleontologist G. F. Eaton.*

are entirely absent and the long sharp bill, which has a distinctly avian appearance, may have been sheathed in a horny beak. There is an obvious resemblance to the pelican, which feeds on fish, and we know that *Pteranodon* also fed on fishes because some specimens have been found with fish remains in the stomach region (Williston, 1902). It has even been suggested that *Pteranodon,* like the pelican, may have had a throat pouch because there is one specimen in which remnants of fishes and crustaceans were found just beneath the mandibles (Brown, 1943). While this makes a nice story, the evidence is entirely circumstantial and should not be given too much weight.

As in all other pterosaurs, most of the length of the wing was formed by the elongated fourth finger, the first three fingers being short and ending in claws. The humerus is relatively short and has a prominent process on its leading edge, which is characteristic of all pterosaurs. As mentioned earlier, this process was probably for the attachment of a muscle that controlled a tendon that kept the wing extended during flight. There is actual evidence for such a tendon in the form of a groove on the leading surface of one of the wrist bones. This groove is about half an inch wide (15mm), indicating that the tendon was the thickness of my little finger (Bramwell and Whitfield, 1974). It appears to have attached, through a very mobile pulley joint, to a prominent knob at the base of the wing finger, just beyond the level of the three free fingers. Tension in this ligament would have kept the wing finger fully extended during flight. But afterward, when the tension was relaxed, the extensive mobility in the pulley joint would have permitted the wing to be folded across the animal's back.

In many vertebrates, ourselves included, the shoulder joint is a ball-and-socket arrangement that allows for a good degree of forelimb mobility. In *Pteranodon,* however, it was essentially a hinge joint that permitted an up-and-down movement and a small amount of twisting. Bramwell and Whitfield (1974) suggested that a bony locking mechanism tended to hold the wing in the normal, outstretched soaring position, but it appears that this was a mistake (Padian, 1985). The albatross achieves such a locking mechanism through a tendinous sheet that passes over the shoulder joint (Pennycuick, 1982), reducing the amount of energy needed to keep the wings outstretched.

As for any flier, light-weight construction was a necessity for *Pteranodon.* This was achieved through the thinness of the bones, and this is most apparent in the wing, where the average thickness of the bones—from humerus to wing tip—is less than one thirty-second of an inch (1 mm), about the thickness of half a dozen pages of this book. How could there possibly have been sufficient strength in such thin bones? First, the bones are hollow, conforming to the well-known engineering principle encountered earlier (Chapter 3). A second strategy for maximizing strength is to concentrate the bone material in regions subjected to the highest stress. This strategy is revealed by examing cross sections of wing

Top: *The wall of the wing bones of* PTERANODON *are not of uniform thickness. Transverse sections across the bones show that they are thickened locally, probably in areas of maximum stress.* Bottom: *The bones of modern birds combine strength and lightness by having thin walls braced by internal struts. A similar strategy is used in aircraft construction.*

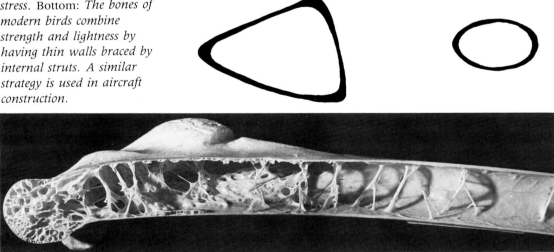

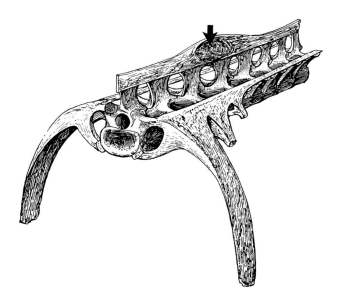

A reconstruction of the notarium of PTERANODON. *The upper end of the pectoral girdle fits into the socket* (arrow).

bones; these are triangular or oval in shape and the walls are obviously not of uniform thickness. Bird bones are strengthened and stiffened by fine internal struts, which are especially abundant toward the ends. Problems of preservation obscure such details in *Pteranodon*, but its bones were probably strengthened in the same way (struts can be seen in some of the pterosaur material from Brazil). Pterosaur bones are relatively thinner than in any other vertebrates, including other fliers, which underscores the importance of weight reduction during their evolution (Currey and Alexander, 1985). Such thin-walled bones would have been at a high risk of being buckled or broken.

The V-shaped shoulder girdle was firmly braced against the rest of the skeleton. The upper end fitted into a socket in the plate-like notarium, formed from the fusion of several dorsal vertebrae, while the lower end located into a socket in the sternum. The sternum has a prominent process, presumably for the attachment of flight muscles, as in birds, though in pterosaurs it has the form of a forwardly projecting process, called the cristospina, rather than a ventral keel.

Birds are unusual in that the hollow spaces of many of their bones are connected to their lungs by a complex system of sacs and ducts. The air-sac system entails a number of openings in the bones, called pneumatic foramina. Similar openings are also seen in the larger pterosaurs, showing that they too probably had an air-sac system. The function of this bizarre arrangement is not altogether clear, but it seems that it maintains a one-way flow of air through the lungs and may be important in shedding excess heat. Bats do not have an air-sac system.

Running Pteranodon *through the computer*

Aeronautics has come a long way since the early days of trial and error. It is now possible to predict the flying characteristics of an aircraft even before it is flown, provided that sufficient information is available regarding its size and shape. The important information includes the area, shape, and profile (cross-sectional shape) of the wings and the total weight of the aircraft. Bramwell and Whitfield (1974) modified a computer program designed for analyzing the performance of man-made sailplanes and used it to deduce the probable flight characteristics of *Pteranodon*. The details of how they estimated the body weight and dimensions need not concern us, but we are interested in two of their assumptions, namely the overall shape and the profile of the wing membrane.

There is no direct evidence for the overall shape of the wing membrane of *Pteranodon*, far less for its profile, so educated guesses had to be made for both. A skeptical reader might conclude that such guesswork invalidates the whole procedure, but an independent analysis by James Brower (1983), using a different wing shape, obtained results that corroborated the methodology. Bramwell and Whitfield (1974) took a traditional approach and assumed that the wing membrane attached to the ankles, but Brower had it attaching to the sides of the body, which gave it a

Estimates of wing loadings and aspect ratios obtained for *Pteranodon* from two computer studies; data for the frigate bird included for comparison

Source	Wing loading (kg/m^2)	Aspect ratio
Pteranodon (Bramwell and Whitfield, 1974)	3.6	10.5
Pteranodon (Brower, 1983)	5.9	19.0
Frigate bird (Pennycuick, 1983)	3.7	12.8

smaller area and therefore a higher wing loading and a higher aspect ratio. He also used a different wing profile, but we need not be concerned with this. As we saw earlier, the outer toe was reduced in the short-tailed pterosaurs, suggesting that the membrane may not have attached to the ankle, so Brower's wing shape may be the closest approximation. However, there is a third possibility. During an investigation into the anatomy of *Pteranodon*, Chris Bennett (1987) discovered that the tail was both longer and more complex than had previously been thought. One of the unusual features was that it ended in a pair of thin, bony rods. The possible function of the tail was puzzling, and he suggested that it may have been attached to the wing membrane, which would have left the hindlegs completely free. This is quite possible, but we will take a more conservative approach and think of the wings as attaching to the legs.

Both computer analyses concluded that *Pteranodon* was specialized for low-speed flying. Estimates for the best flying speed, that is, the speed that gave the slowest sinking speed and therefore the shallowest gliding angle, was about 17 mph (28 kph). At this speed the animal would sink at the rate of about one and a half feet (0.4 m) per second, giving a gliding angle of about 3°. Man-made sailplanes have similar sinking rates, but, because they fly much faster, their gliding angle is shallower than that of *Pteranodon*. The slow flying speeds would have conferred great maneuverability, and Bramwell and Whitfield estimated a minimum turning circle of only about sixteen feet (5 m), banked at an angle of 60°. Such tight turns would have enabled *Pteranodon* to exploit narrow thermals.

Although *Pteranodon* could have soared on weak thermals and slope-soared in weak winds, its slow flying speeds would have prevented its flying in strong winds—it would have been unable to make progress against the force of the wind. Ultra-light aircraft are grounded during windy periods for similar reasons. Because of its low stalling speed, estimated at 15 mph (24 kph) or less, landing approaches would have been slow. Since landing is probably the most hazardous part of any flight, its slow approach would have minimized the risk of injury, an important consideration given the extreme lightness of the skeleton. The low stalling speed resolves the problem of how *Pteranodon* became airborne. All that was required was a wind a little brisker than the stalling speed—a light breeze would do—and it could have taken off merely by facing into the wind with its wings extended.

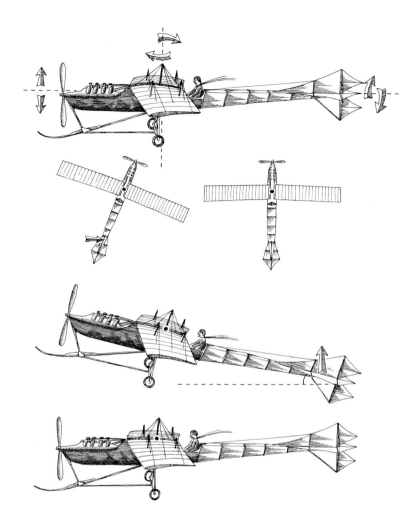

Top: *Movements in the three planes of space are pitch, yaw, and roll.* Second row: *When an aircraft yaws, the rudder becomes an inclined plane and the thrust generated swings the tail back in line again.* Third row: *When the nose pitches up, the horizontal stabilizer becomes an inclined plane and the lift generated raises the tail, returning the aircraft to level flight* (bottom).

Estimates of the maximum power output from the flight muscles, based upon estimates of their mass, suggested that flapping flight was probably possible, but only for very brief periods. *Pteranodon's* muscle power would therefore have permitted bursts of flapping to maintain height during failing wind currents and may also have allowed for flapping takeoffs.

In order to fly successfully an aircraft must be able to compensate for alterations in its flight path caused by air movements, and it must be able to change course at will. There are three basic movements that can occur during flight; yaw, pitch, and roll. Yaw is a side-to-side turning movement about a vertical axis, pitch is an up-and-down movement about a horizontal axis, and roll is a turning movement about a longitudinal axis. All three movements are corrected for during the flight of an aircraft by the action of inclined planes. Yaw is corrected by the action of the vertical rudder. When an airplane is flying on a straight course the rudder cuts through the air in the direction of travel, but when the machine begins to yaw the rudder becomes inclined at a small angle of attack and the deflection force that it generates brings the airplane back onto a straight course. Control of pitch is brought about by the action of the horizontal stabilizer. The flight feathers of a dart or an arrow similarly correct for yaw and pitch; if a dart is thrown with a wobble it can be seen to correct itself very rapidly and to fly straight. In addition to correcting for yaw and pitch, the tail of an airplane has adjustable control surfaces that can *cause* it to yaw (turn) and pitch (climb or dive). Roll is corrected and effected by the ailerons, which are elongated flaps on the trailing edges of the wings, placed toward the wing tips. By increasing the angle of attack of a given aileron, the pilot can give an additional lift to the wing on that side and cause the aircraft to roll toward the other side.

Although birds have tails, these are used mainly during slow flight, as during landing. Most in-flight maneuvers are effected by movements of the wings. These movements are difficult to see and it takes considerable time and experience before an observer can detect the subtle movements that cause changes in the flight path. One such observer was a man named E. H. Hankin (1910), whose painstakingly careful studies of the flight of birds, made in India during the early part of the century, have contributed to the following account. Upward pitch is caused by moving both wings forward so that their lift force is moved farther ahead of the center of mass, thus raising the head. Downward pitch is achieved by the reverse action. Upward pitch can also be achieved by raising the wings above the level of the body, an adjustment that capitalizes on the drag acting on the wings. Raising the point of application of the drag force above the center of mass causes a rotation around the latter, which raises the head. The reverse action is achieved by placing the wings below the level of the body. Yaw can be produced by the partial retraction of one wing, where the resulting reduction in drag causes it to travel faster and turns the bird toward the other wing. Roll is effected by increasing the force of the downbeat of one wing; this generates a greater lift on that side and rolls the body toward the other wing. Turning is probably always accompanied by a roll, the wing on the outside of the turn beating more powerfully than the other. Several of the birds Hankin studied could rotate their wing tips and produce a negative angle of attack. This not only depresses the adjusted wing but also increases its drag: consequently the bird rolls in the direction of the twisted wing. The wing tips therefore function as ailerons, and a similar mechanism probably exists in other birds. The head, which is small relative to the body in most birds, appears to have no role in steering.

Pteranodon, being essentially tailless, had to effect most of its control movements by adjustments of the wings. The long, crested head was probably also used to steer. Roll might have been controlled by changing the profile of one of the wings. If the wing finger on one side were depressed, the leading edge of the wing membrane could probably be lowered, increasing its lift (and drag) and causing the body to roll over toward the opposite side. A roll to the left would therefore be effected by lowering the right wing finger, with or without raising the left wing finger.

Pitch could probably have been controlled by swinging the leading edge of the wings forward, advancing the center of pressure (the point through which the resultant forces of lift and drag act) of the wings in front of the center of mass, thereby raising the head. If the legs were attached to the membrane, or if there were membranes attached along the inside edges of the hindlegs as advocated by Wellnhofer (1978), these additional areas of membrane could have functioned as important control surfaces. Lowering of both legs would have caused a symmetrical lift, pitching the head down, whereas lowering one leg would have produced an asymmetrical lift, causing roll as well as pitch.

Yaw may have been controlled by the partial retraction of one wing, as in birds, or perhaps by movements of the head, which was almost twice the length of the body. The force produced by the inclined head would have caused the body to steer to the same side, because the beak, acting as an inclined plane, would have generated a deflection force around the pivot of the neck. This force would have acted in front of the pivot, so it would tend to rotate the head farther in the same direction. The long crest, which had about the same surface area as the beak, would also have generated a deflection force but, being behind the pivot, it would have rotated the head in the opposite direction, therefore counteracting the turning effect of the beak. Inclination of the

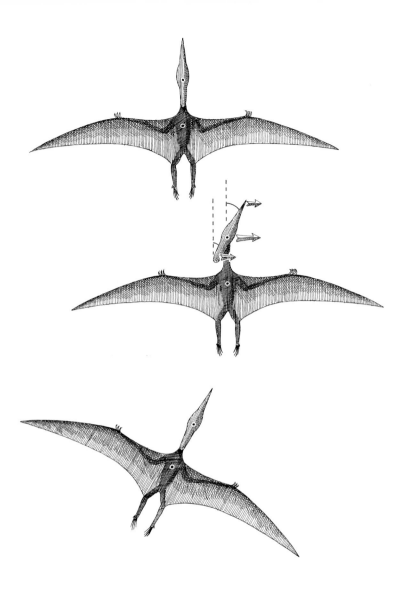

The head as a possible flight-control mechanism in PTERANODON. When PTERANODON flew a straight course (top), a rotation of the head to the right (middle) would have generated a thrust to the head that would have caused the body to yaw onto a new course (bottom). Notice that PTERANODON is depicted as having the wing membrane attached to the knee, but, as noted before, this reconstruction is entirely conjectural.

head during flight would therefore not have resulted in the head being twisted around farther. Instead, the head would have been stabilized, and the resultant deflection force would have turned the whole body toward one side. Without the crest *Pteranodon* would have required large neck muscles to prevent the head from being twisted right around. *Pteranodon*'s crestless contemporaries may not have used their heads for steering.

Pteranodon was probably capable of partially retracting its wings during flight to initiate steeper and faster descents. Many seabirds, like pelicans and gannets, use this strategy—they fall to the sea like stones to catch fishes—but it seems very unlikely that *Pteranodon*'s flimsy skeleton could have withstood such fast dives. A more likely strategy for catching fishes, proposed by Bramwell and Whitfield, was for *Pteranodon* to skim along just above the surface of the sea. By rotating the head down then back toward the feet (the neck vertebrae permitted such movement), it could seize a fish by the tip of the beak and continue soaring with little loss of forward momentum. Bramwell and Whitfield considered that *Pteranodon* may have been able to land and take off again from the water, but given its apparent similarity in wing loading and aspect ratio with the frigate bird, which apparently cannot take off from the sea, this seems unlikely.

Whether *Pteranodon* landed on the ground or on some elevated vantage point is conjectural. It has been suggested that trees would have been avoided because of the vulnerability of the wing membrane to damage. Bramwell and Whitfield pointed out, however, that if the pterosaurian wing membrane was like that of bats it would not have been easily torn. Furthermore, tears in bat wings do heal and bats are still able to fly even with large holes in their wings (Bramwell and Whitfield, 1974). Pterosaurs may well have roosted in trees, especially the smaller ones. Unwin's (1988) study of the Lower Jurassic genus *Dimorphodon* led him to conclude that the hindlimb, with its sharp claws, was specialized for climbing.

Bramwell and Whitfield (1974) compared the clawed feet of *Pteranodon* with those of bats and proposed a similar roosting strategy of hanging upside down, though from cliffs rather than from trees. I cannot see any advantages in hanging upside down from cliffs—in fact the whole idea sounds horribly dangerous to me. Why not simply roost on the tops of cliffs, the same way as many seabirds? As for the clawed toes and the clawed fingers, they might have been very useful in helping to anchor the grounded animal until it was ready for the next flight. One can imagine how cumbersome those folded wings may have been, especially in a breezy location like the top of a cliff—something like trying to control a collapsed umbrella on a windy day. But if *Pteranodon* could have held itself down close to the ground, in the still air of the boundary layer, the task would have been much easier. Takeoff might then have been a simple matter of letting go with the hands and standing up to face into the wind. This, of course, is all very speculative, but that has been the nature of most of this particular chapter.

Out with a Whimper or a Bang?

T H E disappearance of the dinosaurs, the pterosaurs, the plesiosaurs, and all those other magnificent reptiles at the close of the Cretaceous is
one of the great mysteries in the history of the
Earth. Reptiles that had dominated the world for over 150 million
years slipped into oblivion, leaving only a scattering of bones to
bear witness to their former glory. What went wrong? What happened?

There has been no shortage of possible explanations to account
for the demise of the Mesozoic reptiles—everything from changes
in the environment and competition with the mammals to the
spread of disease and the inherent failure of their genetic material.
But no single explanation seemed to fit all of the facts—no all-
embracing solution that would satisfy our quest for order in the
world. And then, in the summer of 1980, a universal explanation
for the great reptilian extinction appeared in the pages of the
journal *Science*. Surprisingly the article was not written by a paleontologist but by a physicist, Professor Luis Alvarez, in association with several others, including his son Walter (1980). Their
solution to the Cretaceous-Tertiary (abbreviated K-T) extinction
problem was startling, and it set off a chain reaction of scientific
enquiry that reverberated around the world for the next several
years.

The story began when Alvarez and his colleagues were analyzing some samples of clay from the K-T boundary in Italy. To their
surprise they detected unusually high levels of the rare element
iridium. Iridium is a metal belonging to the same chemical group
as platinum and, although rare in the Earth's crust, it is more

abundant in extraterrestrial bodies like meteorites. This iridium might have been a localized occurrence and therefore only of limited interest, so they sought samples from other parts of the world. Dale Russell, a paleontologist at Canada's National Museum of Natural Sciences, responded by sending them a boundary clay sample which he had collected in New Zealand. Analysis revealed unusually high amounts of iridium. Similar results were obtained for a boundary clay from Denmark, suggesting that the iridium anomaly at the K-T boundary was a world-wide phenomenon.

The iridium is accompanied by other rare elements and by soot particles and minute globules of minerals, both of which suggest intense heat and explosion. Where could these materials have come from? What event could possibly have covered the whole world in a thin layer of iridium? Because iridium is associated with extraterrestrial bodies, Alvarez and his colleagues concluded that it had originated outside the Earth and had resulted from the impact of an asteroid. On the basis of estimates of the amount of iridium blanketing the world, this group of scientists concluded that the asteroid would have been about six miles (10 km) in diameter. An impact with such a large object, as we will see, would have had severe consequences for the environment. Alvarez correlated this event with the extinction of the dinosaurs and the many other groups of animals and plants that did not survive the end of the Cretaceous. The impacting body was referred to as an asteroid, though it is now more usual to use the general term *bolide* (*bolis* is Greek for a dart or missile).

A six-mile (10 km) bolide hurtling through space and striking the Earth with a velocity of about 55,000 mph (90,000 kph) would have had an explosive force greater than that of all the present stockpiles of nuclear weapons combined (Silver, 1982). The impact force would have melted not only the bolide but also some of the Earth's crust—about ten to one hundred times the mass of the bolide—which would have left a crater about sixty miles (100 km) in diameter.

In the most extreme view of the effects of an impact, forests are thought to have exploded in flames. Smoke and dust thrown into the atmosphere by the explosion are believed to have transformed day into night, plunging the world into a cold darkness that might have lasted for many months, halting photosynthesis and causing plants to become dormant or die. Snow would have fallen where snow had never fallen before and, like the rain, it would have been polluted with acidic gases from the atmosphere. Acid rain would have destroyed lakes and killed trees, just as surely as it does today, but as there would have been so much more, it has been suggested that the seas would have been acidified too (Hsü, McKenzie, and He, 1982). This would have destroyed the myriads of small planktonic organisms with chalky skeletons, wreaking havoc on the food chains in the sea. The effects of all these changes on the living world would have been catastrophic.

This bleak picture of an "impact winter" largely reflects views of the early 1980s, but less extreme environmental effects have since been postulated. The dust and smoke cloud may have been far less extensive, and while some of the world may have experienced cold and darkness, other parts might not.[1]

Alvarez and his colleagues received mixed reviews for their extraterrestrial explanation of extinction. The initial reaction of most paleontologists was skeptical, though some, like Dale Russell, enthusiastically endorsed the idea. Russell, a dinosaur specialist, had been investigating the problem of extinction since the early seventies and had already arrived at the conclusion that an extraterrestrial explanation best fitted the data (1977).

As time passed and more iridium anomalies were reported from around the world, other paleontologists were persuaded that a bolide had collided with the Earth at the end of the Cretaceous. For them, the mass extinctions that heralded the close of the Cretaceous were sudden and therefore readily explicable in terms of a single catastrophic event.

Other paleontologists, however, argued that the decline of the dinosaurs and other organisms was a more gradual process that had been going on for several million years before the end of the Cretaceous. They contended that the extinctions of different groups of organisms were not synchronous and attributed the iridium anomaly to volcanic eruptions, which are also known to eject iridium and other materials into the environment.

Sides were taken—catastrophists on the one hand, gradualists on the other. The media enthusiastically supported the catastrophists—sudden death from outer space made a far more interesting story than a lingering death from earth-bound causes. Magazine articles and television specials alike tended to give the impression that the bolide impact was the accepted view, but an informal vote taken at an international dinosaur meeting held in Drumheller, Alberta, in 1986 showed that most vertebrate paleontologists believed that the extinction of the dinosaurs was a gradual and not a sudden event. In contrast, most of those attending a conference on global catastrophes and mass extinction in Utah in 1988—mainly geologists—appear to have favored an impact event (Kerr, 1988a, 1988b). To weigh the pros and cons of the two sides and thereby identify the most likely culprit for the demise of the Mesozoic reptiles, we will proceed in true Holmesian fashion by asking a series of questions. We will start with the physical evidence, saving the fossils till last.

The physical evidence

What is the likelihood that the Earth has been struck by a large bolide? The Earth is continuously being bombarded by materials from outer space, and we know from our own sightings of shooting stars that encounters with meteorites are quite common. Most burn up in the atmosphere, and an estimated 5–14 million tons of meteoritic dust rains down on our planet every year (Pettersson, 1960). Although this sounds like rather a lot, it does not amount to very much when it is evenly spread over the entire surface of the globe. Sometimes meteorites do impact, and their stony remains, which contain large amounts of iron together with such rare elements as iridium and platinum, may be seen in museums.

Astronomers tell us that the odds against a collision decreases with increasing size of the bolide, and they estimate that the Earth is likely to collide with a six-mile (10 km) bolide once every 40 million years (Wetherill and Shoemaker, 1982). Our planet must therefore have been struck by large bolides many times during its long history. We have direct evidence of this in the craters that were formed, and iridium has been detected for some of these (Gostin, Keays, and Wallace, 1989).

Impact craters are far more obvious on the surface of the moon than on the Earth. They can easily be seen with a pair of binoculars. The reason they are so visible is that there are essentially no weathering processes on the moon's surface, because of the absence of water and an atmosphere, so the craters have remained unchanged since their formation. On Earth the craters become eroded by weathering and although some retain their shape and are therefore readily identified as craters, others are less recognizable. Plate tectonics—the process that causes the movement of continents—also destroys craters.

Just because the Earth *may* have been struck by a large bolide at the end of the Cretaceous does not mean that it *was* struck. If

Direct evidence that the Earth is bombarded with meteorites is provided by impact craters. Left: *This one, Chub Crater, in Quebec, was photographed by the Royal Canadian Air Force. Weathering and other geological processes tend to obliterate craters.* Middle and right: *Since the moon is virtually devoid of such processes, impact craters remain largely unchanged for hundreds of millions of years and the surface is peppered with craters.*

such an event did happen, however, the iridium and other materials released into the atmosphere would have been deposited around the world at the same time—certainly within a few weeks or months—and the deposition would appear in the geological record as a synchronous event.

Is the iridium anomaly synchronous around the world? Let us just consider how we could determine whether the iridium layer in a Late Cretaceous formation in, say, New Zealand was deposited at same instant as an iridium layer in Denmark. If the appropriate minerals were present in both it would be possible to obtain radiometric ages for the bands containing the iridium anomaly. However, radiometric dates for sediments of this age have an inherent error of between two and three million years (Harland et al., 1982), so it would not be possible to tie the ages of the bands together any closer than this interval of time. Fortunately, though, there is a more accurate way of comparing ages, a method that makes use of the phenomenon of magnetic reversals.

At present, the Earth's polarity is said to be normal because a magnetic compass points toward the north. At irregular intervals that vary between about 10,000 years and several tens of millions of years, there is a reversal in polarity, and what was the magnetic north pole becomes the south pole. These reversals in the Earth's magnetic field are imprinted upon the molten rocks spilling out through the Earth's crust, so that a series of magnetic stripes become locked into the new rocks as they cool down. The stripes can be detected using magnetic measuring devices and, because they vary in their widths, specific stripes can be identified. These have been named and numbered in sequence and provide a means of correlating the ages of rocks in different parts of the world. The timing of the iridium layer can therefore be established from the magnetic record.

Alvarez and his colleagues found that the Italian iridium anomaly occurred during the reversed polarity zone of magnetic anomaly number 29. This interval, designated 29R, lasted for about 600,000 years (Officer et al., 1987), or perhaps a little longer (Harland et al., 1982). Therefore, although we may be able to say that the iridium anomaly occurred in different parts of the world during this particular magnetic interval, we are unable to place the timing of the event any more closely than the half-million-year duration of the reversal. Clearly, even if the iridium had been deposited around the globe at the very same instant, we have no way of establishing the fact.

Instant deposition would have resulted in a narrow band of iridium, but Charles Officer and Charles Drake—proponents of a volcanic origin for the iridium layer—found that iridium bands are not always thin. Although there was a single iridium peak at localities in New Mexico, Italy, and Spain, in Denmark and in the South Atlantic (a sample taken from a core drilling) the iridium is spread over a depth of over a foot (30–40 cm; Officer and Drake, 1985). Some of this spreading may be attributable to the action of burrowing organisms, but the evidence suggests that deposition at these two localities took place over a time interval of between 10,000 and 100,000 years (Officer and Drake, 1985). Multiple iridium peaks have subsequently been reported for localities in Germany (Graup and Spettel, 1989) and in Italy (Crocket et al., 1988; actually at the same locality that Officer and Drake had formerly identified as having a single peak). There is also some independent evidence, based on the relative concentrations of other platinum-group elements, to suggest that iridium deposition was not a single occurrence. These analyses showed marked differences between samples collected from the northern hemisphere and those collected from the southern hemisphere, which is incompatible with their having resulted from the same incident (Tredoux et al., 1989).

In summary, we cannot date the deposition of the iridium layer in different parts of the world with greater precision than about half a million years. This does not rule out the possibility of instantaneous deposition, and deposition at localities showing single iridium peaks may in fact have been synchronous. Elsewhere the element was probably accumulated over many thousands of years. We are obviously getting mixed signals from the K-T boundary.

Is there any evidence of an impact crater at the K-T boundary? About 70 percent of the Earth's surface is covered by ocean, so there is a good chance that an impact crater would be underwater. Sea-floor craters are unlikely to be discovered but one has been found, off the east coast of Canada, and it is associated with a slight enrichment of iridium (Jansa and Pe-Piper, 1987). But the crater is too small (about one to two miles; two to three kilometers in diameter) and is geologically too young (Early Eocene) to be of particular interest to us here.

Although the chances are lower that a bolide would strike land than water, the nature of the minerals associated with the iridium suggest an impact on land (Kunk et al., 1989). Furthermore, because these mineral particles are larger in western North America than elsewhere, it is suggested that this is the most likely impact site (Kunk et al., 1989). Are there any likely craters in North America? There is an impact crater in central Iowa, called the Manson Crater, which is Late Cretaceous in age (Kunk et al., 1989). This structure was originally thought to be volcanic in origin, but it was later reinterpreted as an impact feature. It is not a large crater, only about twenty-two miles in diameter (35 km). There are also uncertainties regarding its precise age, but it does appear to be Late Cretaceous and it has been proposed as a likely candidate for the supposed K-T impact event.

The evidence is equivocal, but even if the Manson Crater were taken as documentation of a bolide impact at the close of the Cretaceous, it is too small to have given rise to the large-scale environmental changes that occurred at the close of the Cretaceous. An alternate explanation for the absence of a sufficiently large impact crater is that the Earth may have been struck not by a single bolide but by a shower of them. The search for likely impact craters continues, but a cause-and-effect link between bolide impact and extinction events still has to be established.

Can iridium anomalies be correlated with any other major extinction events in the Earth's history? The end of the Cretaceous was not the only period in geological history marked by major extinctions. There have been about half a dozen others, and if there were a cause-and-effect link between impacts and extinctions we would expect to find iridium anomalies marking these events too. Although there have been some reports to this effect, these have subsequently been dismissed and the K-T extinction event therefore appears to have been unique (McGhee et al., 1984; Orth et al., 1984; Wilde et al., 1986; Clark, Cheng-Yuan, and Orth, 1986; Donovan, 1987a).

Could the iridium anomaly have been produced by volcanic eruptions? Relatively high levels of iridium are found in volcanic ash, together with other rare elements and certain small mineral particles, all of which have been found at the K-T boundary. Officer and Drake (1985) have argued that the concentrations of antimony and arsenic associated with the iridium layer are considerably higher than they are in meteorites but are fairly similar to those in volcanic material, suggesting a volcanic origin.

During volcanic eruptions, fine particles are injected high into the atmosphere, where they become widely distributed. When Krakatau—a small volcanic island in the Sunda Strait between Java and Sumatra—erupted in 1883, dust particles circled the globe for about two years, giving rise to spectacular sunsets seen throughout the world (Alvarez et al., 1980). Eruptions are therefore able to leave a global signature that becomes incorporated into sedimentary rocks.

Among the mineral particles found at the K-T boundary are minute grains of quartz—smaller than the size of the period at the end of this sentence. (Quartz is most commonly found as sand.) In an attempt to determine the origin of these particles, geologists from the U.S. Geological Survey collected some of these grains from a locality in Montana (Bohor et al., 1984). When examined under an electron microscope, about 25 percent of them were found to have multiple hair-line fractures running through them. This suggested that the grains had been subjected to large stresses, consistent with their having been exposed to an impact event. Similar grains, referred to as shocked quartz, have now been found in association with iridium at other K-T boundary sites around the world (Bohor, Modreski and Foord, 1987).

Shocked quartz can also be formed during volcanic eruptions (Carter et al., 1986), but these grains have fewer fracture lines, indicating that they were not subjected to such large stresses. They are also much rarer than the grains in the K-T boundary samples, less than 1 percent compared to 25 percent (Bohor, Modreski, and Foord, 1987).

On balance it seems that shocked quartz grains are marginally supportive of an impact over a volcanic event.

Is there any way of deciding between an impact origin and a volcanic

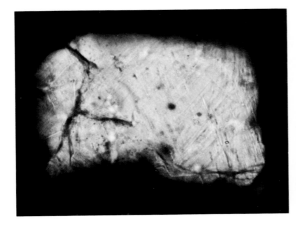

*Shocked quartz grains
showing minute fractures.*

origin for the iridium anomaly? Two lines of evidence may be useful, one involving amino acids the other a mineral called stishovite. But before presenting the evidence I want to emphasize that volcanic eruptions and a bolide impact are not mutually exclusive. Both may have taken place during the Late Cretaceous, and some people have actually suggested that a bolide impact may have triggered a series of volcanic eruptions, a point that will be examined later.

Amino acids, as noted in Chapter 2, are the component parts of protein molecules. Like many other organic molecules most amino acids can exist in two three-dimensional forms, called D, for *dextro* (meaning right-handed), and L, for *laevo* (meaning left-handed). The D and L forms of a molecule are mirror images of each other, like a pair of gloves. A characteristic of living organisms is that, with very few exceptions, they synthesize only the L form of molecules.

Although we normally associate amino acids with living organisms they also occur in nonliving systems, including meteorites, but extraterrestrial amino acids are either completely unknown in living organisms or are very rare. Some meteorites also contain a wide assortment of other organic compounds and these, like the amino acids, are either unknown on Earth or extremely rare. Extraterrestrial bodies therefore carry a unique organic signature, and if this signature were detected at the iridium anomaly it would provide convincing evidence for a bolide impact. With this objective in mind, an analysis was made of organic compounds found in sediments at the K-T boundary at the Danish locality (Zhao and Bada, 1989). Two amino acids were found (isobutyric acid and isovaline), both of which are exceedingly rare on Earth but common in certain meteorites. One of these amino acids, which is also a rare component in some fungi, exists in both the D and L forms. Significantly, there were equal quantities of the D and L forms at the K-T boundary, whereas in the fungi in which this amino acid occurs it is found only in the D form. Neither of these amino acids could have been formed during the breakdown of terrestrial compounds, nor could they have been ejected from a volcanic eruption, so their presence at the K-T boundary is compelling evidence for an extraterrestrial origin. One puzzling aspect of the story is that the alien amino acids were not found within the boundary layer of clay itself, where the iridium concentration is at its highest, but about a foot (30 cm) above and twenty inches (50 cm) below the clay. Perhaps the amino acids had diffused out of the clay into the surrounding chalk. Obviously more work needs to be done. It would be especially interesting to know whether alien amino acids exist at other K-T boundary sites. For the present, all that can be concluded is that foreign amino acids suggest a bolide origin for the iridium layer at the Danish locality.

Another indication of an extraterrestrial origin is the mineral stishovite. Quartz is a common mineral that exists in several crystalline forms, some of which can be obtained by subjecting quartz to stress. One of these minerals, stishovite, requires relatively high stresses for its formation and occurs naturally only in rocks from impact sites (McHone et al., 1989). Significantly, it has not been found at volcanic sites and is therefore regarded as being indicative of bolide impacts. Stishovite has been found in the K-T boundary clay of Raton, New Mexico, and this has been taken as evidence that the iridium layer at this locality originated from a bolide and not from volcanic activity (McHone et al., 1989). Similar investigations are needed for other localities.

Is there any evidence for extensive volcanic activity at the close of the Cretaceous Period? The search for clues of extensive volcanic activity at the close of the Cretaceous has been more successful than the search for impact craters. Huge areas of India and the Indian Ocean are covered by volcanic rocks, bearing testimony to a period of extensive volcanism. These rocks, the Deccan basalts, probably spanned the K-T boundary (there have been doubts regarding their exact age) and the eruptions were probably the largest volcanic event of the entire Mesozoic Era (Duncan and Pyle, 1988). The eruptions appear to have occurred over a period of less than one million years, which is a relatively short interval on the geological time scale, indicating a very intense period of activity.

What can be concluded from the physical evidence so far?

1. The iridium anomaly at the K-T boundary is a well-established fact and was probably global in extent.

2. The date at which the iridium was deposited at the K-T boundary cannot be established with a greater precision than about 500,000 years.

3. There is no reliable evidence of an iridium anomaly at any other extinction boundary.

4. The iridium layer is not always thin, therefore deposition was not always instantaneous.

5. The Manson Crater in Iowa may have been caused by an impact event at the close of the Cretaceous. There are doubts regarding its correct age, however, and it is also too small to account for the amount of iridium deposited.

6. The Deccan basalts of India are incontrovertible evidence of extensive volcanic activity at the time of the K-T boundary.

7. The high concentrations of antimony and arsenic associated with the iridium anomaly at the K-T boundary suggest a volcanic rather than an extraterrestrial origin.

8. Shocked quartz grains, on balance, support a bolide impact event.

9. Atypical amino acids at the Danish K-T boundary site is evidence for a bolide impact at this locality.

10. The occurrence of stishovite in the K-T boundary clay of Raton, New Mexico, is evidence for a bolide impact at this locality.

The only conclusion that appears to fit all the physical evidence is that the Earth underwent extensive volcanic activity *and* was struck by a large bolide at the end of the Cretaceous. These environmentally disturbing events probably continued over a long period of time, probably about one million years. Since much of the iridium probably originated from volcanoes, the bolide may have been smaller than the six-mile estimate given by Alvarez and his coauthors (1980).

Was there a causative link between a bolide collision and volcanic activity? Although this suggestion is not new, an interesting proposal linking the Deccan basalts with a bolide impact has recently been made by D. Alt and his colleagues (1988). The usual interpretation of this plateau of volcanic rock is that it was formed by extensive volcanism, but Alt compared its origin with that of the lunar seas. These large flat areas on the moon's surface are believed to be the result of impacts with very large bolides. The impact energies are thought to be sufficiently high to form deep craters that filled with molten lava from below. The lava eventually solidified as immense lava lakes. The Deccan Plateau, then, is seen as having been formed by a bolide that produced a crater several hundred kilometers in diameter, filled with molten rock from the Earth's interior. Later eruptions broke through the solidifying crust of the lava lake, causing lava flows and injecting volcanic materials into the atmosphere.

Time will tell how well this hypothesis stands up to testing, but it does have the merit of accounting for the occurrence of extensive volcanism and a bolide impact. It also resolves the problem of the location of the crater.

The paleontological evidence

We now have to consider the fossil record to see whether there is evidence for sudden or for gradual extinctions among the various groups of organisms that failed to survive the Cretaceous. There is a surprisingly wide range of opinion on this matter among specialists, even for the same groups of organisms, especially as the K-T boundary is approached. There are several reasons for this, including the problem of recognizing the K-T boundary, sampling errors, and the difficulty of finding good exposures of fossil-bearing sediments that extend right up to the end of the Cretaceous.

How sudden is sudden? If a bolide was responsible for the Cretaceous extinctions, I would imagine that the time interval during which the species died out would be measured in days and months, perhaps in one or two years at the most, but certainly not in thousands of years. But we have already seen that time cannot be measured at this geological period with an accuracy greater than about half a million years. We therefore lack the necessary precision to be able to recognize sudden extinction. If we had a continuous sequence of fossiliferous sediments crossing the K-T boundary, we would be able to see whether the various groups became extinct at the same horizontal level. This would give us some idea as to how suddenly the organisms became extinct at that particular locality. One such sequence occurs at Zumaya, on the northern coast of Spain (Ward, Wiedmann, and Mount, 1986), and this will be referred to later. But just because a particular group of organisms appears to have become extinct at the K-T boundary in one place does not necessarily mean that it became extinct everywhere—the disappearance could be an illusion created by sampling error.

How good are our samples? The Gallup polls are good at forecasting political fortunes because they sample large cross sections of the population. If only a handful of people were asked who they were going to vote for at the next election the polls would be very unreliable. Paleontologists are seldom able to obtain sufficiently large fossil samples, and small samples can lead to some unreliable datings of extinctions. For example, species that are believed to have become extinct sometimes show up again in the fossil record at a later period. Such species are called Lazarus species after the biblical character who was raised from the dead. Sometimes whole groups of organisms fall into this category, as in the case of the coelacanths. This ancient group of fishes was thought to have become extinct at the end of the Cretaceous because they could not be found in post-Cretaceous sediments. Then, to everyone's surprise, a live coelacanth was caught off the African coast in 1938.

Even when sample sizes are large, distribution errors can still be made. Carl Koch (1987), working with samples of fossil invertebrates, each one of which contained at least 1,000 specimens, discovered that as many as 25 percent of the species found in one sample were not found in a second. This means that the chances of wrongly concluding that a particular group of organisms became extinct at a particular time is quite high.

How is the K-T boundary recognized in the field? With great difficulty, would be a quick answer, because the K-T boundary cannot usually be recognized with any degree of certainty. There are some rare instances, as at Zumaya, where the iridium layer can be identified in the field, but this is the exception rather than the

Top: *The end of the Age of the Reptiles is marked by a coal seam, the Z-coal* (arrow), *seen here in Makoshika State Park, eastern Montana.* Bottom: *A thin layer of boundary clay marks the end of the Cretaceous in Berwind Canyon, Colorado.*

rule. The traditional way of recognizing the end of the Cretaceous has been the disappearance of particular fossils. In terrestrial deposits the end of the Cretaceous is frequently taken as the last occurrence of dinosaur bones, whereas in marine sediments the disappearance of large numbers of planktonic species marks the close of the Period. There is an obvious circularity in using these criteria—the last dinosaurs *always* occur at the end of the Cretaceous by definition—and an independent yardstick must be found.

One of the best fossil localities at which to see the last of the dinosaurs is Hell Creek, Montana, where the end of the Cretaceous is conveniently marked by a coal seam, designated the Z-coal. Dinosaur bones are fairly common below the coal seam, but no dinosaurs are found above it. The iridium layer has been found to lay at the base of the coal seam (Smit and Van Der Kaars, 1984), but because the Z-coal often consists of several discontinuous seams, spaced some yards (meters) apart, it does not provide as clear-cut a marker for the end of the Cretaceous as we should like. Dale Russell tells me that there are obvious differences in the appearance of the rocks in the vicinity of the boundary, and he estimates that he can place the K-T boundary to within about a yard (1 m) about 85 percent of the time. But that yard of sediments was deposited over thousands of years, so we are back to the problem of not being able to recognize an instant in geological time. If we were out collecting dinosaur bones in Hell Creek, we could label them according to their distance from the Z-coal and this would give us some idea of their relative ages, but we would not be able to determine their absolute ages with any accuracy. And when we visited other museums and looked at their collections of Hell Creek dinosaurs, especially if they had been collected many years ago, we would probably find even less detailed age data.

Aware of some of the pitfalls, we can now take a census of animals and plants that failed to survive the Cretaceous Period and ask whether their decline was gradual or sudden.

Sudden death and gradual decline during the Cretaceous

The heaviest casualties occurred among the planktonic organisms living in the surface layers of the sea, and the most seriously affected were those with calcareous skeletons. These organisms were so abundant during the Cretaceous Period that the continuous rain of their skeletons down to the seafloor accumulated into vast deposits of chalk, giving rise to the massive chalk cliffs for which the Cretaceous Period was named (*creta* is Latin for "chalk"). But so many of the organisms became extinct at the close of the period that there was an abrupt end to the formation of chalk, which explains why the K-T sediments are clay and not limestone. Their extinction appears to have been sudden in some parts of the world. A study of the Zumaya locality in Spain, for example, showed a large crash in numbers of planktonic species at the K-T boundary (Ward, Wiedmann, and Mount, 1986).

In another study, however, based upon localities in Tunisia and Texas, species extinctions extended over a considerable period (Keller, 1989). The number of species started to drop about 300,000 years before the K-T boundary and continued for another 200,000–300,000 years, during which time there were three episodes of accelerated species extinctions. Iridium was found at the K-T boundary of the Tunisian locality, and 26 percent of the species became extinct at that point, but there were no extinctions at the boundary for the Texan fossils. Extinctions were selective, too. Large, ornate species that were tropical and subtropical were more likely to disappear than the small, simple, temperate forms. Global cooling rather than a catastrophic event was considered the most likely cause of the extinctions.

Mention was made earlier that the extinction of calcareous planktonts (a planktont is a planktonic species) may have been correlated with acidification of the sea. Not only was acid rain unlikely to have caused such an extreme environmental change, but there is evidence that these organisms survived the Cretaceous: a deep-sea core drilled from the floor of the North Pacific reveals a sequence of well-preserved calcareous planktonts spanning the K-T boundary (Zachos, Arthur, and Dean, 1989). Their skeletons had clearly not been dissolved by an acidic sea, but there was a severe decline in the species numbers and evidence of a substantial reduction in the productivity of the sea. These changes occurred over a period of less than 10,000 years and were coincident with iridium and spherules of minerals, all of which were interpreted as evidence for a bolide impact.

Species extinctions were not confined to the plankton. The ammonites, for example, a major group of invertebrates that thrived throughout the Mesozoic seas, failed to survive into the Tertiary. It is generally accepted that these shelled molluscs, related to present-day squids, underwent a slow decline during the Cretaceous and did not extend all the way to the K-T boundary

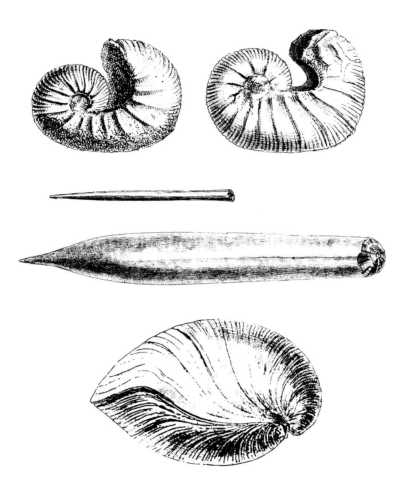

The ammonites (top) *became extinct at the close of the Cretaceous, and so did the belemnites* (middle). *The brachiopods* (bottom) *survive to the present, but they are greatly reduced in numbers. An unusual feature of many Late Cretaceous ammonites* (top) *is that they were loosely coiled or uncoiled.*

(Donovan, 1987b). This view is supported by the fossil sequence at Zumaya, which shows the ammonites stopping short some thirty-two feet (10 m) below the K-T boundary. However, a more recent investigation places them within five inches (13 cm) of the boundary (Kerr, 1988b), and Walter Alvarez and others (1984) reported a fossil sequence in Denmark where ammonites continue right up to the terminal Cretaceous. I conclude from this that ammonites became extinct at the K-T boundary, but the fact remains that they had been in decline for millions of years before, during which species declined in both number and diversity. The same holds true for other groups, including the belemnites (similar to ammonites but with straight shells), the inoceramids (bivalves), and the rudists (reef-building bivalves—Hallam, 1987).

Another marine casualty at the close of the Cretaceous were the brachiopods, a group of invertebrates superficially similar to clams. Although they were not completely eliminated at the close of the Cretaceous—a few stragglers survive to the present—the group did suffer heavy losses, and it seems that their demise was sudden, without any prior decline in the species numbers (Surlyk and Johansen, 1984). But it was not all doom and gloom in the sea—many groups did survive and some, including deep-sea animals, were barely affected (Officer et al., 1987).

There were also large-scale changes occurring on the land at the close of the Cretaceous. Some of these changes involved the terrestrial plants, but there are differences of opinion whether these suggest gradual or catastrophic events. One investigation, based upon pollen and spore fossils across the K-T boundary in the western interior of North America, reported an abrupt disappearance of certain species at the boundary, followed by an immediate and pronounced increase in fern spores, all of which suggested a catastrophic event (Tschudy et al., 1984). A similar pattern was described for a locality in Japan (Saito, Yamanoi, and Kaiho, 1986). Conversely, in another study, which was confined neither to fossil pollen nor to North America, the evidence suggested a gradual change rather than a catastrophic one (Hickey, 1981). Two interesting points emerged: first, more species became extinct in the northern hemisphere than in the south. Second, the extinction of plants and dinosaurs took place at different times. A similar discordance was found in an analysis of fossil pollen and plants collected in North Dakota (Johnson et al., 1989). The K-T boundary was marked by a 30 percent reduction in pollen species (plant species identified on the basis of isolated pollen grains) and this occurred about six feet (2 m) above the level of the highest dinosaur remains. There was a 79 percent change in plant species (species identified on the basis of parts of plants) and although some became extinct right at boundary, others had already disappeared before that time. The results were interpreted as indicating a period of climatic change followed by a bolide impact.

The Mesozoic reptiles are of particular interest to anyone puzzling over the question of Cretaceous extinction. The loss of the dinosaurs and their reptilian contemporaries was the most conspicuous faunistic change at the end of the period. But before reviewing the casualty list we need to remember that many reptilian groups crossed the K-T boundary, seemingly without having been affected. These include the turtles, crocodiles, lizards, and snakes, though the fossil record for snakes is so poor that they probably add little to the story. Furthermore, some of the reptiles that did not survive the Cretaceous had already slipped into oblivion millions of years before its end. The last ichthyosaurs disappeared about 25 million years before the close of the Cretaceous (during the Cenomanian). Plesiosaurs have been found in the Late Maastrichtian (Baird, 1984), which was the very last part of the Cretaceous, but the group was clearly in decline during the latter part of the Cretaceous (Sullivan, 1987). The pterosaurs survived right up till the end of the Cretaceous, but, like the plesiosaurs, they appear to have been much reduced in species numbers and diversity by that time (Wellnhofer, 1978; Langston, 1981).

Most of the major groups of dinosaurs survived to the end of the Cretaceous but some of them, like the sauropods and stegosaurs, had already become much reduced in numbers of species. Whether there was a decline in diversity in the other dinosaurian groups during the Late Cretaceous has been the subject of considerable debate. Most of the discussions have focused upon North American forms, primarily those of Alberta, Montana, and Wyoming. And since Hell Creek, in Montana, is one of the few dinosaur localities where a good sequence of rock can be seen that crosses the K-T boundary, it has figured prominently in the discussions. Robert Sloan and his colleagues (1986) and Robert Sullivan (1987) have argued that there was a decline in the number of dinosaur species well before the K-T boundary, but Dale Russell (1984) and others, including Peter Dodson (in preparation), disagree. Dodson makes the point that small sample sizes make it impossible to determine whether there was a gradual or sudden disappearance of dinosaurs—but there is no evidence that the group as a whole was in decline during the latter part of the Cretaceous. We have already taken note of some of the sampling problems inherent in paleontological studies, even where sample sizes are large, which is certainly not the case for Late Cretaceous dinosaurs of North America. What if we could sample Late Cretaceous dinosaurs from other parts of the world? Would we see evidence of gradual declines there too, or would we find evidence of diversity right up to the K-T boundary? My good friend Dale Russell and I agree to disagree on various issues, including the extinction of the dinosaurs, but our recent discussions on whether there was a real decline in dinosaurian diversity prior to the end of the Cretaceous leaves me feeling less sure of the situation. He has seen more dinosaurs, from all parts of the world, than most other paleontologists and when he tells me he sees diversity rather than decline during the latter part of the Cretaceous, I am inclined to listen.

There have been reports that some dinosaurs did not become extinct at the close of the Cretaceous but survived into the Paleocene, the earliest part of the Tertiary. One of the most widely quoted claims for Paleocene dinosaurs is that of Sloan and his colleagues (1986), who reported finding dinosaur teeth between the upper and lower Z-coal seams in Montana. They presumed that the teeth had been deposited after the iridium layer, and they rejected the idea that they may have been redeposited from older rocks (as sometimes happens) primarily because they showed no signs of abrasion. Aside from the uncertainty of the level of the K-T boundary relative to where the teeth were dug out, it has been shown experimentally that dinosaur teeth are remarkably resistant to abrasion (Argast et al., 1987). I have no difficulty with the idea that a few dinosaurs may have straggled on into the Tertiary, but I believe it has yet to be proven.

The end of an era

So what really did happen during the latter days of the Cretaceous? Were the dinosaurs cut down in their prime by some errant missile from space, or had they and their kind already dwindled into insignificance?

There is little question that the end of the Cretaceous was marked by environmental changes of unprecedented proportions. Aside from the extensive volcanic activity, which seems to have lasted for the final million years or so, sea levels were falling and the vast inland sea that had extended down the length of North America for most of the Mesozoic was drying up. Mountain ranges were being formed, and there was an increasing difference between the seasons, with progressively colder winters. These environmental factors, especially the volcanic activity, could probably have accounted for most, if not all, of the extinctions that occurred during the Late Cretaceous.

The crash in the plankton that occurred during the terminal Cretaceous would have had far-reaching effects upon other organisms. The plankton is the largest source of primary food production (Tait and De Santo, 1972) and we are well aware of the widespread consequences of its depletion. This is graphically demonstrated during the periodic warming that takes place in the eastern Pacific Ocean, a phenomenon referred to as the El Niño. The severe El Niño that occurred during the period 1982–1983 drastically reduced the plankton, hence the fishes and other organisms that feed upon it, and one of the consequences was the disappearance of the entire seabird community from Christmas Island (Schreiber and Schreiber, 1984). The sudden and global changes in the plankton at the close of the Cretaceous would have had far greater consequences and may well have been responsible for some of the other extinctions (Hallam, 1987).

The evidence that the Earth was struck by a large bolide during this time of environmental upheaval is fairly convincing, but I would hazard the opinion that it would probably have made little difference to the final outcome. The reptiles had been the dominant vertebrates throughout most of the Mesozoic—this had been their era—but the best of times were already behind them and many had already slipped quietly away, millions of years before. The turmoil at the end of the Cretaceous was just the *coup de grâce* for those that remained.

In his magnificent book, *An Odyssey in Time: The Dinosaurs of North America*, Dale Russell (1989) attributes the demise of the dinosaurs to the impact of a bolide and asks: "Why did it have to be this way?" The question is an agonizing one because it implies that, but for some quirk of cosmic fate, the dinosaurs would still be with us today. But I believe we can spare ourselves the anguish of asking this question because the great reptiles had already had their day. Bolide or not, the world was ready for something new.

Epilogue

I am watching a pair of chickadees through my window. They have just swooped down upon a bird-feeder hanging among the branches of a silver birch. I know that birds have a well-developed cerebellum—the brain center for balance and coordination. I know that they also have large optic lobes and acute vision and that the movements of each of those wings is controlled by over forty muscles (McGowan, 1986b). But it still amazes me that those tiny birds can flit through the branches without so much as a near-miss. Their maneuvers must be the equivalent of an F–18 fighter-plane—equipped with large-brained pilot and in-board computer—making a landing approach on a downtown street. I would not rate the aircraft's chances very highly. Yes, birds are the most remarkable, the most delightful creatures. How fitting that they should be the living descendants of the dinosaurs.

Making the connection between a songful seed eater and a murderous meat eater requires a certain leap of faith, but the gap narrows considerably if account is taken of *Archaeopteryx*, the earliest bird.

There are seven specimens of *Archaeopteryx*, all collected from the Upper Jurassic of Bavaria, in southern Germany, near the towns of Solnhofen and Eichstätt (Wellnhofer, 1988a). They were all preserved in the fine-grained limestones, once used in the printing process and therefore referred to as lithographic limestones. The nature of the sediments helps explain the remarkable state of preservation of the fossils, where details of the feather impressions can often be seen. (The feathers of *Archaeopteryx* are usually described as being impressions imprinted on the limestone, but it appears that this may be an oversimplification of how these facsimiles have been formed—Rietschel, 1985.) The first specimen,[1] the impression of an isolated feather, was discovered in 1860 and reported the following year by the German paleontologist Hermann von Meyer. Within less than two months von Meyer reported a second specimen from the same locality, this time an almost complete skeleton with well-defined feathers (Meyer, 1861). He described the skeleton and named it *Archaeopteryx lithographica.*

To appreciate the impact that the discovery of *Archaeopteryx* had on the scientific world, we need to remind ourselves of the remarkable timing of the event. This was 1861, just two years after publication of Darwin's controversial book, *On the Origin of Species.*

If the skeleton of *Archaeopteryx* had been like that of a modern bird I doubt it would have attracted very much attention. But here was a feathered animal with an unmistakably reptilian skeleton. It had long been recognized that birds and reptiles were closely related because they share a number of features, including the possession of scales and the laying of shelled eggs, but here was solid fossil evidence of the link between the two major groups. Darwin's proponents, and his opponents, seized every opportunity to use this remarkable discovery to further their cause. The German paleontologist Johann Wagner, who had not actually seen the specimen for himself, dismissed *Archaeopteryx* as nothing more than a feathered reptile. He gave a warning that Darwin and his followers would use its seeming transitional nature between reptiles and birds to support their evolutionary views, but that they would be wrong. Then, with complete disregard for the principle of priority, he attempted to rename the new fossil *Griphosaurus problematicus,* "the puzzling lizard" (Wagner, 1861).

Richard Owen, the final authority on comparative anatomy, made a detailed study of the specimen and declared that *Archaeopteryx* was a bird, albeit a primitive bird possessing certain reptilian characters (1864). He did not discuss the matter further—perhaps because of his ambivalence toward the theory of evolution—but Thomas Henry Huxley, champion of the Darwinian cause, made much of *Archaeopteryx* (1868). He gave a detailed account of its anatomy, emphasizing its transitional features, and in so doing he made a strong case for the theory of evolution. His meticulous observations revealed that *Archaeopteryx* was remarkably similar

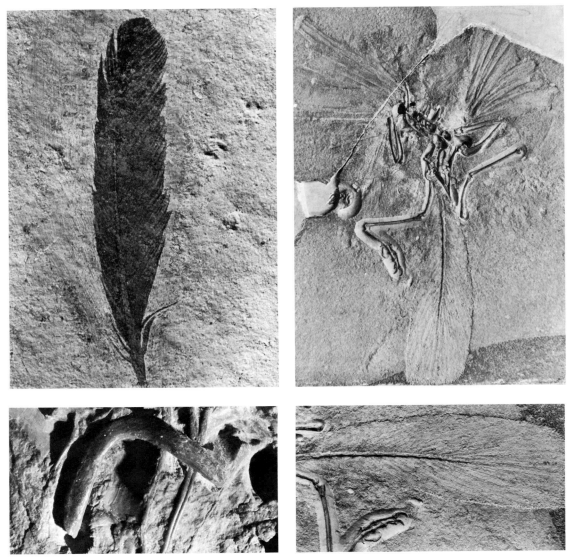

ARCHAEOPTERYX LITHOGRAPHICA *fossils from the Solnhofen limestone of the Upper Jurassic of southern Germany. The first specimen found, shown at upper left, was an isolated feather. Also on this page are views of the London specimen: the entire skeleton at upper right and close-ups of the furcula and the tail in the bottom row. On the facing page are two specimens from Germany, the Berlin specimen at top left and the Eichstätt specimen at top right. In the bottom row the claws and skull of the Berlin specimen are shown in greater detail.*

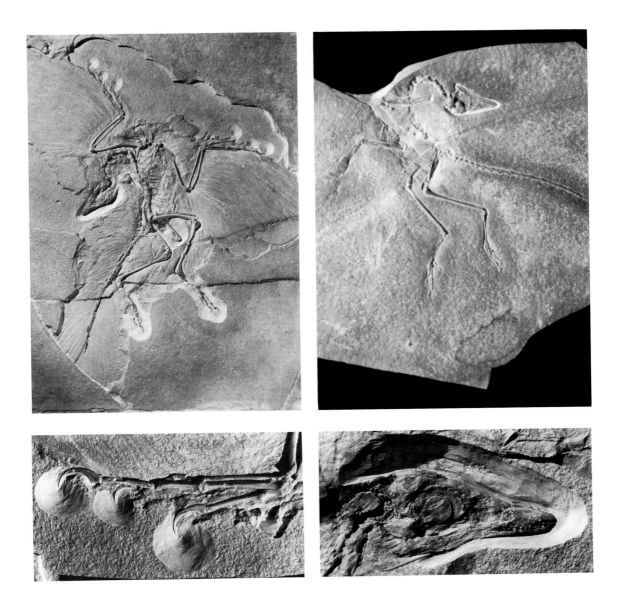

to a dinosaur. Indeed, it has often since been said that if it had not been for the preservation of feathers *Archaeopteryx* would have been classified as a small theropod dinosaur.

Huxley's findings led him to propose that birds had actually evolved from dinosaurs, a view that gained wide acceptance. But the proposal was rejected during the 1920s, following the publication of Gerhard Heilmann's classical study on avian ancestry, *The Origin of Birds* (1926). This authoritative work concluded that although *Archaeopteryx* had much in common with small theropod dinosaurs (the coelurosaurs), a single anatomical feature disqualified them from being ancestral to the birds. The feature in question was the paired clavicles, or collar bones, that form part of the shoulder girdle. *Archaeopteryx,* like most modern birds, has a furcula (the wishbone), which is believed to represent the fused clavicles, but Heilmann pointed out that theropods had already lost their clavicles and were therefore too specialized to have given rise to the birds. However, many new theropods have been discovered since Heilmann's time, and some of these do have clavicles, so this objection no longer holds. Thus argued John Ostrom of Yale University, who, through his painstakingly careful study of the comparative anatomy of *Archaeopteryx* and theropod dinosaurs, made a convincing case that Huxley's hypothesis should be reinstated (1973, 1976b). Although some dissenting voices were raised, mainly by those who believed that crocodiles were the ancestral group, there is now wide and general agreement that birds did evolve from theropod dinosaurs.

Some paleontologists—influenced by the cladistic method of classification, which clusters all organisms sharing the same common ancestor into the same group, or clade—actually classify birds as dinosaurs (Gauthier and Padian, 1985). While this grouping accords with the philosophy of cladistics, I am not alone in thinking that a classification system should, above all else, be practical. It is therefore more useful to separate the birds from their dinosaurian ancestors, rather than classifying them as "avian dinosaurs." Aside from all other considerations, a separate category makes for a more logical organization of specimens in museum collections—imagine the inconvenience if all bird collections had to be amalgamated into the dinosaur collections.

What sort of an animal was *Archaeopteryx*? *Archaeopteryx* was about the size of a raven and if we saw one in our back garden we would certainly recognize it as being a bird. It had feathers, wings, and a long tail, and it would not look very much out of place among the other birds, certainly not from a distance anyway. But a closer look would reveal that it had three long, bony fingers extending from its wings, each ending in a sharp claw, and if it opened its mouth we would catch a glimpse of its small, sharp teeth. Most of its dinosaurian features were worn on the inside, and if we got to dissect one and compare its skeletal features with those of a modern bird, we would notice considerable differences. For example, the long tail of *Archaeopteryx* was largely formed by the bony vertebral column, which supported a fringe of feathers. The tails of modern birds, in contrast, are formed almost entirely of feathers. Modern tail feathers are considerably longer than those of *Archaeopteryx,* and they are attached at their base to a stubby tail, often referred to in culinary circles as the "parson's nose."

The pelvic girdle of *Archaeopteryx* is very much like that of a theropod dinosaur, being a narrow, three-pronged structure. In contrast, modern birds usually have a broad pelvis, and there is such extensive fusion of its component parts that a three-pronged structure cannot be seen.

Extensive bony fusion, mentioned in Chapter 11 with regard to the vertebral column, is a characteristic feature of modern birds. This would be a very obvious difference in any comparisons with the skeleton of *Archaeopteryx.* In the skull, for example, it would be possible to distinguish many of the individual bones of which it is comprised, but these are all fused together in modern birds.

From the outside, then, *Archaeopteryx* and a modern bird would not look so very different, but major differences would be discovered beneath their feathers. This might raise the question as to whether there really was a close relationship between them, but the main differences in the skeleton of the modern bird—the extensive bone-to-bone fusion—arise during the later stages of development. If the skeletons of a bird embryo, *Archaeopteryx,* and a theropod dinosaur are compared, the similarities are striking, even to details of certain small bones in the ankle region (McGowan, 1985). The bird embryo, for example, has a simple, three-pronged pelvis, there are three rudimentary fingers in the wing, and the individual bones of the skull are clearly visible. There is no question that birds are the direct descendants of theropod dinosaurs.

Every year, during the spring school break, the Royal Ontario Museum organizes an extravaganza of special activities to entertain and enlighten young visitors and their parents. There are behind-the-scenes visits, craft activities, film shows, lectures—something for everyone. The Department of Vertebrate Palaeontology is especially popular, and our staff is kept busy for the entire week, but we all have a great time. One year I was told by one of our guards of a distraught little boy who was sobbing his heart out in the dinosaur gallery. When the guard explained what had upset him I dashed off to find the youngster, but by the time

I got to the gallery the boy had gone. What a pity, because I think I could have dried his tears. The reason he was so sad, he had told the guard between sobs, was that all the dinosaurs were dead. I would have put a comforting arm around his shoulder and explained that the dinosaurs were really not dead at all—they live on today as birds.

Notes

1. Material Things

1. Force is the product of mass and acceleration. In the example in the text, the mass is that of the body, measured in pounds or kilograms, and the acceleration is that due to the Earth's gravity, measured in feet per second per second or in meters per second per second. When a mass of 1 kg experiences an acceleration of 1 meter per second per second, the force is 1 newton.
2. Stress cannot be measured directly. Instead, engineers measure strain (see note 3) and use the value for the Young's modulus of the material to calculate stress.
3. Young's modulus is stress/strain. Strain is the change in length divided by the original length and is therefore without units, which means that Young's modulus and stress have the same unit of measure (newtons per square meter, or N/m^2).
4. Work is the product of force (newtons) and distance (meters) and is measured in units of joules.
5. Tensile strength is defined as the stress (force per unit area) at the instant that the material being stretched breaks (in N/m^2). Compressive strength is the stress at the instant when the material being compressed breaks. Because the breaking point is less clearly defined in compressive tests than in tensile tests (the test specimen usually buckles, rather than breaking cleanly as in tensile tests), compressive strength cannot be measured with such accuracy. It is therefore less often used, and many tables of physical constants do not include compressive strengths of materials.
6. The apatite found in bone is specifically calcium hydroxyapatite, which is a crystalline form of calcium phosphate.
7. There is a large literature on bone and bone structure. Two books are recommended for further reading: Currey, 1984, and Evans, 1973.
8. Normal light can be visualized as a wave-form vibrating from side to side as well as up and down. When light passes through a polar-

izer, which is essentially a very fine grid of horizontal or of vertical lines, only the side-to-side or up-and-down vibrations will pass through. Polarized light, then, vibrates only in a single plane.

9. Strain energy = $\frac{1}{2}$ (stress^2/Young's modulus).

10. Because strain energy is proportional to the square of the stress, very large strain energies can be stored in materials, even those with a high Young's modulus, when large stresses are involved. For example, when ships are being towed enormous amounts of strain energy are stored in the steel cables, in spite of the high Young's modulus of steel. Sometimes towing cables break and the loose cables will have so much energy they can cut a man in half.

3. How the Vertebrate Skeleton Works

1. A friend at the riding stable tells me that the training of horses for show jumping includes back-stretching excercises. Horses with more supple backs are better able to clear high fences, presumably because of the storage of some strain energy.

2. A. J. Higgins, of the Department of Athletics and Recreation, University of Toronto, kindly shared his experiences as a trainer with me. He points out that children, being more supple in their limbs than adults, naturally tuck in their lower legs when they run.

4. Reading a Dinosaur Skeleton

1. Sykes (1971) states that the elephant is naturally amphibious or semi-aquatic, a fact well known to "indigenous Africans or Asiatics but apparently almost totally overlooked or ignored by European and American explorers, settlers, and scientists foreign to these countries." The elephants that were observed by Eltringham (1982, p. 113), however, "could have bathed at any time, for they were never more than a few kilometres from a lake or river, yet none of them took the opportunity of doing so during the study period."

2. The relative amounts of work that have to be performed during the power stroke (retraction) and the recovery stroke (protraction) probably vary according to speed and according to species. Quantifying this experimentally would be difficult, but experiments have been conducted to assess the kinetic energy (K.E. = $\frac{1}{2}$ mass $\times$ velocity^2) of limbs in motion, which is probably proportional to the amounts of energy that have been supplied to the limbs by their muscles. Comparisons between an ostrich and a horse of similar body mass, running at the same speeds, have shown that the kinetic energy of the horse's limbs were greater during the power stroke than the recovery stroke, as might be anticipated. In the ostrich, however, the kinetic energies were about the same for each stroke. See Fedak, Heglund, and Taylor, 1982.

3. The degree of development of binocular vision depends on the development of the fovea (a particularly sensitive spot on the retina) as well as on the placement of the eyes. Most animals have a single

fovea, but some birds have two, and these have particularly well developed binocular vision. See Sillman, 1973.

4. In some reptiles and birds the eardrum lies at the bottom of a shallow depression or short canal.

5. A Matter of Scale

1. Mass is a measure of the amount of material in an object and has units such as kilograms, grams, pounds, and ounces. The mass of an object is independent of gravity, which means that an object that has a mass of 40 kg on Earth will have the same mass on the Moon. In everyday usage weight has the same units as mass. Technically, however, weight should be expressed in newtons and not in kilograms, for it is a measure of the force of gravity on an object (weight = mass × gravity). An object that weighs 40 kg on Earth will weigh less than 7 kg on the Moon because the gravity of the Moon is one-sixth that of the Earth. In deep space the object would have no weight, but it would still have mass.

2. These data were obtained from Eltringham, 1982; Hill, 1950; and Spector, 1956.

3. Atmospheric pressure is always present and all pressures are therefore expressed with respect to atmospheric pressure, that is, over and above atmospheric pressure. For example, suppose that in the experiment in the text it was found that the water pressure in the garden hose was equal to a column of water that was 32 feet high—that is, equivalent to atmospheric pressure. This means that the water pressure is actually one atmosphere *above* atmospheric pressure.

  How do we know that atmospheric pressure is equivalent to a 32-foot column of water? A piece of glass pipe, more than 32 feet long and sealed at one end, is filled with water and the open end is temporarily closed. The tube is now inverted over a bucket of water such that the tube is vertical and the temporarily sealed end dips just beneath the water level. If the seal is now removed the water level in the tube would drop down to a height of 32 feet above the water level in the bucket. The water column is held up by the pressure of the atmosphere pressing down on the surface of the water in the bucket. The instrument we have just made is a barometer, but the more usual form uses mercury instead of water. Mercury is about 13 times denser than water; therefore, the barometer tube need be only about 1/13 as long, about 2.5 feet. Atmospheric pressure (at sea level) is about 30.4 inches (760 mm) of mercury.

4. Two numbers are given for blood pressure, for example, 120/80 mm Hg. The first number refers to systolic pressure, the pressure of the blood leaving the heart when the ventricles contract. The second number refers to diastolic pressure, which is the pressure of the blood leaving the heart when the ventricles are relaxed.

5. A Boeing 727–100 has a maximum gross weight of 169,500 pounds (75.6 tons) and the estimated weight of *Brachiosaurus* is 78 tons. Data from Taylor, 1975; and Colbert, 1962.

6. The center of mass (or center of gravity) is the point of an object that moves as though the object's total mass existed at that point and as though all forces (including the force of gravity) were applied at that point. The supports that are closest to the center of mass will obviously carry most of the weight of a body. In most cars, for example, the center of mass is closer to the front than to the back—because of the heavy engine—and therefore the front wheels carry more weight than the back ones. The center of mass of a refrigerator lies closest to the bottom—because the motor and compressor are near the bottom—therefore it is wise to make sure that one is at the top end of a fridge when helping to move one.

7. For example, Alexander (1985) discussed the biomechanical implications of sauropods rearing up on their hindlegs, and a rearing sauropod is illustrated in Norman's book (1985, p. 97).

8. The weight for *Baluchitherium* was estimated by comparing its 18-foot shoulder height with the 11-foot shoulder height of a 6-ton elephant. Thus *Baluchitherium* would have been about $(18/11)^3 =$ 4.4 times heavier, or about 26 tons, which is about the same weight as the sauropod *Apatosaurus*. The fact that *Baluchitherium* was a mammal of sauropod proportions was pointed out in Ricqlès, 1980.

9. The locomotory costs have been found to be 5.3 joules (J) per kg per stride at the trot-gallop transition speed. Incidentally, when 1 J of work is done per second the power is said to be 1 watt.

10. Data for modern predators were taken from Schaller, 1972; Spector, 1956; and Garland, 1983.

11. The value of $\frac{1}{4}$ for the exponent for mammals and birds is only approximate. The actual value given by Lindstedt and Calder (1981) is 0.20, but for our purposes we can use the approximation of $\frac{1}{4}$ (the numerical error that this introduces into extrapolations is relatively small).

12. Calder (1984), citing one of his earlier papers, gives a value of 0.23 for the exponent for reptiles. This is actually closer to the approximation of $\frac{1}{4}$ used in the text.

13. Laws and Parker (1968) suggest that elephants in the wild have an upper age limit of about 60 years. According to age-weight data depicted in their graph (fig. 3), this age corresponds to a weight of about 6 tons (6,000 kg). Tantor, the male African elephant that died at the Toronto Zoo on August 2, 1989, weighed 6.5 tons (14,300 lb), but he was only 21 years old. Zoo animals get less excercise than they would in the wild, and they are also probably better fed; these differences probably explain why this individual was heavier than expected.

6. *What's Hot and What's Not*

1. The idea that dinosaurs were warm-blooded has a long history. The reader is directed to the following sources for an historical perspective: Bakker, 1986; Desmond, 1975; Wieland, 1942; and Russell, 1965.

2. These data, which are very approximate and serve only as a guide, are from Brattstrom, 1970, and Dawson and Hudson, 1970. The data were originally expressed in terms of calories per 100 gm, but for ease of comparison I have converted the units to joules per kilogram.

3. A temperature *change* of three degrees is written as 3 C° (to indicate three centigrade units) and not as 3° C, which is a specific temperature (three units above freezing point).

4. This definition of the term *behavioral temperature regulation* is taken from Bligh and Johnson, 1973.

5. Personal communication from Mike Plyley, School of Physical and Health Education, University of Toronto, January 1990.

6. Metabolism is discussed in many texts but Shephard, 1987, is not only helpful but also very readable.

7. Other factors are also involved. For example, adaptations in the tissues themselves appear to be of considerable importance. Personal communication from Mike Plyley, School of Physical and Health Education, University of Toronto, January 1990.

8. There are a very few reptiles, like the chameleon, that also have an erect posture.

9. Personal communication from Jack Horner, January 1990.

7. Brains and Intellect

1. The logarithm of a number is the exponent to which another number (the base, usually 10) must be raised to get the original number. Tables showing logarithmic values are widely available.

2. The data used by Jerison for plotting the graphs were obtained from Crile and Quiring, 1940.

3. If data for cartilaginous fishes (the sharks and skates) are included in the graphs, they cluster along the upper line with birds and mammals. Although this suggests high intellectual potentials, these fishes are no more intelligent than others. The anomaly is probably attributable to their large olfactory bulbs.

4. When Jerison (1973, p. 61) assessed a broad range of mammals, the best equation for the slope of the line made by the data was found to be: brain weight = $0.12 \times$ (body weight)$^{2/3}$. If this is used to calculate the brain weight of a 1,000-gram monkey, a value of 12 grams is obtained. Some more recent studies, using a larger sample of mammals, show that the gradient of the mammal graph is closer to $\frac{3}{4}$. See Eisenberg, 1981, and also Armstrong, 1983.

5. The ostrich has a fairly prominent sacral swelling, and humans have a slight swelling. See Streeter, 1904.

6. Jerison (1973) cautions that Marsh's data for the primitive toothed birds of the American west are unreliable.

7. Jerison (1973) gives an estimated brain volume of 0.92 cubic centimeters while Hopson (1977) gives almost twice this value (1.76 cm^3).

8. Not Wholly a Fish

1. For a useful account of the early discoveries of ichthyosaurs, see Howe, Sharpe, and Torrens, 1981.

2. There is some doubt as to the date of discovery of this first ichthyosaur. According to Home's (1814) account, most of the specimen was collected in 1812, the remainder the following year. But according to Lang (1959) the dates were 1810 and 1811. The specimen, whose skull is over one yard (one meter) long, is now in the British Museum (Natural History), catalogue number R1158. This was not the very first ichthyosaur to be found because an imperfect skeleton was discovered in 1803 and examined by Home, but it disappeared soon afterward. For more information, see Delair, 1969.

3. Home, 1814, 1816, 1818, 1819a, 1819b; Conybeare, 1822; Hawkins, 1834, 1840; also see Delair, 1969.

4. The name *ichthyosaur* was first proposed in 1818 by Charles Koenig, who was employed by the British Museum (Natural History) until his untimely death from a fall in 1851. But the name was not accompanied by an adequate description, as is required by the rules of nomenclature, so it was not officially acceptable as of that date. Home's (1819) name of *Proteosaurus was* accompanied by an adequate description, and accordingly it has priority over the name *Ichthyosaurus*. Because the latter name has been in such wide use for such a long time, however, the rule of priority is suspended and *Ichthyosaurus* has become the accepted name.

9. The Mechanics of Swimming

1. The relationship between the thickness of the boundary layer and the viscosity of the fluid can be demonstrated by a simple experiment. Syrup or honey is poured into the bottom of one glass, and water is poured into the bottom of a second one. Both surfaces are lightly sprinkled with flour and a toothpick is now slowly drawn across each surface in turn. There will be little disturbance of the flour particles on the surface of the water, but it is almost impossible to move the toothpick through the syrup without disturbing the surface for a considerable distance. This is because of the thickness of the boundary layer imposed by the high viscosity.

2. The units of viscosity can also be expressed as kilograms per meter second (kg/m·s). If this is used in the equation for the Reynolds number you will easily see how the units cancel out.

3. In this personal communication of December 1989, Webb was reporting some new data generated by J. B. Graham.

10. The Sea Dragons

1. The estimate of the number of ichthyosaurs with inclusions is a very approximate one based on my sampling of the major ichthyosaur collections of Germany. At the time of the survey, made during late 1970s, I had seen eleven such ichthyosaurs. This is reported in McGowan, 1979c.

2. Several years ago (1978a) I described an isolated bone from the Upper Cretaceous of New Jersey, which I believed represented the coracoid (part of the pectoral girdle) of an ichthyosaur. The specimen, which was collected during the last century, was from the Maastrichtian—the last stage of the Cretaceous Period—and seemingly represented the geologically youngest ichthyosaur known. My colleague, Don Baird of Princeton University, New Jersey, was never convinced by my findings, and never missed an opportunity to tease me about them! At length a manuscipt arrived on my desk for review, written by Don, in which he systematically demolished my case. Not only did he show that my "ichthyosaur coracoid" was a plesiosaur pubis, but he also showed that an earlier identification, of yet another isolated ichthyosaur coracoid, was also a misidentification (McGowan, 1973a). Embarrassing stuff this, but I had to agree with the findings; the paper was subsequently published. Aside from showing that McGowan is not very good with his coracoids, Don Baird's paper establishes that the geologically youngest ichthyosaur is Cenomanian in age. See Baird, 1984.
3. Camp's latest estimate of ninety presacrals was made in some revisions to the manuscript that he was working on at the time of his death. Joseph Gregory edited the manuscript and it was published posthumously (Camp, 1980). See the footnotes made by Gregory on pp. 149 and 167 of this paper.
4. Massare's estimate is from a personal communication to the author, January 1990.

11. The Winged Phantom

1. Buckland's reference to vampires should be taken to mean vampire bats, not the mythical blood-sucking creatures of fantasy.
2. The usage of the terms *gliding* and *soaring* is consistent with such authorities as Rayner (1981) and Norberg (1981).
3. Most fast aircraft, like modern airliners, are probably flown onto the ground with the wings in an unstalled condition. For further discussion of this and other aspects of man-made flight, see Kermode, 1972.
4. The width of a wing is measured along its chord, that is, a line drawn through the wing section connecting the middle of the leading edge and the middle of the trailing edge. As in aeronautical engineering, the wing length is the wingspan, that is, the distance between the wingtips, measured along the wing surface. The area of the wing also includes the section of the body between the left and right wings. To allow for the varying width of a wing along its length, the aspect ratio is calculated by dividing the square of the wingspan by the wing area. To see how this relationship is arrived at, we need to write out the equation and expand the terms:

$$\text{aspect ratio} = \frac{(\text{wingspan})^2}{\text{wing area}} = \frac{\text{wingspan} \times \text{wingspan}}{\text{wing area}}.$$

But wing area = wingspan × chord, therefore:

$$\text{aspect ratio} = \frac{\text{wingspan} \times \text{wingspan}}{\text{wingspan} \times \text{chord}} = \frac{\text{wingspan}}{\text{chord}}.$$

12. Out with a Whimper or a Bang?

1. See the reports of the 1988 conference on Global Catastrophes in Earth History, held in Snowbird, Utah, and reported by Kerr (1988a, 1988b).

Epilogue

1. Some feathered fossils were reported from the Solnhofen area in 1820 but the material was subsequently lost. For further information, and for an excellent account of the history of *Archaeopteryx*, see Ostrom, 1976.

References

Alexander, R. McN. 1965. The lift produced by the heterocercal tails of Selachii. *Journal of Experimental Biology,* 43:131–138.

——— 1976. Estimates of speeds of dinosaurs. *Nature,* 261:129–130.

——— 1977. Allometry of the limbs of antelopes (Bovidae). *Journal of Zoology,* 183:125–146.

——— 1984. Elastic energy stores in running vertebrates. *American Zoologist,* 24:85–94.

——— 1985. Mechanics of posture and gait of some large dinosaurs. *Zoological Journal of the Linnean Society,* 83:1–25.

——— 1989. *Dynamics of Dinosaurs and Other Extinct Giants.* New York: Columbia University Press.

Alexander, R. McN., M. B. Bennett, and R. F. Ker. 1986. Mechanical properties and function of the paw pads of some mammals. *Journal of Zoology,* 209:405–419.

Alexander, R. McN., N. J. Dimery, and R. F. Ker. 1985. Elastic structures in the back and their role in galloping in some mammals. *Journal of Zoology,* 207:467–482.

Alexander, R. McN., and A. S. Jayes. 1983. A dynamic similarity hypothesis for the gaits of quadrupedal mammals. *Journal of Zoology,* 201:135–152.

Alexander, R. McN., A. S. Jayes, G. M. O. Maloiy, and E. M. Wathuta. 1979. Allometry of the limb bones of mammals from shrews (*Sorex*) to elephant (*Loxodonta*). *Journal of Zoology,* 189:305–314.

Alexander, R. McN., V. A. Langman, and A. S. Jayes. 1977. Fast locomotion of some African ungulates. *Journal of Zoology,* 183:291–300.

Alexander, R. McN., G. M. O. Maloiy, B. Hunter, A. S. Jayes, and J. Nturibi. 1979. Mechanical stresses in fast locomotion of buffalo (*Syncercus caffer*) and elephant (*Loxodonta africana*). *Journal of Zoology,* 189:135–144.

Alt, D., J. M. Sears, and D. W. Hyndman. 1988. Terrestrial maria: the origins of large basalt plateaus, hotspot tracks and spreading ridges. *Journal of Geology,* 96:647–662.

Alvarez, L. W., W. Alvarez, F. Asaro, and H. V. Michel. 1980. Extraterrestrial cause for the Cretaceous-Tertiary extinction. *Science,* 208:1095–1108.

Alvarez, W., E. G. Kauffman, F. Surlyk, L. W. Alvarez, F. Asaro, and H. V. Michel. 1984. Impact theory of mass extinctions and the invertebrate fossil record. *Science,* 223:1135–1141.

Argast, S., J. O. Farlow, R. M. Gabet, and D. L. Brinkman. 1987. Transport-induced abrasion of fossil reptilian teeth: Implications for the existence of Tertiary dinosaurs in the Hell Creek Formation, Montana. *Geology,* 15:927–930.

Armstrong, E. 1983. Relative brain size and metabolism in mammals. *Science,* 220:1302–1304.

Auffenberg, W. 1981. *The Komodo Monitor.* Gainsville: University Presses of Florida.

Augusta, J. 1962. *A Book of Mammoths.* London: Hamlyn.

Axelrod, D. I., and H. P. Bailey. 1968. Cretaceous dinosaur extinction. *Evolution,* 22:595–611.

Baird, D. 1984. No ichthyosaurs in the Upper Cretaceous of New Jersey . . . or Saskatchewan. *Mosasaur,* 2:129–133.

Bakker, R. T. 1971. Dinosaur physiology and the origin of mammals. *Evolution,* 25:636–658.

―――― 1972. Anatomical and ecological evidence of endothermy in dinosaurs. *Nature,* 238:81–85.

―――― 1975a. Dinosaur renaissance. *Scientific American,* 232:58–78.

―――― 1975b. Experimental and fossil evidence for the evolution of tetrapod bioenergetics. In *Evolution of Tetrapod Energetics.* New York: Springer-Verlag.

―――― 1978. Dinosaur feeding behavior and the origin of flowering plants. *Nature,* 274:661–663.

―――― 1980. Dinosaur heresy—dinosaur renaissance; why we need endothermic archosaurs for a comprehensive theory of bioenergetic evolution. In *A Cold Look at the Warm-Blooded Dinosaurs,* ed. R. D. K. Thomas and E. C. Olsen. Boulder: Westview Press.

―――― 1986. *The Dinosaur Heresies.* New York: William Morrow and Co.

Bakker, R. T., M. Williams, and P. Currie. 1988. *Nanotyrannus,* a new genus of pygmy tyrannosaur, from the Latest Cretaceous of Montana. *Hunteria,* 1:1–26.

Ballinger, R. E., J. W. Nietfeldt, and J. J. Krupa. 1979. An experimental analysis of the role of the tail in attaining high running speed in *Cnemidophorus sexlineatus* (Reptilia: Squamata: Lacertilia). *Herpetologica,* 35:114–116.

Baur, G. 1887. On the morphology of ichthyosaurs. *American Naturalist,* 21:837–840.

Beattie, O., and J. Geiger. 1987. *Frozen in Time: Unlocking the Secrets of the Franklin Expedition.* Saskatoon: Western Producer Prairie Books.

Behrensmeyer, A. K. 1978. Taphonomic and ecological information from bone weathering. *Paleobiology* 4:150–162.

Behrensmeyer, A. K., and D. E. Dechant Boaz. 1980. The Recent bones of Amboseli Park, Kenya, in relation to East African paleoecology. In *Fossils in the Making,* ed. A. K. Behrensmeyer and A. P. Hill. Chicago: University of Chicago Press.

Beland, P., and D. A. Russell. 1980. Dinosaur metabolism and predator/ prey ratios in the fossil record. In *A Cold Look at the Warm-Blooded Dinosaurs*, ed. R. D. K. Thomas and E. C. Olsen. Boulder: Westview Press.

Bennett, A. F. 1982. The energetics of reptilian activity. In *Biology of the Reptilia*, vol. 13, ed. C. Gans. London: Academic Press.

Bennett, A. F., and B. Dalzell. 1973. Dinosaur physiology: a critique. *Evolution,* 27:170–174.

Bennett, A. F., and J. A. Ruben. 1979. Endothermy and activity in vertebrates. *Science,* 206:649–654.

Bennett, A. F., R. S. Seymour, D. F. Bradford, and G. J. W. Webb. 1985. Mass-dependence of anaerobic metabolism and acid-base disturbance during activity in the salt-water crocodile, *Crocodylus porosus. Journal of Experimental Biology,* 118:161–171.

Bennett, M. B., and J. A. Stafford. 1988. Tensile properties of calcified and uncalcified avian tendons. *Journal of Zoology,* 214:343–351.

Bennett, S. C. 1987. New evidence on the tail of Pterosaur *Pteranodon* (Archosauria: Pterosauria). In *Fourth Symposium on Mesozoic Terrestrial Ecosystems, Short Papers,* ed. P. M. Currie and E. H. Koster. Drumheller: Tyrrell Museum of Palaeontology.

———— 1990. A pterodactyloid pterosaur pelvis from the Santana Formation of Brazil: Implications for terrestrial locomotion. *Journal of Vertebrate Paleontology,* 10:80–85.

Benton, M. J., and M. A. Taylor. 1984. Marine reptiles from the Upper Lias (Lower Toarcian, Lower Jurassic) of the Yorkshire coast. *Proceedings of the Yorkshire Geological Society,* 44:399–429.

Biewener, A. A. 1983. Allometry of quadrupedal locomotion: The scaling of duty factor, bone curvature and limb orientation to body size. *Journal of Experimental Biology,* 105:147–171.

———— 1989. Scaling body support in mammals: Limb posture and muscle mechanics. *Science,* 245:45–48.

Bird, R. T. 1939. Thunder in his footsteps. *Natural History,* 43:254–261.

———— 1944. Did *Brontosaurus* ever walk on land? *Natural History,* 53:60–67.

Blake, R. W. 1983. *Fish Locomotion.* Cambridge: Cambridge University Press.

Bligh, J., and K. G. Johnson. 1973. Glossary of terms for thermal physiology. *Journal of Applied Physiology,* 35:941–961.

Bock, W. J. 1969. The origin and radiation of birds. *Annals of the New York Academy of Science,* 167:147–155.

Bogert, C. M. 1959. How reptiles regulate their body temperature. *Scientific American,* 200:105–120.

Bohor, B. F., E. E. Foord, P. J. Modreski, and D. M. Triplehorn. 1984. Mineralogic evidence for an impact event at the Cretaceous-Tertiary boundary. *Science,* 224:867–869.

Bohor, B. F., P. J. Modreski, and E. E. Foord. 1987. Shocked quartz in the Cretaceous-Tertiary boundary clays: Evidence for a global distribution. *Science,* 236:705–709.

Bonaparte, J. F., M. R. Franchi, J. E. Powell, and E. G. Sepulveda. 1984. La formacion Los Alamitos (Campanio-Maastrichtiano) del sudeste de Rio Negro, con descripcion de *Kritosaurus australis* n.sp. (Hadrosauridae). *Asociación Geológica Argentina Revista,* 39:284–299.

Bonaparte, J. F., and M. Vince. 1979. El hallazgo del primer nido de dinosaurios Triasicos (Saurischia, Prosauropoda), Triasico superior de Patagonia, Argentina. *Ameghiniana,* 16:173–182.

Bouvier, M. 1977. Dinosaur Haversian bone and endothermy. *Evolution,* 31:449–450.

Bramble, D. M. 1989. Axial-appendicular dynamics and the integration of breathing and gait in mammals. *American Zoologist,* 29:171–186.

Bramwell, C. D., and G. R. Whitfield. 1974. Biomechanics of *Pteranodon. Philosophical Transactions of the Royal Society of London,* 267:503–581.

Branca, H. W. 1907. Sind alle im Innern von Ichthyosauren liegenden Jungen aus nahmslos Embryonen? *Abhandlungen der preussischen Akademie der Wissenschaften,* 1907:1–34.

Brattstrom, B. H. 1970. Amphibia. In *Comparative Physiology of Thermoregulation,* vol. 1, ed. G. Causey Whittow. New York: Academic Press.

Brett-Surman, M. K. 1979. Phylogeny and palaeobiogeography of hadrosaurian dinosaurs. *Nature,* 277:560–562.

Broili, F. 1927. Ein *Rhamphorhynchus* mit Spuren von Haarbedeckung. *Sitzungsberichte der mathematisch-natur Wissenschaftlichen Abteilung der Bayerischen Akademie der Wissenshaften,* 1927:49–65.

Brouwers, E. M., W. A. Clemens, R. A. Spicer, T. A. Ager, L. D. Carter, and W. V. Sliter. 1987. Dinosaurs on the North Slope, Alaska: High latitude, Latest Cretaceous environments. *Science,* 237:1608–1610.

Brower, J. C. 1983. The aerodynamics of *Pteranodon* and *Nyctosaurus,* two large pterosaurs from the Upper Cretaceous of Kansas. *Journal of Vertebrate Paleontology,* 3:84–124.

Brown, B. 1943. Flying reptiles. *Natural History,* 52:104–111.

Brown, J. H., and B. A. Maurer. 1986. Body size, ecological dominance and Cope's rule. *Nature,* 324:248–250.

Buckland, W. 1829. On the discovery of a new species of pterodactyle in the Lias at Lyme Regis. *Transactions of the Geological Society of London,* 3:217–222.

———— 1836. *The Bridgewater Treatise.* Part 6, *Geology and Mineralogy Considered with Reference to Natural Theology.* London: William Pickering.

Buisonjé, P. H. De. 1985. Climatological conditions during deposition of the Solnhofen limestones. In *The Beginnings of Birds,* ed. M. K. Hecht, J. H. Ostrom, G. Viohl, and P. Wellnhofer. Eichstätt: Freunde des Jura-Museums Eichstätt.

Burghardt, G. M. 1977. Learning processes in reptiles. In *Biology of the Reptiles,* ed. C. Gans, London: Academic Press.

Bürgin, T., O. Rieppel, P. M. Sander, and K. Tschanz. 1989. The fossils of Monte San Giorgio. *Scientific American,* 260:74–81.

Burstein, A. H., J. D. Currey, V. H. Frankel, and D. T. Reilly. 1972. The ultimate properties of bone tissue: The effects of yielding. *Journal of Biomechanics,* 5:35–44.

Burstein, A. H., J. M. Zika, K. G. Heiple, and L. Klein. 1975. Contribution of collagen and mineral to the elastic-plastic properties of bone. *Journal of Bone and Joint Surgery,* 57A:956–961.

Cabanac, M. 1987. Glossary of terms for thermal physiology. *Pflügers Archiv (European Journal of Physiology)*, 410:567–587.

Calder, W. A. 1984. *Size, Function, and Life History.* Cambridge: Harvard University Press.

Callaway, J. M. 1989. *Systematics, Phylogeny and Ancestry of Triassic Ichthyosaurs (Reptilia, Ichthyosauria).* Ph.D diss., University of Rochester, Rochester, New York.

Callaway, J. M., and D. R. Brinkman. 1989. Ichthyosaurs (Reptilia, Ichthyosauria) from the Lower and Middle Triassic of Sulphur Mountain Formation, Wapiti Lake area, British Columbia. *Canadian Journal of Earth Sciences,* 26:1491–1500.

Callaway, J. M., and J.A. Massare. 1989a. *Shastasaurus altispinus* (Ichthyosauria, Shastasauridae) from the Upper Triassic of El Antimonio District, Northwestern Sonora, Mexico. *Journal of Paleontology,* 63:930–939.

———— 1989b. Geographic and stratigraphic distribution of the Triassic Ichthyosauria (Reptilia; Diapsida). *Neues Jahrbuch für Geologie und Palaeontologie, Abhandlungen,* 178:37–57.

Camp, C. L. 1976. Vorläufige Mitteilung über grosse Ichthyosaurier aus der oberen Trias von Nevada. *Sitzungsberichte der Österreichischen Akademie der Wissenschaften, Mathematisch-naturwissenshaftliche Klasse, Abteilung I,* 185:125–134.

———— 1980. Large ichthyosaurs from the Upper Triassic of Nevada. *Palaeontographica,* A 170:139–200.

Camp, C. L., and N. Smith. 1942. Phylogeny and functions of the digital ligaments of the horse. *Memoirs of the University of California,* 13:69–124.

Carey, F. G. 1982. A brain heater in the swordfish. *Science,* 216:1327–1329.

Carey, F. G., J. W. Kanwisher, and E. D. Stevens. 1984. Bluefin tuna warm their viscera during digestion. *Journal of Experimental Biology,* 109:1–20.

Carrier, D. R. 1987. The evolution of locomotor stamina in tetrapods: Circumventing a mechanical constraint. *Paleobiology,* 13:326–341.

Carroll, R. L. 1988. *Vertebrate Paleontology and Evolution.* New York: W. H. Freeman.

Carter, N. L., C. B. Officer, C. A. Chesner, and W. I. Rose. 1986. Dynamic deformation of volcanic ejecta from the Toba caldera: Possible relevance to Cretaceous/Tertiary boundary phenomena. *Geology,* 14:380–383.

Case, T. D. 1978. Speculations on the growth rate and reproduction of some dinosaurs. *Paleobiology,* 4:320–328.

Charig, A. J. 1973. Jurassic and Cretaceous dinosaurs. In *Atlas of Palaeobiogeography,* ed. A. Hallam. Amsterdam: Elsevier.

Clark, D. L., W. Cheng-Yuan, and C. J. Orth. 1986. Conodont survival and low iridium abundances across the Permian-Triassic boundary in South China. *Science,* 233:984–986.

Coe, E. 1980. The role of modern ecological studies in the reconstruction of paleoenvironments in Sub-Saharan Africa. In *Fossils in the Making,* ed. A. K. Behrensmeyer and A. P. Hill. Chicago: University of Chicago Press.

Colbert, E. H. 1961. *Dinosaurs: Their Discovery and Their World.* New York: E. P. Dutton.

—— 1962. The weights of dinosaurs. *American Museum Novitates,* 2076:1–16.

Colbert, E. H., R. B. Cowles, and C. M. Bogert. 1946. Temperature tolerances in the American alligator and their bearing on the habits, evolution and extinction of the dinosaurs. *Bulletin of the American Museum of Natural History,* 86:331–373.

Colbert, E. H., and J. H. Ostrom. 1958. Dinosaur stapes. *American Museum Novitates,* 1900:1–20.

Conybeare, W. D. 1822. Additional notices on the fossil genera *Ichthyosaurus* and *Plesiosaurus. Transactions of the Geological Society of London,* 5:103–123.

Coombs, W. P. 1978. Theoretical aspects of cursorial adaptations in dinosaurs. *Quarterly Review of Biology,* 53:393–418.

—— 1975. Sauropod habits and habitats. *Palaeogeography, Palaeoclimatology, Palaeoecology,* 17:1–33.

—— 1982. Juvenile specimens of the ornithischian dinosaur *Psittacosaurus. Palaeontology,* 25:89–107.

Costa, R. L., and W. S. Greaves. 1981. Experimentally produced wear facets and the direction of jaw motion. *Journal of Paleontology,* 55:635–638.

Crile, G., and D. P. Quiring. 1940. A record of the body weight and certain organ and gland weights of 3690 animals. *Ohio Journal of Science,* 40:219–259.

Crocket, J. H., C. B. Officer, F. C. Wezel, and G. D. Johnson. 1988. Distribution of noble metals across the Cretaceous/Tertiary boundary at Gubbio, Italy: Iridium variation as a constraint on the duration and nature of Cretaceous/Tertiary boundary events. *Geology,* 16:77–80.

Currey, J. D. 1960. Differences in the blood-supply of bone of different histological types. *Quarterly Journal of Microscopical Science,* 101:351–370.

—— 1984. *The Mechanical Adaptions of Bones.* Princeton: Princeton University Press.

Currey, J. D., and R. McN. Alexander. 1985. The thickness of the walls of tubular bones. *Journal of Zoology,* 206:453–468.

Cuvier, G. 1809. Mémoire sur le squelette fossile d'un reptile volant des environs d'Aichstedt, que quelques naturalistes ont pris pour un oiseau, et dont nous formons un genre le sauriens, sous le nom de Ptero-Dactyle. *Annales, Musée d'Histoire nauturelle,* 13:424.

Czerkas, S. J., and E. C. Olson, eds. 1987. *Dinosaurs Past and Present.* Los Angeles: Natural History Museum Foundation.

Dawson, W. R., and J. W. Hudson. 1970. Birds. In *Comparative Physiology of Thermoregulation,* vol. 1, ed. G. Causey Whittow. New York: Academic Press.

De La Beche, H. T., and W. D. Conybeare. 1821. Notice of the discovery of a new fossil animal, forming a link between the *Ichthyosaurus* and crocodile, together with general remarks on the osteology of the *Ichthyosaurus. Transactions of the Geological Society of London,* 5:559–594.

Delair, J. B. 1969. A history of the early discoveries of Liassic ichthyosaurs in Dorset and Somerset (1779–1835) and the first record of the occurrence of ichthyosaurs in the Purbeck. *Proceedings of the Dorset Natural History and Archaeological Society,* 90:115–132.

Demment, M. W., and P. J. Van Soest. 1985. A nutritional explanation for body-size patterns of ruminant and nonruminant herbivores. *American Naturalist,* 125:641–672.

DeNiro, M. J., and S. Weiner. 1988. Use of collagenase to purify collagen from prehistoric bones for stable isotopic analysis. *Geochimica et Cosmochimica Acta* 52:2425–2431.

Desmond, A. J. 1975. *The Hot-Blooded Dinosaurs: A Revolution in Palaeontology.* London: Blond and Briggs.

Dimery, N. J., R. McN. Alexander, and K. A. Deyst. 1985. Mechanics of the ligamentum nuchae of some artiodactyls. *Journal of Zoology,* 206:341–351.

Dimery, N. J., R. McN. Alexander, and R. F. Ker. 1986. Elastic extension of leg tendons in the locomotion of horses (*Equus caballus*). *Journal of Zoology,* 210:415–425.

Dodson, P. 1971. Sedimentology and taphonomy of the Oldman Formation (Campanian), Dinosaur Provincial Park, Alberta (Canada). *Palaeogeography, Palaeoclimatology, Palaeoecology,* 10:21–74.

——— 1974. Dinosaurs as dinosaurs. *Evolution,* 28:494–497.

——— 1975. Taxonomic implications of relative growth in Lambeosaurine hadrosaurs. *Systematic Zoology,* 24:37–54.

——— 1980. Vertebrate burials. *Paleobiology* 6:6–8.

——— In preparation. Counting dinosaurs: How many kinds were there?

Donovan, S. K., 1987a. Iridium anomalous no longer? *Nature,* 326:331–332.

——— 1987b. How sudden is sudden? *Nature,* 328:109.

Doran, G. H., D. N. Dickel, W. E. Ballinger, O. F. Agee, P. J. Laipis, and W. W. Hauswirth. 1986. Anatomical, cellular and molecular analysis of 8,000-yr-old human brain tissue from the Windover archaeological site. *Nature* 323:803–806.

Drevermann, F. 1926. Eine neue Ichthyosaura mit Jungen im Senckenberg-Museum. *Bericht der Senckenbergischen Naturforschenden Gesellschaft,* 56:181–186.

Duncan, R. A., and D. G. Pyle. 1988. Rapid eruption of the Deccan flood basalts at the Cretaceous/Tertiary boundary. *Nature,* 333:841–843.

Dunham, A. E., K. L. Overall, W. P. Porter, and C. A. Forster. 1989. Implications of ecological energetics and biophysical and developmental constraints for life-history variation in dinosaurs. In *Paleobiology of the Dinosaurs,* ed. J. O. Farlow. Boulder: Geological Society of America Special Paper 238.

Eaton, G. F. 1910. Osteology of *Pteranodon. Memoirs of the Connecticut Academy of Arts and Science,* 2:1–38.

Edinger, T. 1942. The pituitary body in giant animals, fossil and living: A survey and a suggestion. *Quarterly Review of Biology,* 17:31–45.

Edmund, A. G. 1960. Tooth replacement phenomena in the lower vertebrates. *Life Sciences Contributions of the Royal Ontario Museum,* 52:1–190.

Efremov, J. A. 1940. Taphonomy: A new branch of paleontology. *Pan American Geologist*, 74:81–93.

Eisenberg, J. F. 1981. *The Mammalian Radiations.* Chicago: University of Chicago Press.

Eltringham, S. K. 1982. *Elephants.* Poole: Blandford Press.

Evans, F. G. 1973. *Mechanical Properties of Bone.* Springfield: Charles C. Thomas.

Ewart, J. C. 1921. The nestling feathers of the mallard with observations on the composition, origin and history of feathers. *Proceedings of the Zoological Society of London*, 1921:609–642.

Farlow, J. O. 1976. Speculations about the diet and foraging behaviour of large carnivorous dinosaurs. *Amer. Midland Naturalist*, 95:186–91.

——— 1980. Predator/prey biomass ratios, community food webs and dinosaur physiology. In *A Cold Look at the Warm-Blooded Dinosaurs,* ed. R. D. K. Thomas and E. C. Olsen. Boulder: Westview Press.

Fedak, M. A., N. C. Heglund, and C. R. Taylor. 1982. Energetics and mechanics of terrestrial locomotion. II. Kinetic energy changes of the limbs and body as a function of speed and body size in birds and mammals. *Journal of Experimental Biology*, 97:23–40.

Frair, W., R. G. Ackman, and N. Mrosovsky. 1972. Body temperature of *Dermochelys coriacea:* Warm turtle from cold water. *Science*, 177:791–793.

Galdikas, B. M. F. 1978. Orangutan death and scavenging by pigs. *Science*, 200:68–70.

Galton, P. M. 1970. The posture of hadrosaurian dinosaurs. *Journal of Paleontology*, 44:464–467.

Garland, T. 1983. The relation between maximal running speed and body mass in terrestrial mammals. *Journal of Zoology*, 199:157–170.

Gauthier, J., and K. Padian. 1985. Phylogenetic, functional, and aerodynamic analyses of the origin of birds and their flight. In *The Beginnings of birds,* ed. M. K. Hecht, J. H. Ostrom, G. Viohl and P. Wellnhofer. Eichstätt: Freunde des Jura-Museums Eichstätt.

Giffin, E. B. 1989. Pachycephalosaur paleoneurology (Archosauria: Ornithischia). *Journal of Vertebrate Paleontology*, 9:67–77.

——— 1990. Sacral neural canal enlargement in dinosaurs [Abstract]. *Journal of Vertebrate Paleontology*, 10:26A.

Gooding, R. M., W. H. Neill, and A. E. Dizon. 1981. Respiration rates and low-oxygen tolerance limits in skipjack tuna, *Katsuwonus pelamis. Fisheries Bulletin*, 79:31–48.

Gostin, V. A., R. R. Keays, and M. W. Wallace. 1989. Iridium anomaly from the Acraman impact ejecta horizon: Impacts can produce sedimentary iridium peaks. *Nature*, 340:542–544.

Graham, J. B., F. J. Koehrn, and K. A. Dickson. 1983. Distribution and relative proportions of red muscle in scombrid fishes: Consequences of body size and relationships to locomotion and endothermy. *Canadian Journal of Zoology*, 61:2087–2096.

Graup, G., and B. Spettel. 1989. Mineralogy and phase-chemistry of an Ir-enriched pre-K/T layer from the Lattengebirge, Bavarian Alps, and significance for the KTB problem. *Earth and Planetary Science Letters*, 95:271–290.

Green, W. 1975. *Famous Fighters of the Second World War.* London: MacDonald.

Hallam, A. 1987. End-Cretaceous mass extinction event: Argument for terrestrial causation. *Science,* 238:1237–1242.

Hankin, E. H. 1910. *Animal Flight.* London: Iliffe and Sons Ltd.

Hare, P. E. 1974. Amino acid dating of bone—the influence of water. *Carnegie Institution of Washington Year Book,* 73:576–581.

——— 1980. Organic geochemistry of bone and its relation to the survival of bone in the natural environment. In *Fossils in the Making,* ed. A. K. Behrensmeyer and A. P. Hill. Chicago: University of Chicago Press.

Hargens, A. R., R. W. Millard, K. Pettersson, and K. Johansen. 1987. Gravitational haemodynamics and oedema prevention in the giraffe. *Nature,* 329:59–60.

Harland, W. B., A. V. Cox, P. G. Llewellyn, C. A. G. Pickton, A. G. Smith, and R. Walters. 1982. *A Geologic Time Scale.* Cambridge: Cambridge University Press.

Harlow, H. J., S. S. Hillman, and M. Hoffman. 1976. The effect of temperature on digestive efficiency in the herbivorous lizard, *Dipsosaurus dorsalis. Journal of Comparative Physiology,* 111:1–6.

Hauff, B. 1954. *Das Holzmadenbuch.* Öhringen: F. Rau.

Hawkins, T. 1834. *Memoirs of Ichthyosauri and Plesiosauri, Extinct Monsters of the Ancient Earth.* London: Relfe and Fletcher.

——— 1840. *The Book of the Great Sea-Dragons, Ichthyosauri and Plesiosauri, Gedolim Taninim of Moses. Extinct Monsters of the Ancient Earth.* London: W. Pickering.

Heglund, N.C., and C. R. Taylor. 1988. Speed, stride frequency and energy cost per stride: How do they change with body size and gait? *Journal of Experimental Biology,* 138:301–318.

Heglund, N. C., C. R. Taylor, and T. A. McMahon. 1974. Scaling stride frequency and gait to animal size: Mice to horses. *Science,* 186:1112–1113.

Heilmann, G. 1926. *The Origin of Birds.* London: H. F. and G. Whitherby.

Heinrich, B. 1974. Thermoregulation in endothermic insects. *Science,* 185:747–756.

——— 1987. Thermoregulation by winter-flying endothermic moths. *Journal of Experimental Biology,* 127:313–332.

Heller, W. 1966. Untersuchungen zur sogenannten Hauterhaltung bei Ichthyosauriern aus dem Lias epsilon Holzmadens (Schwaben). *Neues Jahrbuch für Geologie und Paläontologie,* 1966:304–317.

Hickey, L. J. 1981. Land plant evidence compatible with gradual, not catastrophic, change at the end of the Cretaceous. *Nature,* 292:529–531.

Hildebrand, M., and J. P. Hurley. 1985. Energy of the oscillating legs of a fast-moving cheetah, pronghorn, jackrabbit, and elephant. *Journal of Morphology,* 184:23–31.

Hill, A. P. 1980. Early postmortem damage to the remains of some contemporary East African mammals. In *Fossils in the Making,* ed. A. K. Behrensmeyer and A. P. Hill. Chicago: University of Chicago Press.

Hill, A. V. 1950. The dimensions of animals and their muscular dynamics. *Science Progress,* 38:209–230.

Hohnke, L. A. 1973. Haemodynamics in the Sauropoda. *Nature*, 244:309–310.

Home, E. 1814. Some account of the fossil remains of an animal more nearly allied to fishes than any other classes of animals. *Philosophical Transactions of the Royal Society of London*, 101:571–577.

———— 1816. Some further account of the fossil remains of an animal, of which description was given to the Society in 1814. *Philosophical Transactions of the Royal Society of London*, 106:318–321.

———— 1818. Additional facts respecting the fossil remains of an animal, on the subject of which two papers have been printed in the Philosophical Transactions, showing that the bones of the sternum resemble those of *Ornithorhynchus paradoxus*. *Philosophical Transactions of the Royal Society of London*, 108:24–32.

———— 1819a. An account of the fossil skeleton of *Proteosaurus*. *Philosophical Transactions of the Royal Society of London*, 109:209–211.

———— 1819b. Reason for giving the name *Proteosaurus* to the fossil skeleton, which has been described. *Philosophical Transactions of the Royal Society of London*, 109:212–216.

Hopson, J. A. 1975. The evolution of cranial display structures in hadrosaurian dinosaurs. *Paleobiology*, 1:21–43.

———— 1976. Hot, cold-, or lukewarm-blooded dinosaurs? Review of *The Hot-Blooded Dinosaurs: A Revolution in Paleontology*, by Adrian J. Desmond. *Paleobiology*, 2:271–275.

———— 1977. Relative brain size and behavior in archosaurian reptiles. *Annual Review of Ecology and Systematics*, 8:429–448.

Horner, J. R. 1984. The nesting behavior of dinosaurs. *Scientific American*, 250:130–137.

Horner, J. R., and J. Gorman. 1988. *Digging Dinosaurs*. New York: Workman.

Horner, J. R., and R. Makela. 1979. Nest of juveniles provides evidence of family structure among dinosaurs. *Nature*, 282:296–298.

Hotton, N. 1980. An alternative to dinosaur endothermy; the happy wanderers. In *A Cold Look at the Warm-Blooded Dinosaurs*, ed. R. D. K. Thomas and E. C. Olsen. Boulder: Westview Press.

Houston, R. S., H. Toots, and J. C. Kelley. 1966. Iron content of fossil bones of Tertiary Age in Wyoming correlated with climatic change. *Geology*, 5:1–18.

Howe, S. R., T. Sharpe, and H. S. Torrens. 1981. *Ichthyosaurs: A History of Fossil Sea Dragons*. Cardiff: The National Museum of Wales.

Hsü, K. J., J. A. McKenzie, and Q. X. He. 1982. Terminal Cretaceous environmental and evolutionary changes. In *Geological Implications of Impacts of Large Asteroids and Comets on the Earth*, ed. L. T. Silver and P. H. Shultz. Boulder: Geological Society of America.

Huey, R. B., and A. F. Bennett. 1987. Phylogenetic studies of coadaptation: Preferred temperatures versus optimal performance temperatures of lizards. *Evolution*, 41:1098–1115.

Huxley, T. H. 1868. On the animals which are most nearly intermediate between the birds and reptiles. *Annals and Magazine of Natural History*, 2:66–75.

Jansa, L. F., and G. Pe-Piper. 1987. Identification of an underwater extraterrestrial impact crater. *Nature,* 327:612–614.

Jensen, J. A. 1985. Three new sauropod dinosaurs from the Upper Jurassic of Colorado. *Great Basin Naturalist,* 45:697–709.

Jerison, H. J. 1969. Brain evolution and dinosaur brains. *American Naturalist,* 934:575–588.

——— 1973. *Evolution of the Brain and Intelligence.* New York: Academic Press.

Joanen, T., and L. L. McNease. 1989. Ecology and physiology of nesting and early development of the American alligator. *American Zoologist,* 29:987–998.

Johnson, K. R., D. J. Nichols, M. Attrep, and C. J. Orth. 1989. High resolution leaf-fossil record spanning the Cretaceous/Tertiary boundary. *Nature,* 340:708–711.

Jouventin, P., and H. Weimerskirch. 1990. Satellite tracking of Wandering albatrosses. *Nature,* 343:746–748

Keller, G. 1989. Extended period of extinctions across the Cretaceous/Tertiary boundary in planktonic foraminifera of continental-shelf sections: Implications for impact and volcanism theories. *Geological Society of America Bulletin,* 101:1408–1419.

Ker, R. F., M. B. Bennett, S. R. Bibby, R. C. Kester, and R. McN. Alexander. 1987. The spring in the arch of the human foot. *Nature,* 325:147–149.

Kermode, A. C. 1972. *Mechanics of Flight.* London: Pitman Books.

Kerr, R. A. 1988a. Huge impact is favored K-T boundary killer. *Science,* 242:865–867.

——— 1988b. Snowbird II: Clues to Earth's impact history. *Science,* 242:1380–1382.

Koch, C. F. 1987. Prediction of sample size effects on the measured temporal and geographic distribution patterns of species. *Paleobiology,* 13:100–107.

Kosch, B. F. In press. A revision of the skeletal reconstruction of *Shonisaurus popularis* (Reptilia—Ichthyosauria). *Journal of Vertebrate Paleontology,* 10(1990): 512–514.

Kräusel, R. 1922. Die Nahrung von *Trachodon. Palaeontologische Zeitschrift,* 4:80.

Krech, D., R. S. Crutfield, and N. Livson. 1969. *Elements of Psychology.* New York: Alfred A. Knopf.

Kunk, M. J., G. A. Izett, R. A. Haugerud, and J. F. Sutter. 1989. [40]Ar-[39]Ar dating of the Manson impact structure: A Cretaceous-Tertiary boundary crater candidate. *Nature,* 244:1565–1568.

Kurzanov, S. M. 1981. Peculiarities of the structure of the anterior extremities of *Avimimus. Paleontologicheskiy Zhurnal (Akademiya Nauk SSSR),* 24:108–112. [In Russian.]

Lang, W. D. 1959. Mary Anning's escape from lightning. *Proceedings of the Dorset Natural History and Archaeological Society,* 80:91–93.

Langston, W. 1960. The vertebrate fauna of the Selma Formation of Alabama. Part 6: The dinosaurs. *Fieldiana: Geology Memoirs,* 3:315–361.

———— 1981. Pterosaurs. *Scientific American,* 244:122–136.

Laws, R. M., and I. S. C. Parker. 1968. Recent studies on elephant populations in East Africa. *Symposium of the Zoological Society of London,* 21:319-359.

Lawson, D. A. 1975. Pterosaur from the Latest Cretaceous of West Texas: Discovery of the largest flying creature. *Science,* 187:947–948.

Leadbeater, B. S. C., and R. Riding, eds. 1986. *Biomineralization in Lower Plants and Animals.* Oxford: Clarendon Press.

Leidy, J. 1868. Notice of some reptilian remains from Nevada. *Proceedings of the Philadelphia Academy of Science,* 20:177–178.

Liepmann, W. 1926. Leichengeburt bei Ichthyosauriern. *Sitzungsberichte der Heidelberger Akademie der Wissenschaften,* B 6:1–11.

Lindstedt, S. L., and W. A. Calder. 1981. Body size, physiological time, and longevity of homeothermic animals. *Quarterly Review of Biology,* 56:1–16.

Mackay, R. S. 1964. Galápagos tortoise and marine iguana deep body temperatures measured by radio telemetry. *Nature,* 204:355–358.

Manley, G. A. 1971. Some aspects of the evolution of hearing in vertebrates. *Nature,* 230:506–509.

———— 1972. A review of some current concepts of the functional evolution of the ear in terrestrial vertebrates. *Evolution,* 26:608–621.

Mantell, G. A. 1825. Notice on the *Iguanodon,* a newly discovered fossil reptile, from the sandstone of Tilgate Forest, in Sussex. *Philosophical Transactions of the Royal Society of London,* 115:179–186.

Martill, D. M. 1987. Prokaryote mats replacing soft tissues in Mesozoic marine reptiles. *Modern Geology,* 11:265–269.

Martill, D. M., and D. M. Unwin. 1989. Exceptionally well preserved pterosaur wing membrane from the Cretaceous of Brazil. *Nature,* 340:138–140.

Martin, J. 1987. Mobility and feeding of *Cetiosaurus* (Saurischia, Sauropoda)—Why the long neck? In *Fourth Symposium on Mesozoic Ecosystems,* ed. P. J. Currie and E. H. Koster. Drumheller: Tyrrell Museum of Palaeontology.

Massare, J. A. 1987. Tooth morphology and prey preference of Mesozoic marine reptiles. *Journal of Vertebrate Paleontology,* 7:121–137.

———— 1988. Swimming capabilities of Mesozoic marine reptiles: Implications for method of predation. *Paleobiology,* 14:187–205.

Mazin, J. 1981. *Grippia longirostris* Wiman, 1929, un Ichthyopterygia primitif du Trias inférieur du Spitsberg. *Bulletin du Muséum national d'histoire naturelle,* 4:317–340.

McGhee, G. R., J. S. Gilmore, C. J. Orth, and E. Olsen. 1984. No geochemical evidence for an asteroidal impact at late Devonian mass extinction horizon. *Nature,* 308:629–631.

McGowan, C. 1972. The systematics of Cretaceous ichthyosaurs with particular reference to the material from North America. *Contributions to Geology,* 11:9–29.

———— 1973a. A note on the most recent ichthyosaur known: An isolated coracoid from the Upper Campanian of Saskatchewan (Reptilia: Ichthyosauria). *Canadian Journal of Earth Sciences,* 10:1346–1349.

———— 1973b. The cranial morphology of the Lower Liassic latipinnate ichthyosaurs of England. *Bulletin of the British Museum (Natural History), Geology,* 24:1–109.

———— 1973c. Differential growth in three ichthyosaurs: *Ichthyosaurus communis, I. breviceps,* and *Stenopterygius quadriscissus* (Reptilia, Ichthyosauria). *Life Sciences Contributions, Royal Ontario Museum,* 93:1–21.

———— 1974a. A revision of the longipinnate ichthyosaurs of the Lower Jurassic of England with descriptions of two new species (Reptilia: Ichthyosauria). *Life Sciences Contributions, Royal Ontario Museum,* 97:1–37.

———— 1974b. A revision of the latipinnate ichthyosaurs of the Lower Jurassic of England (Reptilia: Ichthyosauria). *Life Sciences Contributions, Royal Ontario Museum,* 100:1–30.

———— 1976. The description and phenetic relationships of a new ichthyosaur genus from the Upper Jurassic of England. *Canadian Journal of Earth Sciences,* 13:668–683.

———— 1978a. An isolated ichthyosaur coracoid from the Maastrichtian of New Jersey. *Canadian Journal of Earth Sciences,* 15:169–171.

———— 1978b. Further evidence for the wide geographical distribution of ichthyosaur taxa (Reptilia: Ichthyosauria). *Journal of Paleontology,* 52:1155–1162.

———— 1979a. The hind limb musculature of the Brown Kiwi, *Apteryx australis mantelli. Journal of Morphology,* 160:33–74.

———— 1979b. Selection pressures for high body temperatures: Implications for dinosaurs. *Paleobiology,* 5:285–295.

———— 1979c. A revision of the Lower Jurassic ichthyosaurs of Germany, with descriptions of two new species. *Palaeontographica,* A 166:93–135.

———— 1982. The wing musculature of the Brown kiwi *Apteryx australis mantelli* and its bearing on ratite affinities. *Journal of Zoology,* 197:173–219.

———— 1983. *The Successful Dragons: A Natural History of Extinct Reptiles.* Toronto: Samuel Stevens.

———— 1985. Tarsal development in birds: Evidence for homology with the theropod condition. *Journal of Zoology,* 206:53–67.

———— 1986a. A putative ancestor for the swordfish-like ichthyosaur *Eurhinosaurus. Nature,* 322:454–456.

———— 1986b. The wing musculature of the Weka (*Gallirallus australis*), a flightless rail endemic to New Zealand. *Journal of Zoology,* 210:305–346.

———— 1988. Differential development of the rostrum and mandible of the swordfish (*Xiphias gladius*) during ontogeny and its possible functional significance. *Canadian Journal of Zoology,* 66:496–503.

———— 1989a. *Leptopterygius tenuirostris,* and other long-snouted ichthyosaurs from the English Lower Lias. *Palaeontology,* 32:409–427.

———— 1989b. Computed tomography reveals further details of *Excalibosaurus,* a putative ancestor for the swordfish-like ichthyosaur *Eurhinosaurus. Journal of Vertebrate Paleontology,* 9:269–281.

———— 1989c. The ichthyosaurian tailbend: A verification problem facilitated by computed tomography. *Paleobiology,* 15:429–436.

———— 1990. Problematic ichthyosaurs from southwest England: A question of authenticity. *Journal of Vertebrate Paleontology,* 10:72–79.

McHone, J. F., R. A. Nieman, C. F. Lewis, and A. M. Yates. 1989. Stishovite at the Cretaceous-Tertiary boundary, Raton, New Mexico. *Science,* 243:1182–1184.

McIntyre, J., ed. 1974. *Mind in the Waters.* Toronto: McLelland and Stewart.

McMahon, T. A. 1975. Allometry and biomechanics: Limb bones of adult ungulates. *American Naturalist,* 109:547–563.

McNab, B. K. 1983. Energetics, body size, and the limits of endothermy. *Journal of Zoology,* 199:1–29.

———— 1986. Food habits, energetics, and the reproduction of marsupials. *Journal of Zoology,* 208:595–614.

Merriam, J. C. 1906. Preliminary note on a new marine reptile from the Middle Triassic of Nevada. *University of California Publications, Bulletin of the Department of Geology,* 5:75–79.

———— 1908. Triassic Ichthyosauria, with special reference to the American forms. *Memoirs of the University of California,* 1:1–196.

Meyer, H. von. 1861. *Archaeopteryx lithographica* (Vogel-Feder) und *Pterodactylus* von Solnhofen. *Neues Jahrbuch für Geologie und Palaeontologie, Monatshefte,* 1861:678–679.

Molnar, R. E. 1987 A pterosaur pelvis from western Queensland, Australia. *Alcheringa,* 11:87–94.

Moodie, R. L. 1923. *Paleopathology: An Introduction to the Study of Ancient Evidences of Disease.* Urbana: University Press.

Morris, W. J. 1970. Hadrosaurian dinosaur bills—Morphology and function. *Los Angeles County Museum Contributions in Science,* 193:1–14.

———— 1981. A new species of hadrosaurian dinosaur from the Upper Cretaceous of Baja California—?*Lambeosaurus laticaudus. Journal of Paleontology,* 55:453–462.

Mrosovsky, N. 1987. Leatherback turtle off scale. *Nature,* 327:286.

Myers, M. J., and K. Steudel. 1985. Effect of limb mass and its distribution on the energetic cost of running. *Journal of Experimental Biology,* 116:363–373.

Nagy, K. A. 1987. Field metabolic rate and food requirement scaling in mammals and birds. *Ecological Monographs,* 57:111–128.

Nopcsa, F. 1929. Sexual differences in orinthopodous dinosaura. *Palaeobiologica,* 2:187–201.

Norberg, U. M. 1981. Flight morphology and the ecological niche in some birds and bats. *Symposium of the Zoological Society of London,* 48:173–197.

Norman, D. 1985. *The Illustrated Encyclopedia of Dinosaurs.* London: Salamander Books Ltd.

Officer, C. B., and C. L. Drake. 1985. Terminal Cretaceous environmental events. *Nature,* 227:1161–1167.

Officer, C. B., A. Hallam, C. L. Drake, and J. D. Devine. 1987. Late Cretaceous and paroxysmal Cretaceous/Tertiary extinctions. *Nature,* 326:143–149.

Oliver, W. R. B. 1955. *New Zealand Birds.* 2d ed. Wellington: A. H. and A. W.Reed.

Orth, C. J., J. D. Knight, L. R. Quintana, J. S. Gilmore, and A. R. Palmer. 1984. A search for iridium abundance anomalies at two Late Cambrian biomere boundaries in western Utah. *Science,* 223:163–165.

Osborn, H. F. 1912. Crania of *Tyrannosaurus* and *Allosaurus. Memoirs of the American Museum of Natural History,* 1:1–30.

Osmolska, H., and E. Roniewicz. 1970. Deinocheiridae, a new family of theropod dinosaurs. *Palaeontologia Polonica,* 21:5–19.

Ostrom, J. H. 1961. Cranial morphology of the hadrosaurian dinosaurs of North America. *Bulletin of the American Museum of Natural History,* 122:37–186.

—— 1969. Osteology of *Deinonychus antirrhopus,* an unusual theropod from the Lower Cretaceous of Montana. *Peabody Museum of Natural History, Yale University Bulletin,* 30:1–165.

—— 1973. The ancestry of birds. *Nature,* 242:136.

—— 1974a. Reply to "Dinosaurs as reptiles." *Evolution,* 28:491–493.

—— 1974b. *Archaeopteryx* and the origin of flight. *Quarterly Review of Biology,* 49:27–47.

—— 1975. The origin of birds. *Annual Review of Earth and Planetary Sciences,* 3:55–77.

—— 1976a. On a new specimen of the Lower Cretaceous theropod dinosaur *Deinonychus antirrhopus. Breviora,* 439:1–21.

—— 1976b. *Archaeopteryx* and the origin of birds. *Biological Journal of the Linnean Society,* 8:91–182.

—— 1978. The osteology of *Compsognathus longipes* Wagner. *Zitteliana Anhandlungen der Bayerischen Staatssammlung für Palaeontologie und historische Geologie,* 4:73–118.

—— 1980. The evidence for endothermy in dinosaurs. In *A Cold Look at the Warm-Blooded Dinosaurs,* ed. R. D. K. Thomas and E. C. Olsen. Boulder: Westview Press.

Owen, R. 1840a. Report on the British fossil reptiles. In *Report of the Ninth Meeting of the British Association for the Advancement of Science, held in Birmingham, August, 1839.* London: John Murray.

—— 1840b. Note on the dislocation of the tail at a certain point observable in the skeleton of many ichthyosauri. *Transactions of the Geological Society of London,* 5:511–514.

—— 1841. A description of some of the soft parts, with the integument, of the hind-fin of the *Ichthyosaurus,* indicating the shape of the fin when recent. *Transactions of the Geological Society of London,* 6:199–201.

—— 1842. Report on British fossil reptiles. Part II. *Report of the British Association for the Advancement of Science, Plymouth,* 11:60–204.

—— 1864. On the *Archaeopteryx* of von Meyer, with a description of the fossil remains of a long-tailed species, from the lithographic stone of Solnhofen. *Philosophical Transactions of the Royal Society of London,* 153:33–47.

Padian, K. 1983. A functional analysis of flying and walking in pterosaurs. *Paleobiology,* 9:218–239.

—— 1987. The case of the bat-winged pterosaur. In *Dinosaurs Past and*

Present, vol. 2, ed. S. J. Czerkas and E. C. Olson. Los Angeles: Natural History Museum of Los Angeles County.

Parker, R. B., and H. Toots. 1980. Trace elements in bones as paleobiological indicators. In *Fossils in the Making,* ed. A. K. Behrensmeyer and A. P. Hill. Chicago: Chicago University Press.

Pearce, J. C. 1846. Notice of what appears to be the embryo of an *Ichthyosaurus (communis ?). Annals and Magazine of Natural History,* 17:44–46.

Pennycuick, C. J. 1972. Soaring behavior and performance of some East African birds, observed from a motor-glider. *Ibis,* 114:178–218.

—— 1982. The flight of petrels and albatrosses (Procellariiformes), observed in South Georgia and its vicinity. *Philosophical Transactions of the Royal Society of London,* 300:75–106.

—— 1988. On the reconstruction of pterosaurs and their manner of flight, with notes on vortex wakes. *Biological Reviews,* 63:299–331.

Peters, R. H. 1983. *The Ecological Implications of Body Size.* Cambridge: Cambridge University Press.

Peters, S. E. 1989. Structure and function in vertebrate skeletal muscle. *American Zoologist,* 29:221–234.

Pettersson, H. 1960. Cosmic spherules and meteoric dust. *Scientific American,* 202:123–132.

Pollard, J. E. 1968. The gastric contents of an ichthyosaur from the Lower Lias of Lyme Regis, Dorset. *Palaeontology,* 11:376–388.

Pooley, A. C., and C. Gans. 1976. The Nile crocodile. *Scientific American,* 234:114–124.

Rayner, J. M. V. 1979. A new approach to animal flight mechanics. *Journal of Experimental Biology,* 80:17–54.

—— 1981. Flight adaptations in vertebrates. *Symposium of the Zoological Society of London,* 48:137–172.

Regal, P. J. 1975. The evolutionary origin of feathers. *Quarterly Review of Biology,* 50:35–66.

Reid, R. E. H. 1981. Lamellar-zonal bone with zones and annuli in the pelvis of a sauropod dinosaur. *Nature,* 292:49–51.

—— 1987. Bone and dinosaurian "endothermy." *Modern Geology,* 11:133–154.

Ricqlès, A. J. de. 1976. On bone histology of fossil and living reptiles, with comments on its functional and evolutionary significance. In *Morphology and Biology of Reptiles,* ed. d'A. Bellairs and C. B. Cox. London: Academic Press.

—— 1980. Tissue structures of dinosaur bone. Functional significance and possible relation to dinosaur physiology. In *A Cold Look at the Warm-Blooded Dinosaurs,* ed. R. D. K. Thomas and E. C. Olson. Boulder: Westview Press.

—— 1983. Cyclical growth in the long limb bones of a sauropod dinosaur. *Acta Palaeontologica Palonica,* 28:225–232.

Riess, J. 1986. Fortbewegungsweise, Schwimmbiophysik und Phylogenie der Ichthyosaurier. *Palaeontographica,* A 192:93–155.

Rietschel, S. 1985. Feathers and wings of *Archaeopteryx,* and the question of her flight ability. In *The Beginnings of Birds,* ed. M. K. Hecht, J. H.

Ostrom, G. Viohl, and P. Wellnhofer. Eichstätt: Freunde des Jura-Museums Eichstätt.

Riggs, E. S. 1903. Structure and relationships of opisthocoelian dinosaurs. Part 1: *Apatosaurus* Marsh. *Field Columbian Museum Publications, Geological Series*, 2:165–196.

Rothschild, B. M. 1987. Diffuse idiopathic skeletal hyperostosis as reflected in the paleontologic record: Dinosaurs and early mammals. *Seminars in Arthritis and Rheumatism*, 17:119–125.

Russell, D. A. 1969. A new specimen of *Stenonychosaurus* from the Oldman Formation (Cretaceous) of Alberta. *Canadian Journal of Earth Sciences*, 6:595–612.

——— 1977. *A Vanished World: The Dinosaurs of Western Canada*. Ottawa: National Museums of Canada.

——— 1984. The gradual decline of the dinosaurs—Fact or fallacy? *Nature*, 307:360–361.

——— 1989. *An Odyssey in Time: The Dinosaurs of North America*. Toronto: University of Toronto Press.

Russell, L. S. 1965. Body temperature of dinosaurs and its relationships to their extinction. *Journal of Paleontology*, 39:497–501.

Saito, T., T. Yamanoi, and K. Kaiho. 1986. End-Cretaceous devastation of terrestrial flora in the boreal Far East. *Nature*, 323:253–255.

Sander, P. M. 1989. The large ichthyosaur *Cymbospondylus buchseri*, sp. nov., from the Middle Triassic of Monte San Giorgio (Switzerland), with a survey of the genus in Europe. *Journal of Vertebrate Paleontology*, 9:163–173.

Sander, P. M., and H. Bucher. 1990. On the presence of *Mixosaurus* (Ichthyopterygia:Reptilia) in the Middle Triassic of Nevada. *Journal of Paleontology*, 64:161–164.

Schaller, G. B. 1972. *The Serengeti Lion*. Chicago: University of Chicago Press.

Schmidt-Nielsen, K. 1972. Locomotion: Energy cost of swimming, flying and running. *Science*, 177:222–228.

Schreiber, R. W., and E. A. Schreiber. 1984. Central Pacific seabirds and the El Niño Southern Oscillation: 1982 to 1983 perspectives. *Science*, 225:713–716.

Scott, E. O. G., and R. H. Green. 1975. Recent whale strandings in Northern Tasmania. *Proceedings of the Royal Society of Tasmania*, 109:91–96.

Seeley, H. G. 1901. *Dragons of the Air: An Account of Extinct Flying Reptiles*. London: Methuen.

Seymour, R. S. 1976. Dinosaurs, endothermy and blood pressure. *Nature*, 262:207–208.

Shapiro, A. H. 1961. *Shape and Flow. The Fluid Dynamics of Drag*. New York: Anchor Books.

Sharov, A. G. 1971. New flying reptiles from the Mesozoic deposits of Kazakhstan and Kirgizia. *Akademiia Nauk SSR, Paleontologicheskii Institut, Trudy*, 130:104–113.

Shephard, R. J. 1984. *Biochemistry of Physical Activity*. Springfield: Charles C. Thomas.

———— 1987. *Exercise Physiology.* Toronto: B. C. Decker Inc.

Shikama, T., K. Tadao, and M. Masafumi. 1978. Early Triassic Ichthyo-saurus, *Utatsusaurus hataii* Gen. et Sp. Nov., from the Kitakami Massif, Northeast Japan. *Tohoku University Science Report (Geology),* 48:77–97.

Sillman, A. S. 1973. Avian vision. In *Avian Biology,* ed. D. S. Farner and J. R. King. New York: Academic Press.

Silver, L. T. 1982. Introduction. In *Geological Implications of Impacts of Large Asteroids and Comets on the Earth,* ed. L. T. Silver and P. H. Shultz. Boulder: Geological Society of America.

Simons, J. R. 1970. The direction of the thrust produced by the hetero-cercal tails of two dissimilar elasmobranchs: the Port Jackson Shark, *Heterodontus portusjacksoni* (Meyer), and the piked dogfish, *Squalus megalops* (Macleay). *Journal of Experimental Biology,* 52:95–107.

Sloan, R. E., J. K. Rigby, L. M. Van Valen, and D. Gabriel. 1986. Gradual dinosaur extinction and simultaneous ungulate radiation in the Hell Creek Formation. *Science,* 232:629–633.

Smit, J., and S. Van Der Kaars. 1984. Terminal Cretaceous extinctions in the Hell Creek area, Montana: Compatible with catastrophic extinction. *Science,* 223:1177–1179.

Somero, G. N., and A. L. De Vries. 1967. Temperature tolerance of some Antarctic fishes. *Science,* 156:257–258.

Spector, W. S., ed. 1956. *Handbook of Biological Data.* Philadelphia: W.B.Saunders.

Spedding, G. R., J. M. V. Rayner, and C. J. Pennycuick. 1984. Momentum and energy in the wake of a pigeon (*Columbia livia*) in slow flight. *Journal of Experimental Biology,* 111:81–102.

Spotila, J. R. 1980. Constraints of body size and environment on temperature regulation of dinosaurs. In *A Cold Look at the Warm-Blooded Dinosaurs,* ed. R. D. K. Thomas and E. C. Olsen. Boulder: Westview Press.

Stanley, S. M. 1973. An explanation for Cope's Rule. *Evolution,* 27:1–26.

Stettner, L. J., and K. A. Matyniak. 1968. The brain of birds. *Scientific American,* 218:64–76.

Stewart, J. M. 1979. A baby that died 40,000 years ago reveals a story. *Smithsonian,* 10:125–126.

Streeter, G. L. 1904. The structure of the spinal cord of the ostrich. *American Journal of Anatomy,* 3:1–27.

Sullivan, R. M. 1987. A reassessment of reptilian diversity across the Cretaceous-Tertiary boundary. *Contributions in Science, Natural History Museum of Los Angeles County,* 391:1–26.

Surlyk, F., and M. B. Johansen. 1984. End-Cretaceous brachiopod extinctions in the chalk of Denmark. *Science,* 223:1174–1177.

Sutcliffe, A. J. 1985. *On the Track of Ice Age Mammals.* Cambridge: Harvard University Press.

Sikes, S. K. 1971. *The Natural History of the African Elephant.* London: Weidenfeld and Nicolson.

Tait, R. V., and R. S. De Santo. 1972. *Elements of Marine Ecology.* New York: Springer-Verlag.

Tarsitano, S. 1983. A case for the diapsid origin of ichthyosaurs. *Neues Jahrbuch für Geologie und Paläontologie, Monatshefte,* 1983:59–64.

Taylor, C. R., N. C. Heglund, and G. M. O. Maloiy. 1982. Energetics and mechanics of terrestrial locomotion. I. Metabolic energy consumption as a function of speed and body size in birds and mammals. *Journal of Experimental Biology,* 97:1–21.

Taylor, C. R., A. Shkolnik, R. Dmi'el, D. Baharav, and A. Borut, 1974. Running in cheetahs, gazelles, and goats: Energy cost and limb configuration. *American Journal of Physiology,* 277:848–850.

Taylor, J. W. R. 1975. *Civil Airliner Recognition.* London: Ian Allan.

Taylor, M. A. 1987. A reinterpretation of ichthyosaur swimming and buoyancy. *Palaeontology,* 30:531–535.

Thomas, R. D. K., and E. C. Olson, eds. 1980. *A Cold Look at the Warm-Blooded Dinosaurs.* Boulder: Westview Press.

Thomson, K. S. 1976. On the heterocercal tail in sharks. *Paleobiology,* 2:19–38.

Thomson, K. S., and D. E. Simanek. 1977. Body form and locomotion in sharks. *American Zoologist,* 17:343–354.

Toit, J. T. du, and N. Owen-Smith. 1989. Body size, population metabolism, and habitat specialization among large African herbivores. *American Naturalist,* 133:736–740.

Toots, H., and M. R. Voorhies. 1965. Strontium in fossil bones and the reconstruction of food chains. *Science,* 149:854–855.

Tredoux, M., M. J. De Wit, R. J. Hart, N. M. Lindsay, B. Verhagen, and J. P. F. Sellschop. 1989. Chemostratigraphy across the Cretaceous-Tertiary boundary and a critical assessment of the iridium anomaly. *Journal of Geology,* 97:585–605.

Tschudy, R. H., C. L. Pillmore, C. J. Orth, J. S. Gilmore, and J. D. Knight. 1984. Disruption of the terrestrial plant ecosystem at the Cretaceous-Tertiary boundary, Western Interior. *Science,* 225:1030–1032.

Tucker, V. A. 1975. The energetic cost of moving about. *American Scientist,* 63:413–419.

Tyne, J. Van, and A. J. Berger. 1976. *Fundamentals of Ornithology.* New York: John Wiley and Sons.

Van Dam, C. P. 1987. Efficiency characteristics of crescent-shaped wings and caudal fins. *Nature,* 325:435–437.

Wade, M. 1984. *Platypterygius australis,* an Australian Cretaceous ichthyosaur. *Lethaia,* 17:99–113.

Wagner, J. A. 1861. Über ein neues, angeblich mit Vogelfedern versehenes Reptil aus dem Solnhofener lithographischen Schiefer. *Sitzungsberichte Bayerische Akademie Wissenschften,* 2:146–154.

Wall, W. P. 1983. The correlation between high limb-bone density and aquatic habits in Recent mammals. *Journal of Paleontology,* 57:197–207.

Walls, G. L. 1942. *The Vertebrate Eye and Its Adaptive Radiation.* Bloomfield Hills: Cranbrook Institute of Science.

Ward, P., J. Wiedmann, and J. F. Mount. 1986. Maastrichtian molluscan biostratigraphy and extinction patterns in a Cretaceous/Tertiary boundary section exposed at Zumaya, Spain. *Geology,* 14:899–903.

Warren, J. V. 1974. The physiology of the giraffe. *Scientific American*, 231:96–105.

Webb, P. W. 1982. Locomotor patterns in the evolution of actinopterygian fishes. *American Zoologist*, 22:329–342.

―――― 1984. Form and function in fish swimming. *Scientific American*, 251:72–82.

―――― 1988. Simple physical principles and vertebrate aquatic locomotion. *American Zoologist*, 28:709–725.

Weihs, D., and P. W. Webb. 1983. Optimization of locomotion. In *Fish Biomechanics*, ed. P. W. Webb and D. Weihs. New York: Praeger Press.

Weishampel, D. B. 1981a. The nasal cavity of lambeosaurian hadrosaurids (Reptilia: Ornithischia): Comparative anatomy and homologies. *Journal of Paleontology*, 55:1046–1057.

―――― 1981b. Acoustic analyses of potential vocalization in lambeosaurine dinosaurs (Reptilia: Ornithischia). *Paleobiology*, 7:252–261.

―――― 1983. Hadrosaurid jaw mechanics. *Acta Palaeontologica Polonica*, 28:271–280.

―――― 1984. Evolution of jaw mechanisms in ornithopod dinosaurs. *Advances in Anatomy, Embryology and Cell Biology*, 87:1–110.

Wellnhofer, P. 1975. Die Rhamphorhynchoidea (Pterosauria) der Oberjura-Plattenkalke Süddeutschlands. *Palaeontographica*, 148:1–33.

―――― 1978. *Handbuch der Paläoherpetologie*. Teil 19. *Pterosauria*. Stuttgart: Gustav Fischer.

―――― 1982. Cuvier and his influence on the interpretation of the first known pterosaur. In *Actes du Symposium Paléontologique G. Cuvier, Montebeliard*, ed. E. Buffetaut, J. M. Mazin, and E. Salmon. Montbéliard.

―――― 1985. Neue Pterosaurier aus der Santana-Formation (Apt) der Chapada do Araripe, Brasilien. *Palaeontographica*, 187:105–182.

―――― 1987a. Die Flughaut von *Pterodactylus* (Reptilia, Pterosauria) am Beispiel des Wiener Exemplares von *Pterodactylus kochi* (Wagner). *Annalen des Naturhistorischen Museums Wien*, 88:149–162.

―――― 1987b. New crested pterosaurs from the Lower Cretaceous of Brazil. *Mitteilungen der Bayerischen Staatssammlung für Paläontologie und historische Geologie*, 27:175–186.

―――― 1988a. A new specimen of *Archaeopteryx*. *Science*, 240:1790–1792.

―――― 1988b. Terrestrial locomotion in pterosaurs. *Historical Biology*, 1:3–16.

Wetherill, G. W., and E. M. Shoemaker. 1982. Collision of astronomically observable bodies with the Earth. In *Geological Implications of Impacts of Large Asteroids and Comets on the Earth*, ed. L. T. Silver and P. H. Shultz. Boulder: Geological Society of America.

Wever, E. G. 1971. Hearing in the Crocodilia. *Proceedings of the National Academy of Science*, 68:1498–1500.

Whitehead, P. J., J. T. Puckridge, C. M. Leigh, and R. S. Seymour. 1989. Effect of temperature on jump performance of the frog *Limnodynastes tasmaniensis*. *Physiological Zoology*, 62:937–949.

Wieland, G. R. 1942. Too hot for the dinosaurs. *Science*, 96:359.

Wild, R. 1978. Die Flugsaurier (Reptilia, Pterosauria) aus der Oberen

Trias von Cene bei Bergama, Italien. *Bollettino della Societa Paleonto-logica Italiana,* 17:176–256.

Wilde, P., W. B. N. Berry, M. S. Quinby-Hunt, C. J. Orth, L. R. Quintana, and J. S. Gilmore. 1986. Iridium abundances across the Ordovician-Silurian stratotype. *Science,* 233:339–341.

Williston, S. W. 1902. Winged reptiles. *Popular Science Monthly,* 60:314–322.

Woodward, A. S. 1899. Exhibition on behalf of Dr. Moreno, of the skull and other specimens of *Neomylodon listai. Proceedings of the Zoological Society of London,* 1899:830.

Wyckoff, R. W. G. 1971. Trace elements and organic constituents in fossil bones and teeth. In *Proceedings of the North American Paleontological Convention for 1969.* Lawrence: Allen Press.

———— 1972. *The Biochemistry of Animal Fossils.* Bristol: Scientechnica.

Young, J. Z. 1962. *The Life of Vertebrates.* Oxford: Clarendon Press.

Zachos, J. C., M. A. Arthur, and W. E. Dean. 1989. Geochemical evidence for suppression of pelagic marine productivity at the Cretaceous/Tertiary boundary. *Nature,* 337:61–64.

Zhao, M., and J. L. Bada. 1989. Extraterrestrial amino acids in Cretaceous/Tertiary sediments at Stevns Klint, Denmark. *Nature,* 339:463–465.

Ziegler, B. 1986. *Der schwäbische Lindwurm.* Stuttgart: Konrad Theiss Verlag.

Zimmerman, L. C., and C. R. Tracy. 1989. Interactions between the environment and ectothermy and herbivory in reptiles. *Physiological Zoology,* 62:374–409.

Zug, G. R. 1974. Crocodilian galloping: An unique gait for reptiles. *Copeia,* 2:550–552.

Credits

Illustrations not listed here appear courtesy of the Photography Department of the Royal Ontario Museum, Toronto, or were provided by the author. At the ROM, Brian Boyle took most of the photographs and Alan McColl prepared the prints.

Prologue

Pages 3, 5, 6–7   Julian Mulock

3. How the Vertebrate Skeleton Works

Page 41   Marg Sansom

Page 42   From R. Owen, *On the Anatomy of Vertebrates,* 3 vols. (London: Longmans, Green and Co., 1866), 2:282

Page 43   O. C. Marsh; published in J. H. Ostrom and J. S. McIntosh, *Marsh's Dinosaurs* (New Haven: Yale University Press, 1966), pl. 37, reproduced here courtesy of the Peabody Museum of Natural History, Yale University

Page 51   From H. A. Nicholson, *Manual of Palaeontology* (Edinburgh: W. Blackwood, 1879), 2:323

4. Reading a Dinosaur Skeleton

Pages 55, 67, 73, 89   Julian Mulock

Pages 64, 79   Marg Sansom

5. A Matter of Scale

Page 94   Julian Mulock

Page 97   From R. Owen, *On the Anatomy of Vertebrates,* 3 vols. (London: Longmans, Green and Co., 1866), 2:282

Page 107   From R. T. Bird, "A dinosaur walks into the museum," *Natural History,* 47(1941):79; neg. 324393 courtesy of the Department of Library Services, American Museum of Natural History, New York

Pages 110–111   O. C. Marsh, *The Dinosaurs of North America,* 16th Annual Report, U. S. Geological Survey (1894), pl. XLII

Pages 112, 115   O. C. Marsh; published in J. H. Ostrom and J. S. McIntosh, *Marsh's Dinosaurs* (New Haven: Yale University Press, 1966), pl. 23 and 72, reproduced here courtesy of the Peabody Museum of Natural History, Yale University

Page 113   From H. F. Osborn and C. C. Mook, *"Camarasaurus, Amphicoelias* and other sauropods of Cope," *Memoirs of the American Museum of Natural History,* n.s., vol. 3, pt. 3 (New York: American Museum of Natural History, 1921), fig. 30(A), courtesy of the American Museum of Natural History, New York

6. What's Hot and What's Not

Page 138   Photo courtesy of A. J. Baker

Page 159   Marg Sansom

7. Brains and Intellect

Page 169   Julian Mulock

Page 171   From R. Owen, *On the Anatomy of Vertebrates,* 3 vols. (London: Longmans, Green and Co., 1866), 3:115, 2:119, 1:292

Pages 180–181   Poem by Bert L. Taylor cited in Alfred Sherwood Romer, *Man and the Vertebrates* (Chicago: University of Chicago Press, 1933, 1941), 92–93

Page 182   O. C. Marsh; published in J. H. Ostrom and J. S. McIntosh, *Marsh's Dinosaurs* (New Haven: Yale University Press, 1966), pl. 24, reproduced here courtesy of the Peabody Museum of Natural History, Yale University

Page 183   From E. T. Newton, "On the skull, brain and auditory organ of a new species of pterosaurian *(Scaphognathus purdoni),* from the Upper Lias near Whitby, Yorkshire," *Proceedings of the Royal Society of London,* 179(1888):521

8. Not Wholly a Fish

Page 187 (bottom)   From J. Phillips, *Illustrations of the Geology of Yorkshire* (London: John Murray, 1875), pl. 6

Pages 188, 189, 194   From T. Hawkins, *Memoirs of Ichthyosauri and Plesiosauri, Extinct Monsters of the Ancient Earth* (London: Relfe and Fletcher, 1834)

Page 191   From R. Owen, *A Monograph of the Fossil Reptilia of the Liassic Formations*. Part III: *Order Ichthyopterygia*, Palaeontographical Society Monographs, 35 (Keyworth, Nottingham: British Geological Survey, 1881), pl. 32

Page 192   From E. Home, "Some account of the fossil remains of an animal more nearly allied to fishes than any other classes of animals," *Philosophical Transactions of the Royal Society of London*, 104(1814): pl. 17

Page 195   Photo kindly supplied by Rupert Wild

9. The Mechanics of Swimming

Pages 201, 202, 205, 206, 207, 208, 216   Julian Mulock

Page 204   J. J. Thomason

10. The Sea Dragons

Pages 220, 221   Julian Mulock

Page 223 (top, middle)   From B. Hauff and R. B. Hauff, *Das Holzmaden Buch* (Fellbach: Repro-Druck Gmbh., 1981), p. 24

Page 223 (bottom)   J. J. Thomason

Page 224 (bottom), 233 (bottom), 236 (bottom), 244 (bottom), 247 (middle), 249 (bottom), 253 (bottom)   Marg Sansom

Page 230   Photos kindly supplied by Rupert Wild

Page 239   From B. F. Kosch, "A Revision of the skeletal reconstruction of *Shonisaurus popularis* (Reptilian Ichthyosauria)," *Journal of Vertebrate Paleontology*, 10(1990): 512–514

Page 242 (top)   Reproduced by permission of the Director, British Geological Survey, Keyworth, Nottingham; NERC copyright reserved

Page 253 (top)   From C. W. Andrews, *A Descriptive Catalogue of the Marine Reptiles of the Oxford Clay* (London: British Museum, Natural History, 1910), fig. 42; by courtesy of the Natural History Museum, London

Page 256 (bottom)   Marg Sansom (modified by Julian Mulock)

11. The Winged Phantom

Page 258 (left)   From S. W. Williston, *The Osteology of the Reptiles* (Cambridge: Harvard University Press, 1925); copyright © 1925 by Harvard University Press, reprinted by permission of the publishers

Pages 258 (right), 266   From M. Cuvier, *Researches sur les ossemens fossiles de quadrupèdes* (Paris: Chez Deterville, Libraire, 1812), vol. 4, vol. 1

Pages 260 (top, bottom left), 261 (top left)   From P. Wellnhofer, *Handbuch der Paläoherpetologie*, vol. 19, *Pterosauria* (Stuttgart: Gustav Fischer, 1978), pp. 8, 15, 9

Page 260 (bottom right)   From P. Wellnhofer, "New crested pterosaurs from the Lower Cretaceous of Brazil," *Mitteilungen Bayerische Staatssammlung für Paläontologie und historische Geologie,* 27(1987): fig. 2a

Pages 261 (top right, bottom right), 264 (top right, bottom right), 265 (bottom), 281, 283   From G. F. Eaton, "Osteology of *Pteranodon,*" *Memoirs of the Connecticut Academy of Arts and Sciences,* 2(1910): pl. 8, 9, 11, 31

Pages 261 (bottom left), 264 (top left, bottom left)   From P. Wellnhofer, "Die Rhamphorhynchoidea (Pterosauria) der Oberjura-Plattenkalke Süddeutschlands," *Palaeontographica,* 148(1975): figs. d, g, i

Page 262   From F. E. Beddard, *Cambridge Natural History* (New York: Macmillan, 1902), figs. 254, 255

Pages 265 (top), 269, 273, 286, 289   Julian Mulock

Pages 274, 275, 282 (top)   J. J. Thomason

12. Out with a Whimper or a Bang?

Page 295 (left)   Photo by the Royal Canadian Air Force

Page 295 (middle, left)   Photos by NASA

Page 299   Photos courtesy of the Geological Survey of Canada, Ottawa

Page 304 (top)   Photo courtesy of D. A. Russell

Page 304 (bottom)   Photo by Harry Foster; reproduced courtesy of the Canadian Museum of Nature, Ottawa

Page 307   From F. Dixon, *The Geology and Fossils of the Tertiary and Cretaceous Formations* (London, 1850)

Index

Achilles tendon and arch of foot, 48
Acid rain: after impact event, 293; effects on calcareous planktonts, 306
ADP (adenosine diphosphate), 141
Aerobic metabolism. *See* Metabolism
Aerobic scope, 144
African elephant, 96, 97, 116
African wild dogs, 87, 123
Age of Reptiles, 3
Ailerons, 287
Air bubbles (thermals), 277
Aircraft: ailerons, 287; high-lift devices, 272; horizontal stabilizer, 287; ice on wings, 206; rivets flush-fitting, 206
Airfoils, 272–273
Air-sac system: birds and pterosaurs, 283; saurischian dinosaurs, 111
Airspeed, 273
Albatrosses: extensive wanderings, 278; flying performance, 275
Albertosaurus, 83–89
Alexander, Annie: American fossil collector, 235
Alexander, R. McN.: on stride lengths and speed, 61
Alligators, 138
Allosaurus, 89, 126
Alula, 274
Alvarez, Luis: on the asteroid impact theory, 291
Ambient temperatures, 130
Amino acids: components of proteins, 17; extraterrestrial forms, 299–300; in fossil bone, 29

Ammonites, 187, 306, 307
Amphicoelous vertebrae, 192
Anaerobic metabolism. *See* Metabolism
Anatosaurus, 82
Angle of attack, 207, 272
Anhanguera, 264
Anning, Mary: pioneer English fossilist, 188
Antarctic fishes, 134, 135
Apatite, 14, 321
Apatosaurus, 109, 110, 114, 116, 117, 118, 120, 126, 127
Aquatic animals: bone density, 219
Archaeopteryx: anatomy, 314–317; dinosaurian skeletal structure, 314; evidence for endothermy, 161–162; natural endocast, 173–174, 183; seven known specimens, 312
Archimedes principle, 104
Area-to-volume ratio, 92, 138
Arteries and arterioles, 102
Arthritic joints in ancient Egyptians, 35
Aspect ratio: defined, 208; frigate birds, 279, 284; gulls and pigeons, 274; measurement, 327; *Pteranodon*, 284
Asteroid. *See* Bolide
ATP (adenosine triphosphate), 141
Avimimus, 161

Bacterial mats and preservation, 196
Bakker, Robert: on dinosaurian energetics, 116; on warm-blooded dinosaurs, 129

Balance perception, 80
Ball-and-socket joints, 111
Baluchitherium, 117, 324
Bandera County, Texas, trackways, 108
Baptanodon, 252
Basal metabolic rate, 131, 140
Basilar membrane, 81
Basking in reptiles, 138
Bastard wing. *See* Alula
Bats: flying and walking, 262, 263; heterothermy, 133
Beam theory, 9, 10
Bears, 133
Bees, 136
Beetles, scavenging by, 25
Behavioral temperature regulation, 136, 325n4
Behrensmeyer, A. K.: extensive field studies in taphonomy, 23
Belemnites, 307
Bernoulli, Daniel: airfoil theory, 272
Biewener, Andrew: on stresses in limb bones, 94
Binocular vision, 78, 183, 322
Binoculars, 77
Biomass, 155
Bird, Roland: on the discovery of sauropod tracks, 106
Birds: air-sac system, 283; as "avian dinosaurs," 316; bipedal balance, 60; embryos resemble dinosaurs, 317; extensive bone fusion late in development, 317; flight, 276; flight control, 287, 288; in fossil record, 28; hearing abilities, 81; instinctive and learned behavior, 171–172; posture and balance, 266; reptilian relationship long recognized, 313
Bitumen, in Holzmaden slate, 196
Blood pressure, 101, 118, 119, 323
Blood remnants suspected in dinosaur remains, 34
Blue Lias, 186, 187
Body temperatures, 131, 133, 136, 137, 139
Body weights: in dinosaurs, 104–105; in selected living animals, 123, 124
Boeing 727: weight, 103, 323n5
Bolide: chances of impact, 294; effects of impact, 292; impact may be linked with volcanism, 302; size, speed, and explosive force, 292

Bone: cancellous, 15; chemical composition, 17; compact, 15; fibrolamellar, 153; Haversian, 16, 153; mechanical properties, 11, 13–15; microscopic structure, 15–17
Bone preservation: destruction rates vary with size and species, 25; mineralization, 29; rapid burial protects bone, 26; reburial, 28; importance of micro-environments, 25; weathering, 25
Bones: significance of being hollow, 45; strength and lightness in fliers, 282–283
Boundary layer: defined, 202; fluid flow within, 205; over land, 278; thickness and Reynolds number, 204; thickness and viscosity, 202, 326n1; turbulence and separation, 206
Brachiopods, 308
Brachiosaurus, 103, 105, 106, 114, 115, 118, 120, 126, 127, 178
Brain: basic anatomy, 169; decomposition after death, 38; mass vs. body mass, 176; reconstructed from cranium, 173; sizes in dinosaurs, 178, 179; sizes in man and other mammals, 168
Brain heater in swordfish, 137
Bramwell, Cherrie: analysis of flight in *Pteranodon*, 280
Breathing rates, 100, 103
Brittle behavior, 13
Brontosaurus. See Apatosaurus
Brown muscle in fishes, 137, 147
Buckland, William: first geology professor at Oxford, 190; on pterosaurs, 257
Buffaloes, 122, 123, 125
Burial, chances of, 26

Californosaurus, 240
Calluses, at site of bone fractures, 35
Camarasaurus, 105, 110, 113
Camber, of airfoil, 272
Camp, Charles: established Ichthyosaur State Park, 238
Camptosaurus, 126
Canaliculi, 16
Cannibalism in ichthyosaurs, 229
Capillaries, 101, 102, 119
Carbohydrate: defined, 140; loading, 144; metabolism, 140

Carbon dioxide, 140
Carbonaceous films, formed during preservation, 196
Carcasses, rapid breakdown, 25
Carrier, David: on the evolution of stamina, 149
Carroll, Robert: on reptilian classification, 5
Cartilage, 8
Caudal peduncle, 212
Cellulose, 69, 100
Center of gravity. *See* Center of mass
Center of mass, 108, 324
Center of pressure, of airfoils, 272
Centra, 42
Cephalopods, 226
Cerebellum, 173
Cerebral cortex, 170
Cerebral hemispheres, 169
Cetiosaurus, 112
Chalk: formation ended at K-T boundary, 306; white cliffs of Dover, 254
Chasmosaurus, bone disease, 36
Cheetahs, 124
Chewing action, 51
China tea cups are not tough, 14
Cladistic method of classification, 316
Clavicle (collarbone), 18
Claws, retractile: cats, 89: dromaeosaurs, 90
Climate, effects on bone destruction, 26
Cochlea, cochlear duct, 81
Coelacanths, 303
Colbert, Edwin: on weight estimates for dinosaurs, 104
Cold-bloodedness, 129
Collagen, 14, 17
Columns, 10, 115
Composite materials, 15
Compression, 9
Compressive strength, 13; defined, 321n5
Compsognathus, 161, 162
Conquest of land, 150
Conybeare, William: early accounts of ichthyosaurs, 192
Coombs, Walter: juvenile dinosaurs and parental care, 164, 165; cursorial adaptations, 66
Cope, Edward Drinker: pioneer American dinosaur collector, 105
Cope's Rule, 121

Coprolites, 225
Core temperatures, 130
Corpora striata, 171
Corythosaurus, 55, 58, 73, 82, 126
Cougars, 136
Cranium, 173
Crests. *See* Hadrosaurs
Cretaceous-Tertiary boundary. *See* K-T boundary
Cristospina, 283
Crocodiles: complex behavior, 178–179; endurance related to body size, 148; galloping speeds, 148; growth rates temperature-dependent, 154; hearing abilities, 81; parental care, 180; physiology and life-style, 152
Crystal Palace, life-sized dinosaur models, 193
CT-scanner, as a research tool, 224
Cunninghamites elegans, 71
Cursorial animals, 48
Cusps, 50
Cuvier, Georges: anatomy of *Pterodactylus,* 260; reconstructing fossils, 53; species extinctions, 190
Cycads, 4
Cymbospondylus, 222, 236, 237

D and L forms of molecules, 299
Darwin, Charles: evolution, 4; publication of *Origin of Species,* 191, 313
Davy, John: discovery that tunas are warm-blooded, 137
De La Beche, Henry: early accounts of ichthyosaurs, 192
Dead space, 103, 120
Deccan basalts, 300, 302
Deep-sea cores, 306
Deer, 125
Deinocheirus, 89
Deinonychus, 90, 158
Delphinosaurus. See Californosaurus
Density, 201, 203
Dental batteries, 68
Dentine, 20, 30, 51
Dermestes beetles, 25
Dextro and laevo. *See* D and L forms of molecules
Diaphragm, 150
Digestion, 100
Digitigrade, 64
Dinosaurs: Alaskan, 157; binocular vision, 88; bone fractures and disease,

Force, 9, 13, 321n1
Forest fires after impact event, 292
Forked neural spines, 113
Fossil record: contemporaneity of species, 27; sampling errors, 27
Fossils: definition and nature, 23–24; bone analyses, 31–34; cell structure preserved, 38; deducing diets from analyses, 32–33; density, 29; human brains preserved, 23; identifying sex, 231; limited information, 54; microscopic details, 29, 30; muscle reconstructions, 65; pterosaur skin preserved, 268
Fraas, Eberhard: studies on German ichthyosaurs, 195
Franklin Expedition: autopsies of bodies, 38; preservation of bodies, 24; skull fragments, 26
Frequency response, 81
Frigate birds, 75, 279
Frogs, 136
Frontal area of swimming bodies, 209
Froude number, 61
Fundamental frequency, 75
Furcula, 316

Gaits, 46, 47
Galápagos tortoises, 138, 139
Gallop, 46
Gallup polls, 303
Galton, Peter: hadrosaur locomotion, 57
Gelatin, 17
Generation times, 121
Geological time scale, 3
Geometric similarity, 94
Gestation periods, 100
Gigantism: advantages, 121; pituitary gland, 172
Gills, ram-jet action, 151
Giraffes: blood pressure, 118; physiology and behavior, 101–102; weight, 123
Glen Rose, Texas, dinosaur trackways, 95, 106
Gliders. *See* Sailplanes
Gliding: defined, 271; angle, 274
Glucose, 140, 144
Glycogen, 144
Glycogen body, 182
Glycolysis, 143
Golf balls, why dimpled, 206

Graviportal posture: defined, 95; features, 97
Grendelius, 253
Greyhounds, 47
Griphosaurus, synonym for *Archaeopteryx*, 313
Grippia, 234
Gulls, compared with pigeons, 274
Gut capacity, 100
Gut contents: hadrosaurs, 71; ichthyosaurs, 225; pterosaurs, 281

Hadrosaurs: bone fractures, 36; crests, 72–76; distribution, 54; ears and hearing, 82; eyes and sight, 78–79; horny bills, 71; inflatable nostrils, 75; life-styles, 67, 82; posture and locomotion, 56–57, 64–66; size and form, 55; teeth and chewing, 68–71
Hammerhead sharks, 218
Hankin, E. H.: observations on bird flight, 287
Harmonics, 75
Hawkins, Thomas: early accounts of ichthyosaurs, 192
Hearing abilities, 81
Heart: anatomy, 151; rates, 100, 102
Heat interchanger of tunas, 137
Heilmann, G.: *Origin of Birds*, 316
Hell Creek, Montana, K-T boundary site, 305
Herbivores: grinding teeth, 68; gut bacteria, 100; large gut, 72
Heterocercal tail, in sharks, 214–218
Heterothermy, 133
Hip girdle. *See* Pelvic girdle
Hippopotamuses: dense bone, 219; diet and life-style, 106; weight, 123
Histology, 153
Holzmaden: ichthyosaur locality, 195; numbers of ichthyosaurs found, 248; pterosaur locality, 259
Home, Sir Everard: first description of ichthyosaurs, 186; perplexed by affinities of ichthyosaurs, 192
Homeothermy, 133
Hopson, James: crest function in hadrosaurs, 74
Horner, Jack: dinosaur nesting sites, 164; growth in *Maiasaura*, 154
Horses: back-stretching exercises, 322n1; backs stiff, 46; hooves,

Horses (*continued*)
snapping action, 48; leg proportions, 67
Huene, Friedrich von: extensive studies on ichthyosaurs, 195
Hummingbirds, 133
Huxley, T. H.: dinosaurian ancestry for birds, 313
Hydrocarbons, 197
Hydroplanes, 197, 211
Hydroxyapatite, 29
Hydroxyproline, 17
Hylaeosaurus, 190
Hypselosaurus, 128

I-beams, 10, 110
Ichthyosaur State Park, Nevada, 238
Ichthyosaurs: ancestry problematic, 234; decline and extinction, 256; density and buoyancy, 219; disappearance before K-T boundary, 309; distinctive vertebrae, 191; diversity reduction during late Jurassic, 253–254; endocast, 184; fat beneath skin, 197; fin functions, 221–222; fin rays, 198; fins serially sectioned, 198; first descriptions, 186; growth changes, 231; Jurassic forms, characterized, 243; largest, 239; latipinnates and longipinnates, 242–243; mothers and embryos, 228–231; nares, position, 191; numbers collected from Holzmaden, 248; oldest, geologically, 232; preservation of body outline, 194, 196–197; senses, 226, 227, 228; skeletal anatomy, 188, 191–192; streamlined profiles, 198; tail function, 220–221; tailbends, 192; tailbends and authenticity, 222–224; tails compared with fishes, 223; teeth and feeding, 225; thermal strategies, 166; thunniform bodies, 213; Triassic, summarized, 241–242; varied body patterns, 222, 224; Victorian misinterpretations, 194; youngest, geologically, 232, 256, 327n2
Ichthyosaurus: history of name, 193, 326n4; *I. breviceps,* 244, 245; *I. communis,* 243, 244; *I. conybeari,* 243, 245
Iguanodon, 190

Impact craters, on Earth and the moon, 294, 295. *See also* K-T boundary
Impact winter, 293
Inclined planes, 207
Incus, 80
Indian elephant, 96, 117
Inertia, 48–49
Inertial homeothermy, 139
Inner ear, 80
Inoceramids, 307
Insects: abundance, 28; muscles and pre-flight warm-up, 135, 136; respiratory system and stamina, 152
Insulation, 133
Intelligence, definitions, 170
Intervertebral disks, 42
Iridium: correlations of deposits with other extinction events, 297; found with antimony and arsenic, 298; metal of the platinum group, 291; single and multiple peaks, 296; synchrony of occurrence around the world, 295; volcanic origin, 298; worldwide occurrence at K-T boundary, 292

Jaw joint, 69
Jerison, Harry: on relative brain sizes, 176–177
Jogging shoes, 48

Kinetic energy, 49
Kiwis, 65, 77
Koch, Carl: errors in sampling fossils, 303
Komodo dragons: hunting strategies, 124; thermal strategies, 139
Krakatau, 298
Krausel, R.: hadrosaur stomach contents, 71
Krebs cycle, 140
K-T boundary: competing views, 293; dinosaur decline gradual or sudden, 309–310; discovery of iridium, 291; evidence for impact crater, 297; extensive environmental changes, 310; extraterrestrial amino acids, 300; iridium anomaly worldwide, 292; mineral spherules, 306; physical evidence, summarized, 301; recognition in the field, 303; size of impact crater, 292; stishovite, at Ra-

ton site, 300; timing of iridium anomaly, 296; volcanic activity extensive, 300

K-T extinctions: ammonites, 306–307; brachiopod decline, 308; deep-sea animals barely affected, 308; differences between hemispheres, 308; plankton seriously affected, 305–306; plants, 308; reptiles, 308–309; selectivity among planktonts, 306; whether sudden, 302

Lactic acid, 143
Lacunae, 16, 29, 30
Lamarck, Jean-Baptiste: species stability discounted, 190
Lambeosaurus, 55, 62, 72, 89
Lamellae, 16
Laminar flow, 200
Lava lakes, 302
Lazarus species, 303
Leading-edge slot, 274
Leatherback turtles, 139
Legs: bone proportions and body weights, 93–95; functions, 44; inclination of bones, 45; proportions and speed, 66; protraction and retraction, 66; segments and speed, 66; solid bones, 97; stresses in bones, 45; stresses in muscles, 95; tuck-in by sprinters, 50
Leopards, 124
Leptopterygius: meaning of name, 248; *L. burgundiae*, 250; *L. tenuirostris*, 222, 241, 242, 243
Liassic: length of interval, 250; lowest division of Jurassic, 186; Upper and Lower divisions, 243
Life spans: dinosaurs, 127, 128; living animals, 100, 126–127
Lift force, of inclined plane, 207
Lift-to-drag ratios, 208, 274
Ligaments: function, 8; mechanical properties, 18
Limestone, 188
Lions, 122, 123
Lipids, 197
Lizards: bipedal running, 60; hearts, 151
Locomotion: body size and costs, 121; endurance versus sprinting, 146; energetic costs and reductions, 48–50, 324n9; performance and muscle types, 146; swimming costs low, 149

Logarithms, 325
Long bones, 97
Lophorhothon, 82
Lunar seas, 302
Lunate tails, 211
Lyme Regis: English seaside resort, 185; ichthyosaur locality, 189, 243; pterosaurs sometimes found, 259

Magnetic reversals, 295
Magnetic stripes in Earth's crust, 295
Magnification factors, 104
Maiasaura, 154, 157
Malleus, 80
Mammals: distinctive teeth, 50; ears and hearing, 79–81; predatory, 122
Mammoths: blood cells preserved, 38; hair, skin, and muscles preserved, 23
Manson Crater, Iowa, 297
Marsh, Othniel Charles: pioneer American dinosaur collector, 105
Martill, David: the authenticity of some ichthyosaurs, 196
Mass, 9, 91, 323n1
Materials and structures, distinctions, 11
Mazin, Jean-Michel: ichthyosaurs from Spitsbergen, 234
McNab, Brian: thermal physiologies of animals, 130
Megalosaurus, 190
Membranous labyrinth, 80
Merriamia, 240, 241
Mesozoic Era: duration, 2; climate and flora, 4; illustrations of selected reptiles, 6, 7
Messerschmitts, 274–276
Metabolism: defined, 131; aerobic, 140, 142–143; aerobic scopes, 144, 147; anaerobic, 143; basal rates in various animals, 131–132; carbohydrate, 140; diets and rates, 130; measurement of rate, 141–143; rates in sharks, 213; rates vary with body size, 99, 132; summary, 140
Meteoritic dust, 294
Middle ear, 80
Mineralization of fossils, 29
Mitochondria, 136, 137
Mixosaurus, 232–234

Predatory dinosaurs, 83
Predentary, 71
Pregnant ichthyosaurs, 231
Pressure, 101, 323n3
Propatagium, 262
Proteins, 17
Proteosaurus, unadopted name for first ichthyosaur, 193
Protoceratops, 163
Psittacosaurus, 164
Pteranodon: anatomy, 280–283; common Niobrara pterosaur, 259; evidence for fish-eating, 281; flying performance, 284–286, 288, 289; large crest, 260; life-style, 290; wing shape, 284; wing tendon, 281
Pterodactyloidea, 258
Pterodactylus, 258
Pterodaustro, 260
Pteroid bone of pterosaurs, 262
Pterosaurs: avian features, 261; classification, 258; decline before K-T boundary, 309; early discoveries, 189, 260; endocasts, 183; evidence for body covering, 166; evidence for endothermy, 165–166, 268; oldest, geologically, 258; posture and walking, 263–267; pterosaurs and pterodactyls, confusion over names, 260; skin preservation, 268; structure of wing, 262; thin bones risk breakage, 283; wing membrane, 267–271; wing tendon, 281
Pulp cavity, 20
Pyruvate, 140

Q_{10} effect, 134
Quartz grains. *See* Shocked quartz
Quetzalcoatlus, 258

Radiometric dates, error margins, 295
Rancho La Brea, California, 31
Raton, New Mexico, K-T boundary site, 300
Rayner, Jeremy: vortex analysis of flight, 271
Regal, P. J.: the origin of feathers, 162
Reptiles: basking in the sun, 133; continuous growth, 104–105, 127; hearing abilities, 81; lack of stamina, 147–149; lactic acid problem, 148; limited binocular vision, 78; living representatives, 3
Reynolds number, Re, 203

Rhamphorhynchoidea, 258
Rhamphorhynchus, 166, 258
Rhinoceroses, 123
Ricqlès, Armand de: bone histology, 153
Roll, 286
Rudders: aircraft, 287; ships, 207
Rudists, 307
Ruminants, 117
Runners, long-distance, aerobic performance, 143
Running costs, 49
Running speeds in living animals, 122–123
Russell, Dale: *An Odyssey in Time: The Dinosaurs of North America,* 311; dinosaur diversity during the Late Cretaceous, 309; proponent of impact theory, 293

Sailplanes, 271
Sampling errors in fossil record, 303
Santana Formation, Brazil, pterosaur locality, 259
Saurischia, 5
Sauropods: blood pressure and heart, 118–119; bone growth, 154; center of mass, 108; decline before K-T boundary, 309; food requirements, 115–118; life spans, 127–128; life-styles, 103–104, 106, 109, 120; neck mobility, 112; size, 103; skeleton and posture, 109–116; teeth, 114; trackways, 106–109
Scavengers, destruction of carcasses, 24
Sclerotic ring, 76, 78, 191; possible function, 226
Scombroid fishes, 211
Seahorse and salmon compared, 204
Segisaurus, 161
Semicircular canals, 80
Sensory organs, 76–82
Shale, 188
Sharks: density and buoyancy, 214; flexibility and swimming costs, 213; lamnid, 212; tail function, 214–218
Shastasaurus, 240
Shearing in fluids, 202
Shocked quartz, 298, 299
Shonisaurus, 238–240
Shoulder girdle. *See* Pectoral girdle
Sinking speeds (rates), 276–277
Sinuses, in elephants, 98